Nitrogen Metabolism of Plants

PROCEEDINGS OF THE PHYTOCHEMICAL SOCIETY OF EUROPE

PROCEEDINGS OF THE
PHYTOCHEMICAL SOCIETY OF EUROPE

Nitrogen Metabolism of Plants

Edited by

K. MENGEL

Institut für Pflanzenernährung
Justus-Liebig-Universität, Giessen, Germany

and

D.J. PILBEAM

Department of Pure and Applied Biology
The University of Leeds, UK

CLARENDON PRESS · OXFORD

1992

Oxford University Press, Walton Street, Oxford OX2 6DP
Oxford New York Toronto
Delhi Bombay Calcutta Madras Karachi
Petaling Jaya Singapore Hong Kong Tokyo
Nairobi Dar es Salaam Cape Town
Melbourne Auckland

and associated companies in
Berlin Ibadan

Oxford is a trade mark of Oxford University Press

Published in the United States
by Oxford University Press, New York

A catalogue record for this book is available from the British Library

Library of Congress Cataloging in Publication Data
Nitrogen metabolism of plants / edited by K. Mengel and D.J. Pilbeam.
(Proceedings of the Phytochemical Society of Europe 33)
Includes index.
1. Nitrogen — Metabolism — Congresses. 2. Plants — Metabolism —
Congresses. I. Mengel, Konrad. II. Pilbeam, D.J. III. Series.
QK898.N6N59 1992   581.1'33 — dc20   91–24151
ISBN 0–19–857752–4

Set by Colset Pte. Ltd., Singapore
Printed in Great Britain by
Bookcraft (Bath) Ltd
Midsomer Norton, Avon

PREFACE

It was exactly 150 years ago that Justus Liebig, Professor of Chemistry at the University of Giessen, published his book *Agrikulturchemie*. Although other workers had shown that inorganic substances could be taken up from the soil by plants, it was Liebig who gradually convinced scientists that the plants take up and assimilate inorganic nutrients rather than preformed organic molecules. Of the inorganic nutrients required for plant growth, the one that is most commonly limiting is nitrogen. The use of nitrogenous fertilizers has probably done more to increase crop yields and improve human nutrition than has anything else.

The improvements in crop yields are largely brought about because nitrogen is a component of proteins, which have both enzymatic and structural properties. However, nitrogen is also a component of many secondary metabolites which have a variety of suggested functions, from plant growth regulators to feeding deterrents and defence compounds. In fact, so many different groups of compounds contain nitrogen that the difficulty in arranging a conference on nitrogen metabolism of plants is knowing what to leave out. Inevitably a choice has to be made between subjects that are essential to an understanding of nitrogen metabolism and subjects which are at first sight less central to the topic but which are developing at an exciting pace.

Many of the reactions of anabolism and catabolism of primary nitrogenous metabolites are common to animals as well as to plants, but the processes of assimilation of inorganic nitrogen are restricted to plants and microorganisms, and are given lengthy consideration here. They are considered from the viewpoints of what happens in the soil, at the soil/root interface, and inside the plant, and from the viewpoints of whole organism physiology, biochemistry, and molecular biology. These three disciplines are also represented in subsequent chapters, where the metabolic fate of assimilated nitrogen within plants is followed. In passing through these topics a wide range of interests are touched upon, from problems of agricultural productivity under nitrogen limitation to the genetic engineering of protein-rich crops, from the regulation of enzyme activity in plants to possible pharmaceutical uses of secondary metabolites.

For a successful conference you require enthusiastic delegates who question what they hear and discuss their work with each other late into the night, and enthusiastic speakers who stimulate the delegates. The animated, sometimes overheated, discussion periods bore witness to the contributions of the delegates, and the high standard of the papers presented here is testimony to the performance of the speakers. The authors of the many posters presented also deserve credit for the part that they played in making the conference a success.

We are very grateful to the numerous staff and students of the Justus-Liebig-Universität, Giessen, who put in so much effort to help run the event efficiently. We also wish to thank Deutsche Forschungsgemeinschaft (Bonn), Licher Privatbrauerei Ihring–Melchior KG (Lich), Gail Architektur-Keramik (Giessen), Bostik–Tucker GmbH (Giessen) and Giessener Hochschulgesellschaft (Giessen) for generous financial support.

Giessen   Konrad Mengel
Leeds   David Pilbeam
April 1991

Contents

Contributors

B. Ahlborn: Fachbereich Biologie, Philipps-Universität Marburg, Karl von Frisch Strasse, D-3550 Marburg, Germany

Nello Bagni: Department of Biology, Institute of Botany, Via Irnerio 42, 40126 Bologna, Italy

S. Bassarab: Fachbereich Biologie, Philipps-Universität Marburg, Karl von Frisch Strasse, D-3550 Marburg, Germany

Friedrich-Wilhelm Bentrup: Botanisches Institut 1 der Justus-Liebig-Universität Giessen, Senckenbergstrasse 17–21, D-6300 Giessen, Germany

Ton Bisseling: Department of Molecular Biology, Agricultural University, Dreijenlaan 3, 6703 HA Wageningen, The Netherlands

Raymond D. Blackwell: Division of Biological Sciences, University of Lancaster, Lancaster LA1 4YQ, UK

Hermann Bothe: Botanisches Institut, Universität zu Köln, Gyrhofstrasse 15, D-5000 Köln 41, Germany

P. Duarte: Department of Botany, Stockholm University, S-106 91 Stockholm, Sweden

Linda E. Fellows: Jodrell Laboratory, Royal Botanic Gardens, Kew TW9 3DS, UK

Anthony D.M. Glass: Department of Botany, University of British Colombia, Vancouver B.C., Canada V6T 2B1

Francine Govers: Department of Phytopathology, Agricultural University, Binnenhaven 9, 6709 PD Wageningen, The Netherlands

Bernd Hoffman: Botanisches Institut 1 der Justus-Liebig-Universität Giessen, Senckenbergstrasse 17–21, D-6300 Giessen, Germany

B. Ingemarsson: Department of Botany, Stockholm University, S-106 91 Stockholm, Sweden

Kenneth W. Joy: Department of Biology, Carleton University, Ottawa K1S 5B6, Canada

R. Kape: Fachbereich Biologie, Philipps-Universität Marburg, Karl von Frisch Strasse, D-3550 Marburg, Germany

A. Kinnback: Fachbereich Biologie, Philipps-Universität Marburg, Karl von Frisch Strasse, D-3550 Marburg, Germany

E.A. Kirkby: Department of Pure and Applied Biology, University of Leeds, Leeds LS2 9JT, UK

Geoffrey C. Kite: Jodrell Laboratory, Royal Botanic Gardens, Kew TW9 3DS, UK

C.M. Larsson: Department of Botany, Stockholm University, S-106 91 Stockholm, Sweden

M. Larsson, Department of Botany, Stockholm University, S-109 91 Stockholm, Sweden

Peter J. Lea: Division of Biological Sciences, University of Lancaster, Lancaster LA1 4YQ, UK

Karl-Heinz Linne von Berg: Botanisches Institut, Universität zu Köln, Gyrhofstrasse 15, D-5000 Köln 41, Germany

Deborah M. Long: Department of Botany, University of Guelph, Guelph, Ontario N1G 2W1, Canada

T. Lundborg: Department of Crop Genetics and Plant Breeding, Swedish Agricultural University, S-268 00 Svalöv, Sweden

M. Mattsson, Department of Crop Genetics and Plant Breeding, Swedish Agricultural University, S-268 00 Svalöv, Sweden

R.B. Mellor: Fachbereich Biologie, Philipps-Universität Marburg, Karl von Frisch Strasse, D-3550 Marburg, Germany

Konrad Mengel: Institut für Pflanzenernährung, Justus-Liebig-Universität Giessen, Südanlage 6, D-6300 Giessen, Germany

E. Mörschel: Fachbereich Biologie, Philipps-Universität Marburg, Karl von Frisch Strasse, D-3550 Marburg, Germany

P. Müller: Fachbereich Biologie, Philipps-Universität Marburg, Karl von Frisch Strasse, D-3550 Marburg, Germany

Adolf Nahrstedt: Institut für Pharmazeutische Biologie und Phytochemie, der Westfälischen Wilhelms-Universität, Hittorfstrasse 56, D-4400 Münster, Germany

Robert J. Nash: Jodrell Laboratory, Royal Botanic Gardens, Kew TW9 3DS, UK

Ann Oaks: Department of Botany, University of Guelph, Guelph, Ontario N1G 2W1, Canada

E. Öhlén: Department of Botany, Stockholm University, S-106 91 Stockholm, Sweden

P. Oscarson: Department of Crop Genetics and Plant Breeding, Swedish Agricultural University, S-268 00 Svalöv, Sweden

E. Pahlich: Institut für Allgemeine Botanik und Pflanzenphysiologie, Heinrich Buff Ring 54–62, D-6300 Giessen, Germany

M. Parniske: Fachbereich Biologie, Philipps-Universität Marburg, Karl von Frisch Strasse, D-3500 Marburg, Germany

D.J. Pilbeam: Department of Pure and Applied Biology, University of Leeds, Leeds LS2 9JT, UK

Rossella Pistocci: Department of Biology, Institute of Botany, Via Irnerio 42, 40126 Bologna, Italy

Thomas W. Rufty, Jr.: USDA-ARS, Department of Crop Science, North Carolina State University, Raleigh, NC 27695–7620, USA

M. Samuelson: Department of Botany, Stockholm University, S-106 91 Stockholm, Sweden

P. Schmidt. Fachbereich Biologie, Philipps-Universität Marburg, Karl von Frisch Strasse, D-3550 Marburg, Germany

A. Schultes: Fachbereich Biologie, Philipps-Universität Marburg, Karl von Frisch Strasse, D-3550 Marburg, Germany

Anthony M. Scofield: University of London, Wye College, Ashford, Kent TN25 5AH, UK

P.R. Shewry: Department of Agricultural Sciences, University of Bristol, AFRC Institute of Arable Crops Research, Long Ashton Research Station, Bristol BS18 9AF, UK

Monique S.J. Simmonds: Jodrell Laboratory, Royal Botanic Gardens, Kew TW9 3DS, UK

Wolfram R. Ullrich: Institut für Botanik, Technisches Hochschule, Schnittspahnstrasse 3, D-6100 Darmstadt, Germany

Richard J. Volk: Department of Soil Sciences, North Carolina State University, Raleigh, NC 27695–7619, USA

Renate Vosswinkel: Botanisches Institut, Universität zu Köln, Gyrhofstrasse 15, D-5000 Köln 41, Germany

D. Werner: Fachbereich Biologie, Philipps-Universität Marburg, Karl von Frisch Strasse, D-3550 Marburg, Germany

1. Nitrogen: agricultural productivity and environmental problems

KONRAD MENGEL

Institut für Pflanzenernährung, Justus-Liebig-Universität, Giessen, Südanlage 6, D-6300 Giessen, Germany

Introduction

On a global scale plants play an essential role in the conversion of inorganic into organic nitrogen, which is a basic process in plant growth and crop production. Crop yields depend much on the supply of inorganic nitrogen and are generally increased by the application of nitrogen fertilizers. Without nitrogen fertilization it would be impossible to feed the world's population.

Of the various fertilizers, nitrogen fertilizers are those which are applied in highest quantities on a global scale. They have a great impact on crop production; nitrogen influences a number of processes which are relevant in determining the quality of plant products, and nitrogen fertilizers also affect ecological processes. In this paper some major aspects of the importance of nitrogen for crop production and its impact on the environment are considered.

Today the annual rate of industrially fixed dinitrogen amounts to 73×10^6 t N, which is about 60 per cent of the dinitrogen fixed biologically. Industrially fixed nitrogen is mainly used as fertilizer and the process of N_2 fixation represents a great interference in the global nitrogen cycle. Most of the fertilizer nitrogen is eventually fed into the environment, in the form of nitrate into aquifers, rivers, lakes, and oceans and in the form of N_2 and N_2O into the atmosphere. Nitrate released into soils and water is a risk to drinking water, while N_2O released into the atmosphere participates in ozone decomposition in the stratosphere. The deposition of NH_4^+ affects natural plant associations and the enrichment of available nitrogen in soils decreases the diversity of plant species.

The release of nitrogen into the environment can be substantially reduced if the nitrogen excreted by farm animals is recycled to fields with a higher efficiency, and if nitrogen fertilizer rates are adjusted to the level of available nitrogen in the soil.

Physiological processes

Nitrogen is involved in numerous physiological processes of which the most important for crop production are N_2 fixation by prokaryotes (in particular those which live in symbiosis with higher plants (Werner 1980)), nitrate–nitrite reduction, assimilation of NH_3 in the physiological source, and the synthesis of proteins and nucleic acids in the physiological sink. According to the terminology of crop physiologists the physiological source represents plant parts, tissues, and organelles in which basic low molecular organic molecules are synthesized. The most important physiological source for nitrogen is the chloroplast (Miflin and Lea 1977). As shown in Fig. 1.1, NO_3^- is reduced in the cytosol and the resulting HNO_2 is transported into the chloroplast, where it is reduced by nitrite reductase to NH_3. In addition, NH_3 taken up directly in the form of NH_4^+-N may be imported into the chloroplast where it is assimilated by glutamine synthetase, and the δ amino-N of the glutamine synthesized is transferred to α-oxoglutarate, forming glutamate. The reducing e^- and ATP required for these steps are directly supplied by photosynthesis. Analogous processes take place in the plastids of roots (Oaks and Hirel 1985).

These reactions are of fundamental importance because they provide the organic nitrogen required for plant growth and crop production. The glutamate produced in the plastids is exported into the cytosol where it may be involved in the synthesis of other amino acids. A substantial amount of amino acids synthesized in fully developed leaves is exported to the

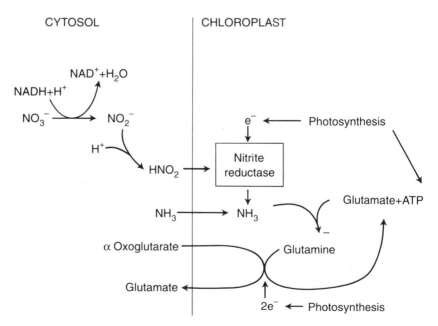

Fig. 1.1. Reduction of NO_3^- in the cytosol, and HNO_2 reduction and NH_3 assimilation in the chloroplast.

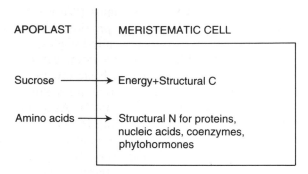

Fig. 1.2. Import of sucrose and amino acids into the meristematic cell required for energy, structural carbon and nitrogen.

physiological sinks. These are tissues in which the amino acids are used for the synthesis of proteins and nucleic acids. Such sinks may be reproductive organs such as seeds and fruits, or meristematic tissues of apices and leaves. Both are important for plant growth, and their supply with amino acids has a direct impact on crop production. Figure 1.2 shows, in a simplified scheme, the most important categories of organic molecules (sucrose and amino acids) which feed the meristematic cells.

The extent of sucrose supply is primarily dependent on photosynthesis, and the extent of amino acid supply depends on the concentration of available nitrogen in the nutrient medium, which under the conditions of practical crop production is usually the soil. Highest growth rates are obtained if the sucrose supply is balanced by an adequate supply of amino acids. From this it follows that under good photosynthetic conditions a high nitrogen supply is also required for optimum growth and vice versa. If the quantity of amino acids is in excess of that of sucrose, the latter is the growth limiting factor. Under such conditions soluble amino acids accumulate in the tissue and may render the tissue more susceptible to fungal attack. If the sucrose supplied is not completely balanced by amino acids the growth rate is reduced and the sugar surplus is converted to non-structural carbohydrates (Hehl and Mengel 1972).

This is the situation in many natural habitats with good photosynthetic conditions but low levels of available nitrogen in the soil. In particular, soils are gradually depleted of available nitrogen if they are cropped, and the nitrogen removed into plant parts (seeds, fruits, leaves, roots) is not restored. In rural societies substantial amounts of nitrogen are recycled back to the field in the form of urine and faeces of animals and human beings. With the beginning of industrialization in Europe in the nineteenth century, a strong nitrogen drain commenced from the farmers' fields. Nitrogen was drained directly, in the form of plant produce, and indirectly as animal

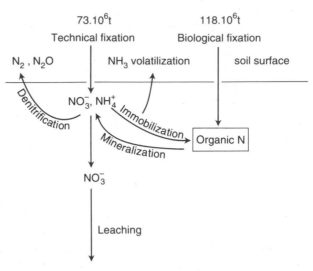

Fig. 1.3. Principal processes in global nitrogen turnover. (Technical N_2 fixation per year according to Brändlein 1987, biological N_2 fixation per year according to Delwiche 1983.)

produce sent to the towns; and from there, nitrogen in the form of urine and faeces drained into rivers, lakes and oceans via sewage systems (where they existed). Farmers' fields became more and more infertile by this process, exacerbated by population increase.

The situation improved when the technical fixation of dinitrogen became possible (see Fig. 1.3). Today the quantity of industrially fixed N_2 amounts to 73×10^6 t N per year (Brändlein 1987), which is about 60 per cent of the global biological fixation rate. This nitrogen input represents a tremendous interference in the natural nitrogen cycle, the consequences of which are as yet little understood. By far the greatest fraction of industrially fixed nitrogen is used as fertilizer nitrogen, and this gives rise to questions of whether such huge quantities of nitrogen fertilizers are required, whether alternatives exist, and about the fate of the fertilizer nitrogen incorporated into soils.

The biological productivity of nitrogen fertilizers

Two hundred years ago, before nitrogen fertilizers were known and cultivation of leguminous species with N_2-fixing bacteria was sparse in Central

Europe, one hectare of arable land produced hardly sufficient food for one person.

Today, on the same soils and under the same climatic conditions, food for almost five people is harvested from one hectare (Siemes 1979). This improvement is not solely due to nitrogen fertilizer application, but nitrogen does play a key role in this development. As outlined above, optimum crop growth is only attained if the CO_2 assimilation rate is balanced by an adequate supply of nitrogen.

If farmers were to stop using nitrogen fertilizers now, crop yields would drop during some years to a very low level, and many countries which today use substantial amounts of fertilizers and export agricultural produce would barely be in a position to feed their own populations. The data shown in Table 1.1 may help to illustrate these relationships (Odell *et al.* 1984). The results shown here originate from a long-term field experiment carried out on a fertile soil in Illinois ('Morrow Plots'). In the treatment without fertilizer, the yield was much lower than in the fertilizer treatments. Particularly low yields were obtained when maize was grown in monoculture without fertilizer. The incorporation of a leguminous species (such as soya or clover) in the rotation increased the yield to above that recorded in the treatment without fertilizer, showing that nitrogen was an important limiting factor.

Table 1.1. Effect of fertilizer and rotation on the grain yield of maize. Long-term field trial in Illinois. Yields (in t ha^{-1}) are average grain yields from 1967–1978 (after Odell *et al.* 1984)

Fertilizer	Corn monoculture	Corn/soya	Corn/oats/soya
None	1.54	2.38	2.97
Mineral NPK since 1975	4.05	4.69	4.76
Farmyard since 1904, and mineral NPK since 1955	4.12	4.64	4.79

This finding may raise the question of whether the demand for fertilizer nitrogen can be completely covered by growing leguminous species. Theoretically, this should be possible. It would, however, require more than 40 per cent of arable land to be cultivated with leguminous species, and thus a substantial proportion of arable land would be used primarily for the production of fertilizer nitrogen (Mengel 1979). This kind of biological nitrogen fertilizer production would be much more expensive than industrial production, and its environmental hazard would be at least as high as that of technically produced nitrogen fertilizer, since nitrate can be leached in large quantities from soils covered with deteriorating leguminous crops (Low and Armitage 1970). 'Biological' production of nitrogen fertilizers is also of no major significance for global energy conservation since the energy consumption for the production of fertilizers amounts to only about 1 per cent of the total energy consumption of the industrialized countries (White 1976).

The productive effect of nitrogen fertilizer is mainly based on two of its effects: nitrogen improves both plant growth and protein production in kernels of seeds and other plant parts. The first effect is a general one and is true for all crop species, while the second is particulary important for crops grown mainly for protein production such as seed legumes, forage crops, and to some degree also cereals.

The general growth effect of nitrogen is particularly important for the growth of leaves. Plants undersupplied with nitrogen have small leaves, hence the leaf area is reduced, with negative consequences for CO_2 assimilation and yield production (Evans *et al.* 1975). Ample nitrogen supply delays senescence, the leaves remain green longer, and the 'leaf area duration' is extended, with positive effects on grain yield (Woolhouse 1981). The delay of senescence under the conditions of ample nitrogen nutrition presumably results from the promoting effect of nitrogen on the synthesis of cytokinins (Wagner and Michael 1971). This example shows that nitrogen influences plant growth not only by being an elemental constituent of essential molecules such as protein and nucleic acids, but also by influencing the phytohormonal status of plants.

The fate of fertilizer nitrogen

The amount of nitrogen fertilizer which is generally applied to a crop and required for optimum crop production is not completely taken up by the crop stand. Between 30 and 80 per cent is taken up, the exact amount being dependent on various factors (Craswell and Godwin 1984). The nitrogen remaining in the soil is partially incorporated into the biomass of soils (Olfs and Werner 1989) and may be provided to following crops (Teske and Matzel 1976) after mineralization. Some fertilizer nitrogen may be leached in the form of nitrate into deeper soil layers, and some nitrate may be denitrified with the resulting N_2 and N_2O escaping into the atmosphere. Gaseous losses of fertilizer nitrogen may also occur from the volatilization of NH_3. Nitrate leaching, denitrification of NO_3^-, and volatilization of NH_3 represent a hazard to the environment, as will be discussed below in more detail.

If the crops are produced for their protein content, most of the nitrogen taken up by the crop is removed from the field with the harvested plant parts. This is the case for forage crops, numerous vegetables, seed legumes (soya, beans, and peas) and also for cereals. Olson and Swallow (1984), experimenting with ^{15}N labelled fertilizer in field trials for a period of 5 years, found that 50 per cent of fertilizer nitrogen was taken up by the wheat and 30 per cent of the fertilizer nitrogen was removed from the field by the grains. The loss of fertilizer nitrogen by leaching and/or denitrification was 15–20 per cent of the nitrogen fertilizer rate.

The situation is quite different in cases in which the plant parts are harvested for ingredients other than protein, such as for oil or sugar. A good sugar beet stand needs about 250 kg N ha$^-$ to develop an adequate leaf surface area; the roots harvested contain hardly 20 per cent of this nitrogen quantity. The residual nitrogen, mainly present in the leaves, remains on the field. The situation is similar for oil crops, such as rape and sunflowers. The nitrogen present in the crop residues may be partially mineralized and the nitrate finally produced can be leached by winter rainfall. The hazard of NO_3^- leaching is particularly true for sandy, pervious soils. On clay soils water-logging may occur and nitrate can be denitrified. Generally nitrogen losses are higher on sandy than on loamy and clay soils. Soils with particular clay minerals (vermiculite, illite) can adsorb the NH_4^+ produced by the mineralization of organic nitrogen and thus protect the nitrogen from leaching and denitrification. The NH_4^+ so fixed is available to the plant (Mengel and Scherer 1981; Li Chang-wei *et al.* 1990).

Proteins present in forage, seeds, and grains are used for animal and human nutrition. In the case of animal nutrition they are converted to animal proteins with efficiencies ranging between 6 and 30 per cent (Cooke 1975). This means that from the total protein nitrogen fed to the animal in the best case (milk production) 30 per cent is present in the animal protein, the rest is excreted by the animal, mainly in the form of urea or uric acid. The excreted nitrogen can be recycled to the field and serve as nitrogen fertilizer for the crop. In this nitrogen recycling process heavy nitrogen losses may occur, depending on farming systems. The problem will be discussed below.

Animal and plant proteins not used for animal feeding are mainly used in human nutrition. Of the nitrogen taken up by humans, on average 99 per cent is excreted in urine and faeces (Isermann 1990). In rural societies the excretions were recycled to the fields and the nitrogen in them was used by crops. Today nitrogen excreted by humans is frequently released directly into rivers or into sewage systems. Even in efficient sewage plants about 50 per cent of the soluble nitrogen delivered to the sewage plant leaves it, mainly in the form of NH_4^+ and NO_3^- and is fed into the rivers and eventually reaches the sea (Hähnel 1986).

Because of the poor efficiency of animal protein production, the total amount of nitrogen required for the production of 1 unit of plant protein is much less. This is demonstrated by a simple example: assuming a fertilizer nitrogen utilization efficiency for soya production of 50 per cent, the production of 1 tonne of soya protein requires about 400 kg nitrogen. Assuming an efficiency of about 17 per cent for pork protein production, 6 tonnes of soya protein are required for the production of 1 tonne pork protein. These 6 tonnes of soya protein require about 2400 kg nitrogen. From this quantity, about 2200 kg nitrogen remain somewhere in the environment (soil, drainage water, atmosphere) during the various steps of production.

Environmental hazards of nitrogenous compounds

Excessive nitrogen fertilizer application may lead directly to the leaching of NO_3^- into deep soil layers where eventually it may reach the ground water and aquifers which feed drinking water wells (Müller *et al.* 1985). Nitrate leaching rates are particularly high on pervious (sandy) soils which play a major role in the regeneration of drinking water (Strebel *et al.* 1985). In loamy soils excessive nitrogen fertilizer rates result in a build-up of organic nitrogen compounds in the upper soil layers which later are gradually mineralized, and the resulting NO_3^- is leached by winter rainfall (Weller 1983). In addition, regular high rates of slurry application over several years result in the enrichment of organic nitrogen in the soil profile (Scherer *et al.* 1988). Excessive fertilizer rates may lead to high NO_3^- concentrations in the plant tissue. This problem is of particular importance for parts of plants such as vegetables which are directly consumed by people.

Nitrate taken up in food or drinking water is not noxious. It can, however, be reduced to NO_2^- in the human body and thus give rise to methaemoglobinaemia. This risk is particularly high for babies which still lack a reductase which can reduce the Fe^{III} of the methaemoglobin to Fe^{II}. Until now the question of whether NO_2^- may form carcinogenic nitrosamines by reaction with amines in the human body has not been clearly answered (Owen and Jürgens-Gschwind 1986). Sander (1987) suggests that synthesis of nitrosamines is possible in the human stomach. As yet, however, no clear relationship between the intake of NO_3^- and the occurrence of cancer has been found (Forman *et al.* 1985). Until this problem is resolved NO_3^- concentrations in human food and drinking water need to be examined critically.

Nitrate-containing water percolating through the soil profile may reach anaerobic zones where denitrification of NO_3^- may occur. Lind and Pedersen (1976) reported that at a soil depth of 2.5 m there is a clear-cut border between the aerobic and anaerobic zones. In the anaerobic zone, NO_3^- is reduced to N_2 and N_2O by denitrifying bacteria which take the reducing electrons from Fe^{II} compounds and/or organic carbon (Kölle *et al.* 1983). Waters heavily charged with NO_3^- may induce an enhanced oxidation of organic carbon and/or Fe^{II} compounds so that the border between the aerobic and anaerobic zones gradually moves deeper and finally reaches the aquifer (Obermann 1984). At this stage the soil profile has lost its power to regenerate NO_3-free drinking water irreversibly.

Denitrification may occur in various environments such as soils, lakes, rivers, and also sewage plants. The denitrification products are N_2 and N_2O and hence denitrification represents a kind of recycling of gaseous nitrogen into the atmosphere. Dinitrogen (N_2) is harmless. Dinitrogen oxide (N_2O) is rather stable, having a mean time to decomposition of 20 to 100 years. It therefore reaches the stratosphere where it is involved in the decomposition of ozone (Crutzen 1991). The main reactions in this process are:

$$N_2O + O^{\cdot} \longrightarrow 2NO \tag{1.1}$$

$$NO + O_3 \longrightarrow NO_2 + O_2 \tag{1.2}$$

$$2O_3 + NO + NO_2 \xrightarrow{h\nu} 3O_2 + NO_2 + NO \tag{1.3}$$

Reactions (1.2) and (1.3) are ozone-consuming. Nitrogen dioxide (NO_2) has a short life since it easily forms HNO_3 with hydroxyl radicals:

$$NO_2 + OH^{\cdot} \longrightarrow HNO_3$$

This reaction removes NO_2 from the system and may therefore reduce the rate of ozone decomposition. Removal of NO_2 is particularly high if water droplets are present since they are a strong sink for HNO_3.

The ozone-decomposing reactions proceed at the surface of ice crystals at a low temperature, a condition which occurs in the antarctic stratosphere during spring and which is said to be responsible for the 'ozone hole' (Tolbert *et al.* 1987). It should be emphazied that it is not only NO which is involved in the ozone decomposition; radicals such as H, OH, and Cl decompose O_3 in an analogous way, the radical Cl being said to have the strongest effect. Denitrification is a biological process which has obviously occurred for millions or even billions of years and, therefore, it is unlikely that it plays the dominant role in the recent decomposition of the ozone layer. Nevertheless, critical attention is required because the atmospheric rate of increase of N_2O is currently three times higher than it was 25 years ago (Crutzen *et al.* 1985). Dinitrogen and N_2O are not produced in a constant ratio during denitrification. The amount of N_2O released by denitrifying bacteria is generally less than one tenth the amount of N_2 produced. Dinitrogen oxide is not only produced by denitrification, but also in the process of nitrification. This means that it can also be produced under aerobic conditions (Goodroad and Keeney 1985). The source for this N_2O can be organic soil nitrogen, urea, and NH_4^+-containing fertilizers.

Substantial amounts of nitrogen can be released from the soil in the form of NH_3. The release rate depends greatly on soil pH and is favoured by alkaline conditions. High release rates have been found in areas with intensive animal production (Fuhrer 1986), where large amounts of slurry are deposited on fields, with nitrogen values greater than 300 kg N ha^{-1} per year. Such quantities can hardly be absorbed by the soil and metabolized by soil microbes. Under such conditions nitrogen losses due to NO_3^- leaching, and particularly as volatile NH_3, are severe. High nitrogen losses result from the urine of grazing animals on pastures. Here the nitrogen utilization by grass is only about 20 per cent (Whitehead *et al.* 1986). During the process of slurry application NH_3 losses may occur, which may amount to 90 per cent of the NH_4^+ present in the slurry if it is not incorporated into the soil (Amberger *et al.* 1987; Döhler and Wiechmann 1987). High losses of volatile NH_3 also occur after the application of urea on flooded soils in the production of paddy rice (Schnier *et al.* 1988).

The NH_3 released from the soil into the atmosphere is brought back to the soil and vegetation by rain and fog (Sigg *et al.* 1987). The 'agricultural emission density' of NH_3 varies between 10 and 70 kg N ha^{-1} per year in western European countries. Similar NH_3 rates are recycled to the soil by precipitation (Isermann 1987). This kind of 'nitrogen fertilizer' is more or less evenly distributed on highly differing regions such as urban areas, rivers, lakes, wasteland, forests, grassland, and arable land, with a strong impact on their ecology. The NH_4^+ brought into the soil can be oxidized by soil microbes to NO_3, a process that is associated with a decrease in soil pH and which may therefore give rise to toxic levels of soluble aluminium in the soil (Roelofs *et al.* 1988). Ammonium deposition is involved in forest decline, as was reported by Roelofs *et al.* (1988) from The Netherlands and by Temmermann *et al.* (1988) from Belgium. The high rates of NH_4^+ deposition have a strong impact on the botanical composition of natural vegetation. Steubing and Buchwald (1989) reported that typical heath species such as *Calluna vulgaris* were replaced by grasses. According to Ellenberg (1985) the enrichment of soils with available nitrogen decreases the diversity of plant species.

Conclusions and consequences

From the total amount of industrially fixed nitrogen, by far the greatest part is eventually fed into the environment as NO_3^- via percolating soil water, aquifers, sewage water, rivers, and lakes. The final sink is the ocean. Some of the NO_3^- is denitrified to N_2 and N_2O. The hazards of NO_3^- and N_2O were considered above. The question of whether the current level of input of nitrogen into the environment can be reduced is answered in the affirmative. Some important measurements for attaining this target are considered below.

There is no doubt that industrially produced nitrogen fertilizer is required in order to feed the world population. As discussed above, very high nitrogen losses associated with environmental hazards occur in animal production systems with high stocking rates. The modern trend to specialized farms, i.e. farms with only animal production or only crop (plant) production, may have economic advantages, but from an environmental point of view is very destructive. In animal production systems, nitrogen is released into the environment in high quantities, while crop production systems lack nitrogen and therefore have to use mineral nitrogen in high quantities. The integration of animal production with crop production as in traditional farming is therefore required so that the nitrogen excreted by the animals can be used for plant growth. The stocking rate (number of animals per hectare of available land) should be reduced to a certain level so that the quantity of nitrogen recycled to the field can be absorbed by the soil and utilized to the optimum extent by the crop.

Plant protein production requires much less fertilizer nitrogen than animal protein production, as was shown above. The question therefore arises of whether animal proteins can be substituted by plant proteins to a greater extent. Most plant proteins have a poorer biological quality (concentration of essential amino acids) than animal proteins. With gene technology, however, it should be possible to develop cultivars of beans and peas of which the seed protein quality is as high as that of soya, which has a high protein quality. A major shift from animal proteins to high value plant proteins, however, would represent a tremendous interference in today's farming with serious consequences, particularly for small farmers. It should also be mentioned that the average protein consumption in industrialized countries is very high and a reduction is advisable for health reasons. A reduction in human protein consumption would also lead to reduced nitrogen fertilizer consumption.

A very important development for obtaining a higher efficiency of use of fertilizer nitrogen and thus for reducing the entry of nitrogen into the environment would be the controlled use of fertilizer nitrogen, where fertilizer rates are adjusted to the concentration of available nitrogen already in the soil. This demand becomes more urgent the more the soils have been enriched with available nitrogen by earlier excessive nitrogen administration. In a recent field trial carried out over several years in Hessia, Heyn and Brüne (1990) found that on average 50 kg fertilizer N ha^{-1} were required to attain the optimum sugar beet yield, a crop which needs about 200–250 kg N ha^{-1}. This example shows that on average about 70 per cent of the nitrogen required came from the soil. Steffens *et al.* (1990) reported that on wheat-growing land around Giessen in Germany the amount of available nitrogen in the soil profile in autumn varied between 25 and 220 kg N ha^{-1}. This tremendous variation shows that in most cases farmers have no idea of the amount of available nitrogen in their soils.

The adjustment of nitrogen fertilizer rates to the level of available nitrogen in the soil demands a reliable and practicable soil test method. In Germany the 'Nmin-method' has been widely encouraged, but its acceptance by farmers was poor because the soil sampling required a great deal of work. It is for this reason that of the total arable land in the former Federal Republic of Germany less than 5 per cent has been analysed for available nitrogen. In other countries the situation is similar or even worse. In recent years the electro-ultra-filtration (EUF) method has gained importance (Nemeth 1988) and the number of farmers who allowed their soils to be analyzed by EUF has increased steadily during recent years. The soil sampling technique is simple, and soil extraction by EUF not only yields values for the quantity of inorganic nitrogen present but also a value for organic nitrogen, which provides information about the quantity of organic nitrogen mineralized during a growing period (Nemeth *et al.* 1987). Horn (1990), examining the EUF soil test over several years on various sites used for maize in Southern Germany found a close correlation between the nitrogen fertilizer rates based on the EUF test and the optimum nitrogen fertilizer rates found in the field trials ($R^2 = 0.69$***). The optimum nitrogen fertilizer rates varied considerably; the average optimum fertilizer rate was 120 kg N ha^{-1}, which is about two-thirds of what is usually applied to maize.

Nitrogen fertilizer rates should on one hand be low enough for the available soil nitrogen to be exhausted by the end of the growth period and for the NO_3^- in the rooting soil in autumn not to exceed 50 kg N ha^{-1}. On the other hand, fertilizer rates must be high enough to attain a satisfactory yield.

Recent unpublished experiments carried out by the Institute of Plant Nutrition, Justus-Liebig-Universität, Giessen have shown that based on an EUF-based nitrogen fertilizer recommendation, 87–97 per cent of the optimum grain yield levels were attained and the residual NO_3^- in the soil was low in autumn in accordance with the above-mentioned level. On average the fertilizer rates recommended according to EUF were 40 kg ha^{-1} lower than the rates required for the maximum yield. This investigation shows that both ecological and economic aspects can be accommodated in nitrogen fertilization provided that a reliable soil nitrogen test is available.

References

Amberger, A., Huber J., and Rank M. (1987). Gülleausbringung: Vorsicht Ammoniakverluste. *DLG-Mitteilungen,* **20**, 1084–6.
Brändlein, W. (1987). Fertilizers, consumption, production, world trade. In *Ullmann's Encyclopedia of Industrial Chemistry*, Vol. 1. A10, pp. 414–20. VHC Verlagsgesellschaft, Weinheim.
Cooke, G.W. (1975). Sources of protein for people and livestock; the amounts now available and future prospects. In *Fertilizer use and protein production*, pp. 29–51, Proceedings of 11th Colloquium Internernational Potash Institute, Berne.

Craswell, E. T. and Godwin, D. C. (1984). The efficiency of nitrogen fertilizers applied to cereals in different climates. *Advances in Plant Nutrition*, 1, 1–55.

Crutzen, P. J., Delany, A. C., Greenberg, J., Haageson, P., Heidt, L., Lueb, R., Pollock, W., Seiler, W., Wartburg, A., and Zimmermann, P. (1985). Tropospheric chemical composition measurements in Brazil during the dry season. *Journal of Atmopheric Chemistry*, 2, 233–56.

Crutzen, P. J. (1991). Global changes in atmospheric chemistry. *Tellus* (in press).

Delwiche, C. C. (1983). Cycling of elements in the biosphere. In *Inorganic plant nutrition* (ed. A. Läuchli and R. L. Bieleski). In *Encyclopedia of Plant Physiology*, New Series, Vol 15A, pp. 212–38. Springer Verlag, Berlin.

Döhler, H. and Wiechmann, M. (1987). *Ammonia volatilization from liquid manure after application in the field*. 4th International CIEC-Congress (ed. E. Welte and I. Scabolcs), pp. 305–13. Belgrade, Göttingen, Vienna.

Ellenberg, H. (1985). Veränderungen der Flora Mitteleuropas unter dem Einfluß von Düngung und Emmissionen. *Schweizer Zeitung Forstwesen*, 136, 19–39.

Evans, L. T., Wardlaw, I. F., and Fischer, R. A. (1975). Wheat. In *Crop physiology* (ed. L. T. Evans), pp. 101–49. Cambridge University Press.

Forman, D., Al-Dabbagh, S., and Doll, R. (1985). Nitrates, nitrites and gastric cancer in Great Britain. *Nature*, 313, 620–25.

Fuhrer, J. (1986). Chemistry of fogwater and estimated rates of occult deposition in an agricultural area of Central Switzerland. *Agriculture, Ecosystems and Environment*, 17, 153–64.

Goodroad, L. L. and Keeney, D. R. (1985). Site of nitrous oxide production in field soils. *Biology and Fertility of Soils*, 1, 3–7.

Hähnel, K. (1986). *Biologische Abwasserreinigung mit Belebtschlamm*. Volkseigener Betrieb G. Fischer Verlag, Jena.

Hehl, G. and Mengel, K. (1972). Der Einfluß einer variierten Kalium- und Stickstoffdüngung auf den Kohlenhydratgehalt verschiedener Futterpflanzen. *Landwirtschaftliche Forschung, Sonderheft*, 27/II, 117–29.

Heyn, J. and Brüne, H. (1990). Ein Vergleich zwischen N-Düngeempfehlungen nach Nmin und EUF-Bodenuntersuchungen anhand hessischer Feldversuche. *Verband Deutscher Landwirtschaftliche Untersuchungs-Schriftenreihe Kongressband* 1989, 30, 195–200.

Horn, D. (1990). Bedeutung des EUF-extrahierbaren Stickstoffs im Boden für die Ermittlung des Stickstoffdüngebedarfs von Mais. Unpublished Ph.D. thesis. Agricultural Faculty of the Justus-Liebing-University, Giessen.

Isermann, K. (1987). Environmental aspects of fertilizer application. In *Ullmann's Encyclopedia of Industrial Chemistry*, Vol A 10, pp. 400–9. VHC Verlagsgesellschaft, Weinheim.

Isermann, K. (1990). Share of agriculture in nitrogen and phosphorus emissions into the surface waters of Western Europe against the background of their eutrophication. *Fertilizer Research*, 26, 253–69.

Kölle, W., Werner, P., Strebel, O., and Böttcher, J. (1983). Denitrifikation in einem reduzierenden Grundwasserleiter. *Vom Wasser*, 61, 125–47.

Li Chang-wei, Fan Xiao-lin, and Mengel, K. (1990). Turnover of interlayer ammonium in loess derived soil grown with winter wheat in the Shaanxi Province of China. *Biology and Fertility of Soils*, 9, 211–14.

Lind, A. M. and Pedersen, M. B. (1976). Nitrate reduction in the subsoil. II. General description of boring profiles, and chemical investigations on the profile cores. *Tidsskrift Planteavl,* **80,** 82–99.

Low, A. J. and Armitage, E. R. (1970). The composition of the leachate through cropped and uncropped soils in lysimeters compared with that of the rain. *Plant and Soil,* **33,** 393–411.

Mengel, K. (1979). Pflanzenbau ohne Mineraldüngung, eine Alternative? *Kali-Briefe (Büntehof),* **14** (10), 707–11.

Mengel, K. and Scherer, H. W. (1981). Release of nonexchangeable (fixed) soil ammonium under field conditions during the growing season. *Soil Science,* **131,** 226–32.

Miflin, B. J. and Lea, P. J. (1977). Amino acid metabolism. *Annual Review of Plant Physiology,* **28,** 299–329.

Müller, W., Gärtel, W., and Zakosek, H. (1985). Nährstoffauswaschung aus Weinbergsböden an der Mittelmosel. *Zeitschrift für Pflanzenernährung und Bodenkunde,* **148,** 417–28.

Nemeth, K. (1988). Wissenschaftliche Grundlagen der EUF-Stickstoffempfehlung zu Getreide und Hackfrüchten. EUF-Symposium, Vol 1, pp. 11–46. Mannheim.

Nemeth, K., Maier, J., and Mengel, K. (1987). EUF-extrahierbarer Stickstoff und dessen Beziehung zu Stickstoffaufnahme und Ertrag von Weizen. *Zeitschrift für Pflanzenernährung und Bodenkunde,* **150,** 369–74.

Oaks, A. and Hirel, B. (1985). Nitrogen metabolism in roots. *Annual Review of Plant Physiology,* **36,** 345–65.

Obermann, P. (1984). Möglichkeiten des Nitratabbaues im Sicker- und Grundwasser. *Gewässerschutz, Wasser, Abwasser,* **65,** 577–91.

Odell, R. T., Melsted, S. W., and Walker, W. M. (1984). Changes in organic carbon and nitrogen of Morrow plot soils under different treatments, 1904–1973. *Soil Science,* **137,** 160–71.

Olfs, H. W. and Werner, W. (1989). Veränderungen extrahierbarer 'Norg'-Mengen unter dem Einfluß variierter C/N-Verhältnisse und Biomasse. *VDLUFA-Schriftenreihe,* **28,** 15–26.

Olson, R. V. and Swallow, C. W. (1984). Fate of labeled nitrogen fertilizer applied to winter wheat for five years. *Soil Science Society of America Journal,* **48,** 583–6.

Owen, T. R. and Jürgens-Gschwind, S. (1986). Nitrates in drinking water: a review. *Fertilizer Research,* **10,** 3–25.

Roelofs, J. G. M., Boxma, A. W., and van Dijk, H. F. G. (1988). Effects of airborne ammonium on natural vegetation and forests. In *Air pollution and ecosystems* (ed. P. Mathy), pp. 876–80. D. Reidel Publishing Company, Dordrecht.

Sander, J. (1987). Endogene Nitrosaminentstehung. In *DFG-Mitteilungen,* **III,** 149–56.

Scherer, H. W., Werner, W., and Kohl, A. (1988). Einfluß langjähriger Gülledüngung auf den Nährstoffhaushalt des Bodens. 1. Mitteilung: N-Akkumulation und N-Nachlieferungsvermögen. *Zeitschrift für Pflanzenernährung und Bodenkunde,* **151,** 57–61.

Schnier, H. F., DeDatta, S. K., Mengel, K., Marqueses, E. P., and Faronilo, J. E. (1988). Nitrogen use efficiency, floodwater properties, and nitrogen-15 balance in transplanted lowland rice as affected by liquid urea band placement. *Fertilizer Research*, **16**, 241–55.

Siemes, J. (1979). Ernährungssicherung durch Mineraldüngung. *Pflug und Spaten*, **6**, 2.

Sigg, L., Stumm, W., Zobrist, J., and Zürcher, F. (1987). The chemistry of fog; factors regulating its composition. *Chimica*, **41**, 159–65.

Steffens, D., Barekzai, A., Bohring, J., and Poos, F. (1990). Die EUF—löslichen Stickstoffgehalte in Ackerböden des Landkreises Giessen. *Agribiological Research*, **43**, 319–29.

Steubing, L. and Buchwald, K. (1989). Analyse der Artenverschiebungen in der Sand-Ginsterheide des Naturschutzgebietes Lüneburger Heide. *Natur und Landschaft*, **64**, 100–5.

Strebel, O., Böttcher, J., and Duynisveld, W. H. M. (1985). Einfluß von Standortbedingungen und Bodennutzung auf Nitratauswaschung und Nitratkonzentration des Grundwassers. *Landwirtschaftliche Forschung Kongressband*, 1984, **37**, 34–44.

Temmermann, L. de, Ronse, A., van den Cruys, K., and Meeus-Verdine, K. (1988). Ammonia and pine tree dieback in Belgium. In *Air pollution and ecosystems* (ed. P. Mathy), pp. 774–9. D. Reidel Publishing Company, Dordrecht.

Teske, W. and Matzel, W. (1976). Stickstoffauswaschung und Stickstoffausnutzung durch die Pflanzen in Feldlysimetern bei Anwendung von 15-N-markiertem Harnstoff. *Archiv für Acker- und Pflanzenbau und Bodenkunde*, **20**, 489–502.

Tolbert, M. A., Rossi, M. J., Malhotra, R., and Golden, D. M. (1987). Reaction of chlorine, nitrate, hydrogen, chloride and water at antarctic stratospheric temperature. *Science*, **238**, 1258–60.

Wagner, H. and Michael, G. (1971). Der Einfluß unterschiedlicher Stickstoffversorgung auf die Cytokininbildung in Wurzeln von Sonnenblumenpflanzen. *Biochemie und Physiologie der Pflanzen (BPP)*, **162**, 147–58.

Weller, F. (1983). Stickstoffumsatz in einigen obstbaulich genutzten Böden Südwestdeutschlands. *Zeitschrift für Pflanzenernährung und Bodenkunde*, **146**, 261–70.

Werner, D. (1980). Stickstoff (N_2)-Fixierung und Produktionsbiobiologie. *Angewandte Botanik*, **54**, 67–75.

White, D. J. (1976). Energy use in agriculture. In *Aspects of energy conversion* (ed. B. Jones and van Horn), pp. 141–76. Pergamon Press, Oxford.

Whitehead, D. C., Pain, B. F., and Ryden, J. C. (1986). Nitrogen in UK grassland agriculture. *Journal of the Royal Agricultural Society, England*, **147**, 190–201.

Woolhouse, H. W. (1981). Crop physiology in relation to agricultural production: the genetic link. In *Physiological processes limiting plant productivity* (ed. C. B. Johnson), pp. 1–21. Butterworth, London.

2. Nodule development and nitrogen fixation in the *Rhizobium/Bradyrhizobium* system

D. WERNER, B. AHLBORN, S. BASSARAB,
R. KAPE, A. KINNBACK, R.B. MELLOR,
E. MÖRSCHEL, P. MÜLLER, M. PARNISKE,
P. SCHMIDT, and A. SCHULTES

Fachbereich Biologie, Philipps-Universität Marburg, Karl von Frisch Strasse, D-3550 Marburg, Germany

Introduction

Various aspects of legume nodule development and function have recently been reviewed by Rolfe and Gresshoff (1988), Dazzo *et al.* (1988), Hennecke *et al.* (1988), Long and Cooper (1988), Lugtenberg *et al.* (1989), Appelbaum (1990), Mellor and Werner (1990), and Surin *et al.* (1990). The subject of nodulins in root nodule development will be covered in detail in this chapter and in Chapter 3, this volume.

Some highlights and new concepts resulting from the intensive research of several laboratories are:

1. Flavonoids have been found to possess an essential role in developmental biology, in the communication between host plants and microsymbionts.

2. In the morphogenesis of the new plant organ 'nodules' the synthesis of more than 30 new specific proteins (nodulins) is induced. At least 7 of these nodulins are associated with a symbiotic structure, the peribacteroid membrane.

3. In later phases of the symbiosis, the symbiotic interaction can be reversed to a parasitic interaction by changes in the genotype of the host plant as well as specific mutations in the microsymbiont.

4. The ecological significance of symbiotic nitrogen fixation has been further increased by the characterization of new genera of the microsymbionts: *Sinorhizobium*, fast growing soybean nodule-inducing bacteria from China; and *Azorhizobium*, stem nodulating bacteria from Africa.

5. New nitrogenases in *Azotobacter*: a vanadium nitrogenase and a Fe nitrogenase have re-stimulated biochemistry and also the genetics of N_2-fixation in general.

The development of nodules — an overview

The most obvious result of an effective legume nodule development can be demonstrated under field conditions with a very low endogenous effective *Rhizobium/Bradyrhizobium* population in the soil. In general, inoculation experiments to demonstrate this effect are more successful in subtropical and tropical soils than in temperate regions.

The two cell types communicating during the first steps of nodule development are rhizodermis cells, especially root hairs from the host plant and the free-living *Rhizobium/Bradyrhizobium* cells. The major cytological events are well documented. However, the biochemical and genetical basis of the communication is still rather incomplete. The four major steps are:

(1) recognition and root hair curling;

(2) infection sack and infection thread development together with meristem formation in the root cortex;

(3) differentiation of all major host cell organelles together with the symbiosome formation;

(4) establishment of the major biochemical functions of effective nodules: carbon supply, nitrogenase activity, oxygen protection mechanisms, ammonia assimilation and transport systems.

Nodule development begins with the infection of a root hair or, less often, some other root surface component. The first morphologically recognizable reaction of the plant is the curling of the root hair (Turgeon and Bauer 1985). When soybean is used as a model, the next steps are that the bacteria break through the slime layer and are then pressed against the root hair cell wall or the cell wall of a nearby epidermis cell through the curling action of the root hair. The cell wall is then subjected to local dissolution and curls in upon itself bringing the bacterium with it. This is the beginning of the infection thread formation.

The infection sack then grows intracellularly in the root hair. At this point the bacteria are embedded in an electron translucent matrix material (Turgeon and Bauer 1985). Around this is a net of cell wall fibrils which are in continuum with the regularly ordered fibrils of the cell wall. These are divided from the cell cytoplasm of the root hair cell by a membrane which is in continuum with the plasma membrane.

In the surrounding cell cytoplasm a massive increase in the amount of Golgi, ER, and microtubules is evident. The infection sack branches out in the form of tubes or threads. The cortical cells at this stage have a meristematic character, with a small per cent cell volume being taken up with vacuole, a large cell nucleus and clear nucleoli. Host cells further away from the infection thread have much larger central vacuoles. The root hair cell itself remains vacuolated and the increase in dense cytoplasm and cellular organelles is localized around the infection sack. The tips of the growing infection threads are surrounded only by a very thin layer of cell wall material.

Seventy two hours after inoculation, successful infection threads enter the cortex. The favoured path leads between two newly divided cells. Twelve hours after inoculation the first cell divisions can be observed in the hypodermis: 4–8 daughter cells are formed through anticline division. After 48 h cell division in the cortex is discernible. At this time a small group of the darker-coloured meristematic cells are seen.

Three to four days after inoculation the meristematic zone is so well developed that a morphologically recognizable swelling is noticeable on the root surface. The connection between the nodule meristem and the vascular system of the root is made six days after inoculation (Calvert *et al.* 1984).

Parallel to this typical sequence of infection events is the occurence of many 'pseudo-infections'. In these cases initiation of cortical cell division is observed without the formation of an infection thread. Cell division is restricted to the outermost two or three layers of cells. This does not lead to a macroscopically observable nodule. Sometimes in the case of successful infection thread formation the growth of the thread preceeds cell division. That infection thread growth and cortical cell division are not coupled is also proven by experiments with sym-plasmid-containing *Agrobacterium tumefaciens* or *Escherichia coli*, where cell division without infection is observed. In these cases, however, the cell division does result in macroscopically observable nodule swellings, even though no bacteria are inside. Thus signal substances have been postulated to occur, although it is unknown if this is really the case.

Three days after inoculation with *Bradyrhizobium japonicum* on a 7 cm-long piece of root 80 pseudo-infections take place, along with 50 root hair infections. Such roots normally exhibit five nodules. Consequently, where infection threads are initiated, over 90% of these are abortive.

When successful infection threads reach the meristematic cortex, bacteria bud off into these cells. This first invagination is bordered by a membrane in continuum with the host cell plasma membrane. The individual bacteria are later enclosed in the 'peribacteroid membrane'.

This process is influenced by bacterial genes. Some mutant *Bradyrhizobium japonicum* strains are not set free from the infection thread. At the time of release the meristematic cells are not polyploid, since cells in nodules where bacteria are not intracellularly released are not polyploid. Thus the question arises must cells be: polyploidy for infection to occur or do infected cells become polyploid? For soybean and clover the latter may be the case.

The degree of ploidy rises in *Lathyrus, Medicago, Trifolium,* and *Vicia* from 2 n in the cortex cells to tetraploidy in infected cells. In *Pisum sativum* in nodule cells the levels of ploidy are increased to 8 or 16 n.

Legume nodules contain high concentrations of auxins, kinins and gibberellins. Auxin (indoleacetic acid, IAA) can reach levels in pea and lucerne nodules 40–60 times higher than in roots. *Rhizobium* can metabolize exogenous tryptophan to auxin.

Whether these high auxin levels are plant-or bacteria-derived is not known. IAA oxidase activity in nodules is dramatically reduced after infection. Thus the high IAA concentrations (10^{-7}–10^{-6} g IAA per g nodule fresh weight) may simply be due to reduced degradation. Cytokinin in nodules of *Phaseolus vulgaris* reaches 1.5×10^{-7} g kinetin equivalents per g nodule fresh weight. In *Vicia faba* nodules this is 10–15 times higher than in roots. Pure cultures of *Rhizobium leguminosarum* produce cytokinin to an extracellular concentration of 10^{-9} g ml^{-1}. LPS from *Rhizobium* also contains cytokinin (up to 2×10^{-7} g per g LPS). It is, however, not known if cytokinin production by bacteria in the infection thread stimulates cell division. The concentration of gibberellin in nodules is also 40–100 times higher than in root tissue. *Rhizobium leguminosarum* can also export various gibberellins.

The involvement of these three phytohormones in the various steps of nodule differentiation can only be explained when the location of the biosynthetic and degradative pathways are known. The effects of abscissic acid and ethylene on nodules is not known apart from that they inhibit nodule development in concentrations too low to inhibit secondary root formation. Nodule form and morphology falls basically into two categories. Type A has an apical meristem without a determined growth and a cylindrical nodule. These occur on, for example *Vicia* and *Trifolium*. These nodules have a branched vascular system, the uppermost end being connected to the meristematic zone. These nodules are biochemically characterized by the N-transport form being predominantly amides (asparagine and also glutamine). The nodules also have transport cells in the pericyclic tissue.

The second type of nodule (Pate and Atkins 1983) is spherical, with limited growth. The vascular branches encircle the infected tissues and are connected to the root vascular bundle at the base. Transfer cells have not been reported here. The form of nitrogen transport in such nodules is ureides such as allantoin and allantoic acid. This type of nodule is characteristic of *Phaseolus* and *Glycine*. It is postulated that a relationship between nodule type and N-transport form exists. One explanation could be that since ureides are less soluble than amino acids they can be better transported in the closed vascular system of a type B nodule. Nodules of other legumes, e.g. *Aotus ericoides*, lie between the two types. Cross-inoculation experiments prove nodule morphology to be plant-determined (Allen and Allen 1981).

Nodules developing on soybean root fall into several discontinuous size classes (Stripf and Werner 1980). Independently of their age some size classes are favoured over others (classes 2 mm, 2.6 mm, and 3.3 mm). Size classes between these are badly represented. It is not known whether in these classes optimal nutritional conditions are reached. It is, however, evident that nodule size and nodule age are independent factors. Young, big nodules can be present along with old, small ones. The number of nodules developing on a root system is influenced by nitrogen. In ineffective (no N_2-fixation) symbioses over 100 nodules per root system can be observed. In effective (N_2-fixing) symbioses 20–30 nodules in 50 days is the average (Werner 1990).

The phenyl–propane communication concept

Based on the results of several laboratories in the UK, The Netherlands, France, Switzerland, the United States, Australia, Germany, and other countries, the development of the symbiosis has been transferred into a signal-exchange concept between microsymbiont and host plant.

On the biochemical basis of flavonoids and related compounds the host plant turns on specific genes (common nodulation genes and host specific nodulation genes) and the bacteria on the other side turn on the synthesis of other flavonoids in the plant cells. This means that a large group of secondary plant metabolites has come into the main focus of molecular developmental biology. The other stages, such as attachment to the plant root hair surface, root hair curling and branching, and infection thread formation are still black boxes as far as biochemistry and cell genetics are concerned. The later stages of communication in nodule development and nitrogen fixation in the nodule are much better studied and known in much more detail. The orientation and the sizes of the nodulation genes in *Rhizobium leguminosarum* are given in Fig. 2.1.

We see the central role of the *nod*D gene product which in the presence of flavonoids such as naringenin induces the operons of all the other nod genes. The mechanism is, that all other operons are preceeded by a conserved

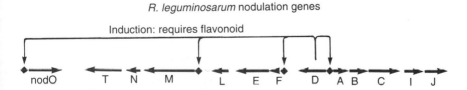

Fig. 2.1. *Rhizobium leguminosarum* nodulation genes. (from Economou *et al.* 1990 and J. A. Downie, personal communication 1990).

DNA sequence, the 'nod box' (Kondorosi *et al.* 1988).

The communication between plants and microsymbionts by flavonoids exceeds the specific nodulation effects if we look to the precursors and to the following compounds (Fig. 2.2). A very rapid and significant accumulation of phytoalexins (glyceollin I) after symbiotic infection has been demonstrated by Parniske *et al.* (1988). If we look now for the effects of

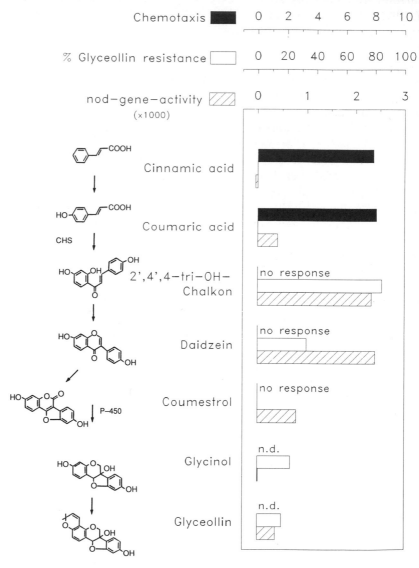

Fig. 2.2. Effects of intermediates in glyceollin biosynthesis on symbiotic *Bradyrhizobium* markers.

the produced glyceollin on the compatible *Bradyrhizobium japonicum*, we observe a resistance to glyceollin (Parniske *et al.* 1991). If we look for the precursors of flavonoids and phytoalexins, we can observe that the cinnamic acid derivatives have a strong chemotaxic effect on *Bradyrhizobium japonicum*, whereas flavonoids such as coumestrol, daidzein, and genistein, nodulation gene-inducing substances, have no chemotactic effect (Kape *et al.* 1991). When we summarize the results concerning comprehensive communication between host plant and microsymbiont and those of the biosynthetic pathway of phenylpropane compounds (Fig. 2.2), we can see that from chemotaxis to nodulation to phytoalexin resistance the specific substances in the pathway have very distinct effects in the communication. Genistein, a major nodulation gene-inducing substance was also found as another major flavonoid in soybean cell suspension cultures, but only in trace amounts in intact plants (Zacharius and Kalan 1990).

On the bacterial side, besides the nodulation genes other gene families are important for the development of symbiosis: these are the exopolysaccharide genes (Müller *et al.* 1988a, b), the DCT (Dicarboxylic transport) genes (Ronson 1988), the *hup* genes (Evans *et al.* 1987) and of course the *nif* genes, homologous to the *nif* gene cluster in *Klebsiella pneumoniae* and the *fix* genes, additional genes in rhizobia without homology to the *nif* gene cluster in *Klebsiella pneumoniae* (Gubler and Hennecke 1988; Gubler *et al.* 1989; Kullik *et al.* 1989; Martinez *et al.* 1990).

The symbiosome

The concept of the symbiosome as a new cell organelle-like compartment in infected cells was proposed by G. Stacey (Evans *et al.* 1989). It includes the peribacteroid membrane, the peribacteroid space and the bacteroid. The complete structure is essential for effective symbiotic N_2-fixation. The concept of communication between host plant and symbiont is also essential for this stage of interaction. The evidence that the peribacteroid membrane is a symbiotic structure can be summarized as follows:

Table 2.1. Stimulation of some membrane-building enzyme activities

Nodules infected with *Rhizobium japonicum* strain	Choline kinase	Choline phospho-transferase	GDP–DMP mannosyl-transferase	UDP–ASGF galactosyl-transferase	UDP–ASGF Nacetyl-galactosamine transferase
61-A-101	350	300	300	780	1600
RH 31-Marburg	80	120	280	800	800
61-A-24	50	90	180	555	570

Figures given are per cent stimulation over root tissue (= %). From Mellor *et al.* (1986).

To characterize the specific proteins of the peribacteroid membrane it was essential to improve the isolation procedures in separating this membrane from other membrane systems in the host cells (Bassarab *et al.* 1986; Mellor and Werner 1986, 1987). With these improved methods the isolation of intact symbiosomes on a preparative scale as well as the isolated peribacteroid membrane in soybeans was achieved. The evidence that specific genes in the microsymbionts are essential for biosynthesis stability, and protein (nodulin) composition was established with several mutants in the system of *Bradyrhizobium japonicum/Glycine max* (Werner *et al.* 1988, Mellor *et al.* 1989). Results with these mutants indicate that at least four different signals (genes) are essential for a complete set of proteins in the peribacteroid

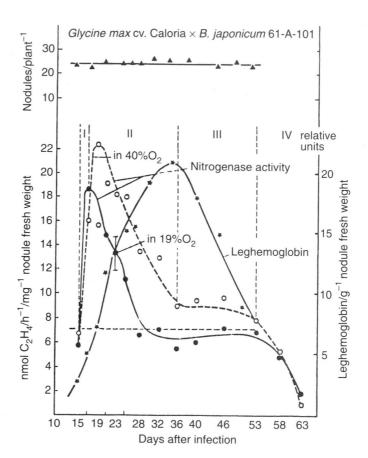

Fig. 2.3. Nitrogenase activity under 19 and 40 per cent O_2 and leghaemoglobin concentration in synchronous nodules of *Glycine max* infected with *Bradyrhizobium japonicum* 61-A-101 (From Werner and Krotzky 1983; Werner 1990).

membrane and the stability of the system. The plant response to these specific signals results in the stimulation of several membrane-building enzyme activities. Effective (N_2-fixing) symbiosis can be discriminated from ineffective (non N_2-fixing) symbiosis, especially by the activities of choline kinase, choline phosphotransferase and UDP-ASGF-N-acetyl-galactosaminetransferase (Table 2.1).

The decisive physiological function of the peribacteroid membrane was also proven by the experimental evidence that the instability of this membrane system lead to a hypersensitive reaction and phytoalexin accumulation in high concentrations also in symbiotic organs (Werner *er al.* 1985; Parniske *et al.* 1990).

N_2-fixation: activity and regulation

The oxygen limitation in effective legume nodules can be easily demonstrated within synchronous nodules (a constant number of nodules per plant during development) by incubating the root system under 19 and 40 per cent oxygen respectively (Fig. 2.3). The oxygen enhancement effect is very obvious in Stages II and III of the nodule development, that is 15–53 days

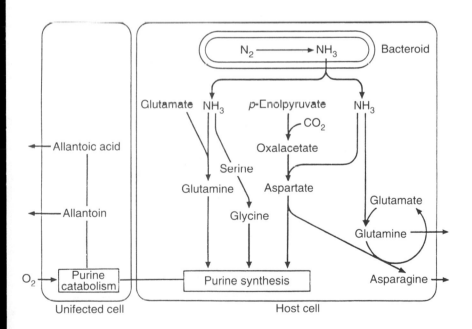

Fig. 2.4. Carbon and nitrogen metabolism in the host cell and non-infected cells of legume in nodules (modified from Dilworth and Glenn 1984).

after infection. Within synchronous nodules it is also obvious that nitrogenase activity and leghaemoglobin concentration in the nodules can be clearly separated. The results for the host plant nitrogen metabolism are summarized in Fig. 2.4. Exported from the host cell are glutamine or asparagine or, for example, in the case of soybeans, purines which are catabolized in the non-infected cell to allantoin and allantoic acid.

The essential role of oxygen concentration for *nif* gene regulation can be summarized by comparing the current models for *nif* regulation in *Klebsiella pneumoniae* (Fig. 2.5) and *Bradyrhizobium japonicum* (Fig. 2.6). In *Klebsiella* we have two pairs of regulatory proteins for the control of *nif* transcription: the NTRBC protein and the *nif*A and *nif*L proteins. Transcription of *nif*A and *nif*L are activated by phosphorylated NTRC protein. High levels of oxygen inhibit the expression of the *nif*A and *nif*L promoter. Under low oxygen concentrations and under nitrogen limitation *nif*A activates the other *nif* genes, e.g. *nif* KDH. Already at 0.1 μM concentrations we find an inhibition of de-repression of the other *nif* operons via the *nif*L product. At 6 μM concentrations of oxygen, nitrogenase activity is inhibited. This means we have at least three levels of oxygen concentration involved in gene regulation and nitrogenase activity regulation.

In *Bradyrhizobium japonicum* the regulation of oxygen is different from *Klebsiella* and also apparently different from *Rhizobium meliloti* (Thöny

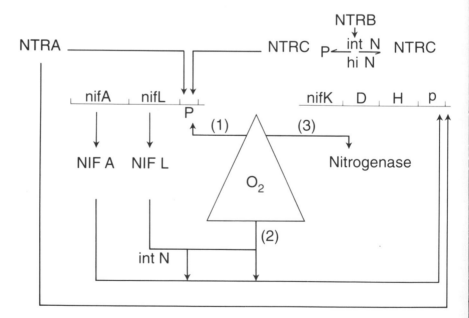

Fig. 2.5. Current model for *nif* regulation in *Klebsiella pneumoniae* (from Hill *et al.* 1988).

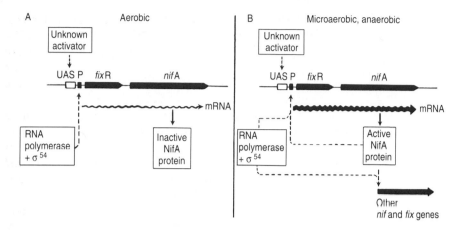

Fig. 2.6. *nif* and *fix* gene regulation under (a) aerobic and (b) microaerobic and anaerobic conditions in *Bradyrhizobium japonicum* (from Thöny *et al.* 1989; H. Hennecke, personal communication 1989).

et al. 1989). The *fix*R–*nif*A operon is activated by a sigma factor and an additional unknown activator. Under aerobic conditions it is expressed at moderate levels. However, the gene product is inactive. Under microaerobic or anaerobic conditions, the *nif*A protein remains active and serves itself as a transacting factor, enhancing its own synthesis to high levels of mRNA and activity. It follows the activation of the *nif* and *fix* genes by the active *nif*A protein.

References

Allen, O.N. and Allen, E.K. (1981). *The leguminosae*. Macmillan, London.

Appelbaum, E. (1990). In *Molecular biology of symbiotic nitrogen fixation* (ed. P.M. Gresshoff), p. 131. CRC Press, Boca Raton, Florida.

Bassarab, S., Mellor, R.B., and Werner, D. (1986). Evidence for two types of Mg^{++}-ATPase in the peribacteroid membrane from *Glycine max* root nodules. *Endocytobiosis and Cell Research*, **3**, 189–96.

Calvert, H.E., Pence, M.K., Pierce, M., Malik, N.S.A., and Bauer, W.D. (1984). Anatomical analysis of the development and distribution of *Rhizobium* infections in soybean roots. *Canadian Journal of Botany*, **62**, 2375–84.

Dazzo, F., Hollingsworth, R., Philip-Hollingsworth, S., Robeles, M., Olen, T., Salzwedel, J., Djordjevic, M., and Rolfe, B. (1988). In *Nitrogen fixation: hundred years after* (ed. H. Bothe, F.J. de Bruijn, and W.E. Newton), p. 431. Gustav Fischer, Stuttgart.

Dilworth, M. and Glenn, A. (1984). How does a legume nodule work? *Trends in Biochemical Sciences*, **9**, 519–23.

Economou, A., Hamilton, W. D. O., Johnston, A. W. B., and Downie, J. A. (1990). The *Rhizobium* nodulation gene nodO encodes a Ca^{2+}-binding protein that is exported without N-terminal cleavage and is homologous to haemolysin and related proteins. *The EMBO Journal*, 9, 349–54.

Evans, H. J., Harker, A. R., Papen, H., Russel, S. A., Hanus, F. J., and Zuber, M. (1987). Physiology, biochemistry and genetics of the uptake hydrogenase in rhizobia. *Annual Review of Microbiology*, 41, 335–61.

Evans Roth, L. and Stacey, G. (1989). Bacterium release into host cells nitrogen-fixing soybean nodules: the symbiosome membrane comes from three sources. *European Journal of Cell Biology*, 49, 13–23.

Gubler, M. and Hennecke, H. (1988). Regulation of the fixA gene and fixBC operon in *Bradyrhizobium japonicum*. *Journal of Bacteriology*, 170, 1205–14.

Gubler, M., Zürcher, T., and Hennecke, H. (1989). The *Bradyrhizobium japonicum* fixBCX operon: identification of fixX and of a 5′ mRNA region affecting the level of the fixBCX transcript. *Molecular Microbiology*, 3, 141–8.

Hennecke, H., Fischer, H.-M., Gubler, M., Thöny, B., Anthamatten, D., Kullik, I., Ebeling, S., Fritsche, S., and Zürcher, T. (1988). In *Nitrogen fixation: hundred years after* (ed. H. Bothe, F. J. de Bruijn, and W. E. Newton), p. 339. Gustav Fischer, Stuttgart.

Hill, S., Kavanagh, E. P., Arnott, M., Sidoti, C., Coppard, J. R., Merrick, M. J., Henderson, N. C., Austin, S. A., Dixon, R. A., Smith, A. T., and Anthony, C. (1988). *Annual report*, p. 114. AFRC Institute of Plant Science Research, Norwich.

Kape, R., Parniske, M., and Werner, D. (1991). Chemotaxis and *nod* gene activity of *Bradyrhizobium japonicum* in response to hydroxycinnamic acids and isoflavonoids. *Applied and Environmental Microbiology*, 57, 316–19.

Kondorosi, E., Gyuris, J. Schmidt, J., John, M., Duda, E., Schell, J. and Kondorosi, A. (1988). In *Molecular genetics of plant-microbe interactions* (ed. R. Palacios and D. P. S. Verma), p. 73. APS Press, St. Paul, Minnesota.

Kullik, I., Hennecke, H., and Fischer H.-M. (1989). Inhibition of *Bradyrhizobium japonicum* nifA-dependent nif gene activation by oxygen occurs at the nifA protein level and is irreversible. *Archives of Microbiology*, 151, 191–7.

Long, S. R. and Cooper, J. (1988). In *Molecular genetics of plant-microbe interactions* (ed. R. Palacios and D. P. S. Verma), p. 163. APS Press, St. Paul, Minnesota.

Lugtenberg, B. J. J. (1989). *Signal molecules in plants and plant-microbe interactions*, NATO ASI Series H, Cell biology 36. Springer Verlag, Berlin.

Martinez, E., Romero, D., and Palacios, R. (1990). The *Rhizobium* genome. *Plant Sciences*, 9, 59–93.

Mellor, R. B. and Werner, D. (1986). The fractionation of *Glycine max* root nodule cells: A methodological overview. *Endocytobiosis and Cell Research*, 3, 317–36.

Mellor, R. B. and Werner, D. (1987). Peribacteroid membrane biogenesis in mature legume root nodules. *Symbiosis*, 3, 75–100.

Mellor, R. B., Christensen, T. M. I. E., and Werner, D. (1986). Choline kinase II is present only in nodules that synthesize stable peribacteroid membranes. *Proceedings of the National Academy of Sciences, USA*, 83, 659–63.

Mellor, R. B., Garbers, C., and Werner, D. (1989). Peribacteroid membrane nodulin gene induction by *Bradyrhizobium japonicum* mutants. *Plant Molecular Biology*, **12**, 307–15.

Mellor, R. B. and Werner, D. (1990). In *Molecular biology of symbiotic nitrogen fixation* (ed. P. M. Gresshoff), p. 111. C R C Press, Boca Raton, Florida.

Müller, P., Hynes, M., Kapp, D., Niehaus, K., and Pühler, A. (1988a). Two classes of *Rhizobium meliloti* infection mutants differ in exopolysaccharide production and in coinoculation properties with nodulation mutants. *Molecular and General Genetics*, **211**, 17–26.

Müller, P., Enenkel, B., Hillemann, A., Kapp, D., Keller, M., Quandt, J., and Pühler, A. (1988b). In *Molecular genetics of plant-microbe interactions* (ed. R. Palacios and D. P. S. Verma), p. 26. A P S Press, St. Paul, Minnesota.

Pate, J. S. and Atkins, C. A. (1983). In *Nitrogen fixation*, Vol. 3 (ed. W. J. Broughton), p. 245. Clarendon, Oxford.

Parniske, M., Pausch, G., and Werner, D. (1988). In *Nitrogen fixation: hundred years after* (ed. H. Bothe, J. F. de Bruijn, and W. E. Newton), p. 466. Gustav Fischer, Stuttgart.

Parniske, M., Zimmermann, C., Cregan, P., and Werner, D. (1990). Hypersensitive reaction of nodule cells in *Glycine max* x *Bradyrhizobium japonicum* symbiosis occurs at the genotype specific level. *Botanica Acta*, **104**, 143–8.

Parniske, M., Ahlborn, B., and Werner, D. (1991). Isoflavonoid-inducible resistance to the phytoalexin glyceollin in soybean rhizobia. *Bacteriology*, **173**, 3432–9.

Rolfe, B. and Gresshoff, P. M. (1988). Genetic Analysis of legume nodule initiation. *Annual Review of Plant Physiology*, **39**, 297–319.

Ronson, C. W. (1988). In *Nitrogen fixation: hundred years after* (ed. H. Bothe, J. F. de Bruijn, and W. E. Newton), p. 547. Gustav Fischer, Stuttgart.

Stripf, R. and Werner, D. (1980) Development of discontinuous size classes of nodules of *Glycine max*. *Zeitschrift für Naturforschung*, **35c**, 776–82.

Surin, B. P., Watson, J. M., Hamilton, W. D. O., Economou, A., and Downie, J. A. (1990). Molecular characterization of the nodulation gene, *nod*T, from two biovars of *Rhizobium leguminosarum*. *Molecular Microbiology*, **3**, 245–52.

Thöny, B., Anthamatten, D., and Hennecke, H. (1989). Dual control of the *Bradyrhizobium japonicum* symbiotic nitrogen fixation regulatory operon *fix*R *nif*A: analysis of *cis-* and *trans-*acting elements. *Journal of Bacteriology*, **171**, 4162–69.

Turgeon, B. G. and Bauer, W. D. (1985). Ultrastructure of infection-thread development during the infection of soybean by *Rhizobium japonicum*. *Planta*, **163**, 328–49.

Werner, D. (1990). *Plant and microbial symbiosis*. Chapman and Hall, London.

Werner, D. and Krotzky, A. (1983). Die symbiontische Stickstoff-Fixierung der Leguminosen. *Funktionelle Biologie und Medizin*, **2**, 31–9.

Werner, D., Mellor, R. B., Hahn, M. G., and Grisebach, H. (1985). Soybean root response to symbiotic infection. Glyceollin I accumulation in an ineffective type of soybean nodules with an early loss of the peribacteroid membrane. *Zeitschrift für Naturforschung*, **40c**, 179–81.

Werner, D., Mörschel, E., Garbers, C., Bassarab, S., and Mellor, R. B. (1988).

Particle density and protein composition of the peribacteroid membrane from soybean root nodules is affected by mutation in the microsymbiont *Brady-rhizobium japonicum. Planta*, **174**, 263–79.

Zacharius, R.M. and Kalan, E.B. (1990). Isoflavonoid changes in soybean cell suspensions when challenged with intact bacteria or fungal elicitors. *Journal of Plant Physiology*, **135**, 732–6.

3. Nodulins in root nodule development: function and gene regulation

FRANCINE GOVERS and TON BISSELING*

*Department of Molecular Biology, Agricultural University,
Dreijenlaan 3, 6703 HA Wageningen, The Netherlands, and
*Department of Phytopathology, Agricultural University,
Binnenhaven 9, 6709 PD Wageningen, The Netherlands*

Root nodule formation

The symbiotic interaction between bacteria of the genus *Rhizobium* and leguminous plants leads to the establishment of nodules, specialized organs on the roots of the host plant in which the *Rhizobium* bacteria fix atmospheric nitrogen. The formation of root nodules is known to proceed through a series of characteristic stages (Vincent 1980). Prior to the actual infection, the symbiotic partners recognize each other and the bacteria become attached to the root hairs which then start to curl. The bacteria are entrapped in the curls and they induce the plant cells to deposit cell wall-like material from which a tubular structure is formed, the infection thread. Within the growing infection thread the bacteria multiply, but in front of the growing infection thread tip, in the cortex, mitosis is induced and a nodule primordium is formed. Once the infection thread tip has reached the primordium the bacteria are released in the plant cells and they are enclosed by a membrane. In plants which have indeterminate nodules, for example pea and alfalfa, a meristem is formed at the apical site of the nodule primordium. Continuous cell division, cell differentiation and cell enlargement finally result in a specialized plant organ.

A mature root nodule is made up of central tissue surrounded by peripheral tissue. The latter is traversed by vascular strands which are connected to the central cylinder of the root. In indeterminate nodules the central tissue can be divided into different zones. Most distal is the apical meristem. Adjacent to this is the invasion zone, which is the site where bacteria are released from the infection threads. In the early symbiotic zone cells differentiate into infected and uninfected cells. In the infected cells the bacteria differentiate into bacteroids and these are capable of fixing nitrogen. The late symbiotic zone consists of uninfected and infected cells, fully packed with bacteroids. In this zone nitrogen fixation and ammonia assimilation occur. As the infected cells age the nitrogen fixing capacity decreases. Plant cells and bacteroids degenerate and they make up the most proximal zone of the nodule, the senescent zone. Consequently all tissues of indeterminate nodules are of graded age from the meristem to the root attachment point.

Nodulin genes and nodulin function

The establishment and functioning of a root nodule depends on complex interactions between the two symbiotic partners. Both the bacterium and the host plant have genes which are expressed in a highly developmentally regulated manner and whose products are essential for nodule formation and symbiotic nitrogen fixation (reviewed by Nap and Bisseling 1990). The plant genes in question, the nodulin genes are differentially expressed. The early nodulin genes are expressed in stages preceding the actual nitrogen fixation whereas expression of late nodulin genes is first detectable around the onset of nitrogen fixation. Since expression coincides with particular stages in the development of the symbiosis, functions of early and late nodulins must be related to specific events that occur during these stages. Thus early nodulins can be involved in nodule formation or in the infection process and late nodulins will most probably function in establishing and maintaining a proper environment within the nodule that allows nitrogen fixation to occur.

Late nodulins

Indeed, late nodulins for which the functions have been established fulfil these criteria. Leghaemoglubin, for example, is an oxygen carrier and it transports sufficient oxygen to the bacteroids at a low free oxygen concentration for nitrogenase to function properly. *n*-Uricase and glutamine synthetase are involved in the assimilation of ammonia whereas sucrose synthase is a nodulin that apparently is required for a nodule specific pathway in the catabolism of sucrose. Several nodulins are located in the membrane that surrounds the bacteroids and they may function in the transport of metabolites from the bacteroids to the plant cytoplasm or vice versa. For several other late nodulins the genes or corresponding cDNAs have been isolated but functions remain to be established. For more detailed information on late nodulins and on the regulation of nodulin gene expression see Nap and Bisseling (1990).

Early nodulins

From the early nodulins that have been described so far only two, ENOD2 and ENOD12, have been characterized in detail (Franssen *et al.* 1987; Van de Wiel *et al.* 1990; Scheres *et al.* 1990). ENOD2 and ENOD12 have been classified as early nodulins because expression of their genes is detectable as early as six and two days respectively after inoculation of the plants with rhizobia, whereas mRNAs from late nodulins are first detectable 12 days after inoculation. Within the legumes ENOD2 is highly conserved and cDNA clones have been isolated from several leguminous plant species. ENOD12 is much less conserved. The only cDNA clone that is available is isolated from a pea cDNA library. From the deduced amino acid sequence it appears that ENOD2 and ENOD12 have a similar structure. The N-terminal parts have typical features of a signal peptide. Adjacent to this is a stretch of pentapeptide repeats in which the first two amino acids in each pentapeptide are prolines. The homology of ENOD2 and ENOD12 with other plant proteins strongly suggests that they both belong to a group of hydroxyproline-rich glycoproteins which are found in cell walls. In order to gain further insight into the functions of ENOD2 and ENOD12, it is important to know in which tissues in the nodule these nodulins are localized.

Using *in situ* hybridization it has been demonstrated that ENOD2 mRNA is located exclusively in the nodule parenchyma ('inner cortex'), one of the tissues located in the peripheral tissue that surrounds the central tissue of the nodule (Fig. 3.1a,b) (Van de Wiel *et al.* 1990). The nodule parenchyma has a morphology which differs from the surrounding cell layers in that it contains relatively few and small intracellular spaces. The nodule parenchyma is thought to be the site where the barrier against oxygen diffusion from the atmosphere to the central tissue is situated, and measurements of the O_2 concentration in different nodule compartments are in full agreement with this (Witty *et al.* 1986). The presence of the putative cell wall protein ENOD2 may contribute to the morphology of the nodule parenchyma and thus indirectly, to establishment of the oxygen barrier. Information concerning the regulation of ENOD2 gene expression has been summarized recently (Govers *et al.* 1990).

In contrast to ENOD2, ENOD12 is located in a completely different cell type in the nodule. *In situ* hybridization has shown that in mature pea nodules, ENOD12 mRNA is produced in the invasion zone, the site where infection thread growth occurs and where the bacteria are released from the infection threads (Fig. 3.1c,d) (Scheres *et al.* 1990). In infected roots, two days after inoculation infection threads were visible in the outer cortical cell layers. In the inner cortex, a few cells had divided and a nodule primordium was being formed. At that stage ENOD12 mRNA was detectable at two sites in the root. Firstly, in the outer cortical cells through which the infection threads were migrating and in a few cells that lay in front of the threads towards the central cylinder. Secondly, in the nodule primordium that was formed in the root inner cortex. These are the cells that would be penetrated by the infection threads a few days later. By then the two sites of expression had become one site; a few days later, expression is detectable exclusively in the invasion zone of the nodule. These experiments suggest that the ENOD12 gene product is in one way or another involved in the infection process. All the cells in which ENOD12 transcripts have been found are

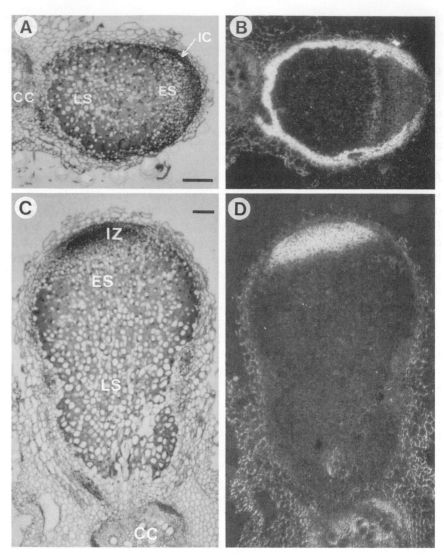

Fig. 3.1. Localization of ENOD2 (A, B) and ENOD12 (C,D) transcripts by *in situ* hybridization. Brightfield micrographs of longitudinal sections through a nodule from a 16 day old pea plant (A) and one from a 21-day-old pea plant (C). (B) and (D) Darkfield micrographs of the sections shown in (A) and (C) respectively. Silver grains representing hybridization signals are visible as white spots. ENOD2 mRNA is located in the inner cortex (IC) and ENOD12 mRNA in the invasion zone (IZ). ES: early symbiotic zone; LS: late symbiotic zone; CC: central cylinder of the root. Bar represents 200 μm.

sites of new cell wall synthesis and, therefore, possible sites of incorporation
of the ENOD12 protein.

Regulation of ENOD12 gene expression

ENOD12 gene expression is one of the first recognizable and detectable
responses of the host plant upon infection with *Rhizobium* and this in itself
makes an ENOD12 probe a convenient tool for investigating the signals that
are exchanged between host plant and *Rhizobium* during the infection.
Scheres *et al*. (1990) made a first attempt to unravel the pathway by which
Rhizobium induces expression of ENOD12 in pea. They showed that
soluble compounds from *Rhizobium* which are excreted upon activation
of the *Rhizobium* nodulation genes trigger the induction of PsENOD12.
Induction not only occurs at the site where the bacteria are, and so in the cells
containing infection threads, but also in cells at a significant distance from
the bacteria. Therefore the factors involved must be capable of diffusing
through several cell layers. Identification of the factors involved is currently
in progress.

   An additional reason why PsENOD12 is an interesting subject for investigating gene regulation is that ENOD12 transcripts are found in two other
plant tissues, that is in flower and stem. *In situ* hybridization of stem sections
has shown that, as in the nodule, the presence of ENOD12 mRNA is
restricted to certain cell types (Scheres *et al*. 1990). The tissue specific expression of ENOD12 in three completely different tissues raised the question
whether the ENOD12 mRNAs in nodule, stem, and flower are derived from
the same gene and if so, how its expression is regulated in the different
tissues. To answer these questions we started to analyse the pea genome
and it appeared that pea has two ENOD12 genes which have been designated
PsENOD12A and PsENOD12B. For each of these we now have probes
available. The cDNA clone described by Scheres *et al*. (1990) is a representative of gene PsENOD12A whereas gene PsENOD12B has been isolated
from a genomic library (Govers *et al*. 1991). Both genes have large stretches
of homology in their coding regions. However in comparison with gene A,
gene B has two deletions, one of 60 basepairs (bp) and a smaller deletion
of 12 bp. The 60 bp deletion is located in the region coding for the pentapeptide repeats and thus the number of repeat units in the product of gene
B is reduced from 12 to 8. The occurrence of deletions in gene PsENOD12B
gave us the opportunity to analyse expression of gene PsENOD12A and
gene PsENOD12B separately. We made use of the polymerase chain reaction to amplify specifically the ENOD12 mRNA molecules present in the
total RNA population isolated form nodule, stem and flower (PCR expression analyses as described in Scheres *et al*. 1990). By choosing one oligonucleotide complementary to DNA located upstream of the 60 bp deletion

and the other one complementary to DNA located downstream of the deletion, the DNA amplified by PCR from the mRNA of gene B will be 60 bp smaller than the DNA amplified from gene A transcripts. It appeared that in all RNA samples from nodule, stem and flower two different ENOD12 mRNAs are amplified indicating that in all three tissues expression of both genes, PsENOD12A and PsENOD12B, is induced. It is evident that both in infected roots and in the nodule *Rhizobium* produces the factors that finally trigger ENOD12 gene expression but in flower and stem these signals are absent. The question remains, whether the same mechanism is involved in the induction of ENOD12 gene expression in the different tissues. So does *Rhizobium* in fact mimic signals that are operational in other plant organs? Or are the promotors of the ENOD12 genes inducable by two, or maybe three, different signals. Now that several tools have become available to attack this problem we may be in the position to find an answer to these questions in the future.

Acknowledgements

We thank ir. E. M. Kesaulya-Monster for typing the manuscript and Wei Cai Yang for help with the figure. We are grateful to our colleagues for their support and helpful discussions. F. G. was supported by The Netherlands Foundation for Biological Research (BION), with financial aid from the Netherlands Organization for Scientific Research (NWO).

References

Franssen, H. J., Nap, J. P., Gloudemans, T., Stiekema, W., Van Dam, H., Govers, F., Louwerse, J., Van Kammen, A., and Bisseling, T. (1987). Characterization of cDNA for nodulin-75 of soybean: a gene product involved in early stages of root nodule development. *Proceedings of the National Academy of Sciences, USA*, **84**, 4495–9.

Govers, F., Franssen, H. J., Pieterse, C., Wilmer, J., and Bisseling, T. (1990). Function and regulation of the early nodulin gene ENOD2. In *Genetic engineering of crop plants* (ed. G. W. Lycett and D. Grierson), pp. 259–69. Butterworths, London.

Govers, F., Harmsen, H., Heidstra, R., Michielsen, P., Prins, M., Van Kammen, A., and Bisseling, T. (1991). Characterization of the pea ENOD12B gene and expression analyses of the two ENOD12 genes in nodule, stem and flower tissue. *Molecular and General Genetics*. (In press.)

Nap, J.-P. and Bisseling, T. (1990). Nodulin function and nodulin gene regulation in root nodule development. In *Molecular biology of symbiotic nitrogen fixation* (ed. P. M. Gresshoff), pp. 181–229. CRC Press, Boca Raton, Florida.

Scheres, B., Van de Wiel, C., Zalensky, A., Horvath, B., Spaink, H., Van Eck, H., Zwartkruis, F., Wolters, A.-M., Gloudemans, T., Van Kammen, A., and

Bisseling, T. (1990). The ENOD12 gene product is involved in the infection process during the pea-Rhizobium interaction. *Cell*, **60**, 281-94.

Van de Wiel, C., Scheres, B., Franssen, H., Van Lierop, M.J., Van Lammeren, A., and Bisseling, T. (1990). The early nodulin ENOD2 transcript is located in the nodule specific parenchyma (inner cortex) in pea and soybean root nodules. *The EMBO Journal* **9**, 1-7.

Vincent, J.M. (1980). Factors controlling the legume-Rhizobium symbiosis. In *Nitrogen Fixation II* (ed. N.E. Newton and W.H. Orme-Johnson), pp. 103-29. University Park Press, Baltimore.

Witty, J.F., Minchin, F.R., Skøt, L., and Sheehy, J.E. (1986). Nitrogen fixation and oxygen in legume root nodules. *Oxford surveys of plant and cellular biology*, **3**, 275-315.

4. Denitrification: the denitrifying bacteria and their interactions with plants

KARL-HEINZ LINNE VON BERG, RENATE
VOSSWINKEL, and HERMANN BOTHE

*Botanisches Institut, Universität zu Köln, Gyrhofstr.15, D-5000
Köln 41, Germany*

Plant physiologists are well aware that plants and many micro-organisms assimilate nitrate to meet their nitrogen requirements. Dissimilatory nitrate reduction, on the contrary, is only performed by prokaryotes and can clearly be differentiated from nitrate assimilation. The synthesis of the enzymes of nitrate assimilation, nitrate and nitrite reductases, is repressed by NH_4^+ or organic nitrogen compounds and is unaffected by O_2. The expression of dissimilatory nitrate reduction is not dependent on NH_4^+ or organic nitrogen but requires O_2-exclusion or low O_2-tensions. Some exceptional bacteria, such as *Aquaspirillum magnetotacticum* (Bazylinski and Blakemoore 1983) or *Thiosphaera pantotropha* (Robertson and Kuenen 1984) perform dissimilatory nitrate reduction even in air. These isolates are of particular interest, because they may be useful for the removal of nitrate in waste water treatment.

The capability for dissimilatory nitrate reduction is widespread among systematically unrelated bacteria. Relatively few organisms, however, for example, *Paracoccus denitrificans*, *Pseudomonas* spp., *E. coli*, and some photosynthetic bacteria have been investigated extensively. Several recent detailed reviews covering all aspects of the field are available (Payne 1985; Ferguson 1987; Soerensen 1987; Stewart 1988; Stouthamer 1988; Hochstein and Tomlinson 1988; Zumft *et al.* 1988*b*). Therefore this article will concentrate on more recent developments.

Bacteria utilize NO_3^- instead of O_2 under anaerobic conditions as a respiratory electron acceptor when degrading organic carbon (Fig. 4.1). The utilization of nitrate is energetically less favourable. Therefore O_2 is the preferred respiratory electron acceptor in air. Nitrate dissimilation and nitrate respiration are thus two terms for one and the same microbiological process. The term denitrification implies the formation of gaseous N-compounds (N_2, N_2O, NO) in this process. The microorganisms largely differ with respect to the formation of the end product of nitrate respiration which can be either NO_2^-, N_2O, or N_2. Bacteria of the enteric group, for example *Escherichia coli*, excrete ammonia. This process is called nitrate-

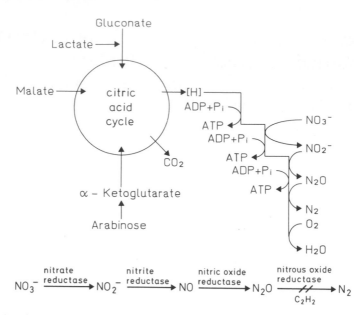

Fig. 4.1. The degradation of organic carbon compounds by dinitrifying bacteria such as *Azospirillum*.

ammonification. Nitrite is an intermediate in this process, and the reduction of nitrite to ammonia is accompanied by the formation of small but significant amounts of N_2O and NO in the Enterobacteriaceae (Stouthamer 1988; Ji and Hollocher 1988; and unpublished observations from our laboratory).

In denitrification, nitrite and N_2O are well established intermediates in the evolution of N_2 from NO_3^-. The role of NO in the process is less clear. NO may be a direct intermediate of a linear reaction sequence:

$$NO_3^- \rightarrow NO_2^- \rightarrow NO \rightarrow N_2O \rightarrow N_2$$

or may be component of a branched pathway:

$$NO_3^- \rightarrow NO_2^- \nearrow^{NO}_{\rightarrow} \searrow N_2O \rightarrow N_2$$

The conversion of NO to N_2O implies the build up of a dinitrogen bond with two mesomeric structures, which is mechanistically interesting but not understood (see Hochstein and Tomlinson 1988).

The properties of the enzymes involved in denitrification will be mentioned briefly here. Dissimilatory nitrate reductase is characterized best in *E. coli*. The enzyme catalyses the reduction of NO_3^- to NO_2^- and contains molybdopterin-like assimilatory reductase, and, in addition, three to four

4Fe-4S and one 3Fe-4S centres. Its molecular weight is 500 kDa with a probable subunit structure $\alpha_2\beta_2\gamma_4$. The α-subunit is the active site and contains the molybdopterin and the Fe-S centres. The reduction of NO_3^- to NO_2^- involves a valence change of Mo^{4+} to Mo^{6+} or 2 Mo^{4+} to 2 Mo^{5+}. The β-subunit contains no metal. It might be involved in membrane association, otherwise its role is uncertain. The γ-subunit contains cytochrome b_{556} as the electron donor for NO_3^- reduction. The structural genes are encoded by the nar C H J I-operon. In addition, several genes involved in the regulation of the enzyme and in the synthesis of the molybdenum cofactor are known (see Stewart 1988). The genetic system of dissimilatory nitrate reductase resembles that of *Klebsiella* nitrogenase which is, however, much better understood currently. Dissimilatory nitrate reductase of other organisms is smaller in size, but not so well investigated as the *E. coli* enzyme (see Payne 1985; Stewart, 1988).

Two different types of dissimilatory nitrite reductases are known for denitrifiers. The cytochrome cd_1-containing enzyme occurs in *Pseudomonas aeruginosa*, *Pseudomonas stutzeri*, *Paracoccus denitrificans*, and *Azospirillum brasilense*, whereas the Cu-containing protein is found in *Alcaligenes faecalis*, *Pseudomonas aureofaciens*, and in the phototroph *Rhodopseudomonas sphaeroides*. Thus one and the same genus (*Pseudomonas*) contains both forms, and bacterial systematics do not allow any prediction of the enzyme present in one bacterium. The end product of nitrite reduction may be either N_2O or NO. When the enzyme is membrane-bound, cell free systems mainly form N_2O from NO_2^-. The solubilized protein catalyses the reduction of NO_2^- to NO. Nitric oxide is the main product with phenazine methosulphate and ascorbate as reductant, whereas N_2O is formed with $Na_2S_2O_4$ and methylviologen in cell free systems. In the latter case, the NO formed enzymatically may be reduced chemically to N_2O by $Na_2S_2O_4$. It was stated that NO is not liberated from denitrifying cells except under a few, and sometimes peculiar, conditions (Zumft *et al.* 1988*b*), but the situation can be more complex. In our experience *Azospirillum brasilense* and *A. lipoferum* evolve at best low amounts of NO from NO_2^-, in contrast to a new isolate (*Pseudomonas* spp.) from soils near Braunschweig, Germany. The isolate produces mainly N_2O and small amounts of NO when the concentration of NO_2^- in the medium is low. At higher amounts of NO_2^- (10 mM), the cells evolve more NO than N_2O (Vosswinkel and Bothe 1990). The gas formation by this *Pseudomonas* isolate is also very much affected by the pH in the medium (Fig. 4.2). The isolate produces only N_2O at pH > 7.5 and mainly NO in the pH range 6.5–7.0. Controls with boiled bacteria indicated that these gases are not formed chemically.

Dissimilatory nitrite reductase from *E. coli* and other Enterobacteriaceae is remarkably different from the enzymes just described. The *E. coli* protein catalyses the reduction of NO_2^- to NH_4^+, uses either NADH or formate as electron donor, and contains FAD, Fe-S clusters, and siroheme. It resembles the assimilatory nitrite reductase of *Neurospora crassa* but serves dissimilatory purposes. *E. coli* excretes NH_4^+ when grown anaerobically with NO_3^-.

After a period of controversy it has now become clear that nitrous oxide reductases from all organisms investigated so far are multi-Cu-containing enzymes. The enzyme from *Pseudomonas stutzeri* contains eight Cu atoms per molecule, and can exist in at least three differently coloured forms with different specific activities when isolated (Coyle *et al.* 1985; Riester *et al.* 1989). Nitrous oxide reductase from all organisms is blocked by sulphide, rhodanide, and low concentrations of C_2H_2. Denitrification in microorganisms is very often tested in the presence of C_2H_2, because N_2O can be

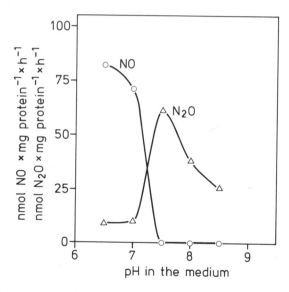

Fig. 4.2. The dependence of the formation of NO and N_2O by E 8/2 on the pH of the medium.
The cells were grown anaerobically in a medium containing 10 mM malate, 10 mM KNO_3, and mineral salts for 48 h at pH 7.0. The bacteria were harvested by centrifugation and suspended in the same medium which, however, did not contain KNO_3. The pH was adjusted as indicated in the abscissa of the figure. The experiment was performed in 7.0 ml Fernbach flasks under argon in a shaking water bath at 30 °C for 4 h. The flasks contained cells with 4.46 mg protein in 1.9 ml medium. The experiment was started by injecting 0.1 ml of an anaerobic solution of KNO_2 (final concentration in the flasks = 10 mM). NO was determined in a Tecan NO_x-analyser and N_2O by gas chromatography.

more easily monitored by gas chromatography than can N_2. In addition, the measurement of N_2 requires strict removal of air from the assay flasks.

As mentioned, the role of NO in denitrification is uncertain. Many bacteria utilize NO. A mutant of *Pseudomonas stutzeri* defective in cytochrome cd_1 showed an unimpaired utilization of NO (Zumft *et al.* 1988*a*), indicating that NO_2^--reductase and NO-reductase are different entities. Homogenous NO-reductase from *Pseudomonas stutzeri* consists of two polypeptides with each one containing either cytochrome *b* or cytochrome *c* (Heiss *et al.* 1989).

Another enzyme purified from vesicles of *Pa. denitrificans* reduced NO with much lower activity than the *Ps. stutzeri* protein and was claimed not to contain any metal (Hoglen and Hollocher 1989). This preparation probably does not represent the true NO-reductase from this bacterium. A NADH-NO reductase in particles from *Pa. denitrificans* catalysed ATP-formation with a P/2e ratio of 0.75 (Carr *et al.* 1989). Respiration with NO as sole respiratory electron acceptor was shown to generate a proton gradient in intact cells of *Pa. denitrificans* (Garber *et al.* 1982) and other denitrifiers (Shapleigh and Payne 1985). All the evidence mentioned above strongly indicates the existence of a NO-reductase different from N_2O-reductase. However, all attempts so far to grow bacteria solely with NO have failed.

The situation is illustrated by the experiments of Table 4.1 performed with our own isolate E 8/2 (*Pseudomonas* spp.) which readily reduces NO to N_2 via N_2O. As indicated by the increase in protein content, the isolate grows when being allowed to reduce NO to N_2. In the presence of C_2H_2, NO reduction stops at N_2O, and no growth is observed under various conditions employed although CO_2 is produced. Similar findings were also obtained with *A. brasilense* Sp 7 (Table 4.1). If the reduction of NO to N_2O, indeed, results in the generation of a proton motive force and ATP, growth is to be expected. This point needs clarification. It cannot be ruled out that the NO-reductase described reduces NO artificially and utilizes an electron acceptor other than NO in intact cells.

The orientation of the enzymes involved in denitrification at the cytoplasmic membrane has a strong impact on the size of the proton motive force to be generated. The cytoplasmic membrane is *per se* impermeable to protons. A reduction of NO_3^- or NO_2^- at the periplasmic side requires utilization of the H^+ translocated. When these reductions occur at the cytoplasmic face of the membrane, the cytoplasm should compensate for the H^+-requirement. Detailed studies, particularly by Stouthamer's group (see Stouthamer 1988) indicate that NO_3^- is reduced at the cytoplasmic side and NO_2^- and N_2O at the periplasmic face of the membrane in *Pa. denitrificans*. Such a distribution requires import of NO_3^- into the cytoplasm and export of NO_2^-. Two transport systems seem to exist in *Pa. denitrificans*: a H^+/NO_3^- active symport process in the low NO_3^--concentration range and a

Table 4.1. Growth of *Azospirillum brasilense* Sp 7 and isolate E8/2 (*Pseudomonas* spp.) with NO as sole respiratory electron acceptor

	Azospirillum brasilense Sp 7		E 8/2	
	$-C_2H_2$	$+C_2H_2$	$-C_2H_2$	$+C_2H_2$
1. NO concentration				
at the start	357	357	357	357
after 18 h	84	84	80	80
after 4 days	0	0	0	0
2. N_2O concentration				
after 18 h	27	124	24	126
after 4 days	0	181	0	206
3. N_2 concentration				
after 18 h	110	0	102	0
after 4 days	179	0	207	0
4. Per cent recovery N	100	101	116	115
5. Protein content (mg/ml)				
at the start	55.9	55.9	43.1	43.1
after 4 days	73.1	53.9	62.8	46.0
6. CO_2 concentration				
after 18 h	539	309	758	618
after 4 days	984	821	988	767

Data are given in μmol/assay except for protein. (For determining the gases see Danneberg *et al.* 1989.)

NO_3^-/NO_2^- antiport system when higher amounts of NO_3^- and NO_2^- are accumulated (Stouthamer 1988).

The distribution of the enzymes of dissimilatory NO_3^--reduction across the cytoplasmic membrane apparently differs from bacterium to bacterium. In photosynthetic bacteria all three enzymes appear to be orientated outwards (Satoh 1981; McEwan *et al.* 1984). In *A. brasilense* Sp 7, the energy transformation efficiencies were recently determined in anaerobic respiration with either NO_3^-, NO_2^-, or N_2O as respiratory electron acceptor by measuring the H^+-translocation or the maximal molar growth yields (Y_s^{max} values) in cells grown in continuous cultures (Danneberg *et al.* 1989). The Y_s^{max} values were approximately the same with O_2 and N_2O and were one-third and two-thirds lower with NO_2^- or NO_3^-, respectively, as respiratory electron acceptors. The reduction of NO_2^- to N_2O, but not the conversions of NO_3^- to NO_2^- and of N_2O to N_2, were accompanied by consumption of H^+, when the utilization of pulses of these oxidants was followed polarographically. Such observations allow us to conclude that there is periplasmic orientation of NO_2^--reductase and cytoplasmic orientation of NO_3^- and N_2O-reductases in *A. brasilense* (Fig. 4.3). Experiments with ionophores such as valinomycin/K^+ or $TPMP^+$ only partly support this conclusion (Danneberg *et al.* 1989).

The switch from aerobic to anaerobic mode of life and vice versa requires an extensive synthesis of proteins. It was shown for *E. coli* that up to 20 per

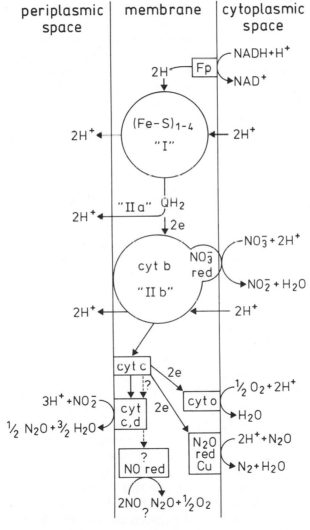

Fig. 4.3. The distribution of the enzymes involved in denitrification in the cytoplasmic membrane of *Azospirillum*.

cent of the protein pattern changes at that time (Sawers *et al.* 1988). Synthesis of enzymes of the tricarboxylic acid cycle, of the glyoxylate cycle, of the Mn-dependent superoxide dismutase and of components of the aerobic respiratory electron transport chain is induced by O_2 (see Stewart 1988). Such synthesis is genetically controlled at the transcriptional level by the gene *arc* A. The gene product, ARC A, probably functions as a repressor protein under anaerobic conditions (Iuchi and Lin 1988). A positive regulator gene under O_2-exclusion is *fnr* (Lambden and Guest 1976; Shaw and Guest 1982; Unden and Guest 1985; Stewart 1988, Unden and Trageser 1990). Its product FNR is a homodimer with a molecular weight of approximately 30 kDa. It acts at the transcriptional level on the synthesis of several but not all enzymes expressed only anaerobically. Examples of FNR-controlled proteins in *E. coli* or *Salmonella typhimurium* are fumarate reductase, nitrate reductase, nitrite reductase, pyruvate:formate lyase and one of the three hydrogenases. The sequence of FNR shares typical homologies with other DNA-binding regulatory proteins (Shaw *et al.* 1983). Similarities with the cyclo-AMP receptor protein which regulates glucose catabolite repression in *E. coli* are pronounced, indicating that both proteins have evolved by gene duplication. Genes which are activated by FNR (e.g. those for nitrate, nitrite, and fumarate reductases) possess a dyad consensus sequence of 22 bp which is probably the FNR binding site (Spiro and Guest 1987; Li and deMoss 1988; Eiglmeier *et al.* 1989). The concentration of FNR in the cells does not change at the switch from aerobic to anaerobic conditions (Unden and Duchêne 1987; Spiro and Guest 1987) and FNR is not modified during this transition. As metal binding protein, FNR possesses four characteristic cysteine residues at the N-terminus (Unden and Guest 1985). Deletion of the cysteine-containing region or mutation of one of the cysteine residues (= cysteine 19) to serine results in an inactive FNR protein (Spiro and Guest 1988). The cysteine residues can be alkylated to different degrees when isolated from aerobically or anaerobically grown cells (Trageser and Unden 1989). All these findings suggest that FNR is a metal binding protein and exists in different forms in aerobic and anaerobic cells. The conversion of the aerobic to the anaerobic form would be regulated by the amount of O_2 available in the cells and might involve metal (Fe?) binding. The oxidized form of FNR could interact with the oxidized form of a metal (Fe^{3+}?) in contrast to the reduced form, which could bind a more reduced state of a metal (Fe^{2+}?). Gene activation would be achieved only by the metal-activated reduced form (Fig. 4.4). Such an interpretation is substantiated by the recent observation that fumarate reductase and other enzymes of anaerobic respiration can be expressed by setting the redox potential in the medium to values more electronegative than $+300\,mV$ (Unden *et al.* 1990).

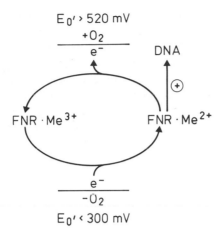

Fig. 4.4. The function of the FNR protein in *E. coli*. From Unden and Trageser (1990), modified.

FNR apparently functions as a coarse regulator. Additional factors regulate the differential expression of enzymes under anaerobic conditions. It is known for *E. coli* that nitrate induces the synthesis of nitrate reductase under anaerobic conditions and represses that of fumarate reductase, pyruvate:formate lyase and formate:hydrogen lyase (Iuchi and Lin 1987; Kalman and Gunsalus 1988; Sawers *et al.* 1988; Stewart and Berg 1988). This fine regulation is controlled by the two genes *nar* L and *nar* X. The nar L gene product, NAR L, is the DNA-binding regulator of the nitrate reductase operon (nar CHJI) and of some other regulatory genes (Stewart 1982), whereas NAR X may be the sensor component of the NAR X – NAR L two-component regulator system (Stewart *et al.* 1989).

The statement at the beginning of this chapter that denitrification is restricted to prokaryotes needs some modification. Assimilatory NO_3^--reductase of soybean leaves forms NO and trace amounts of N_2O in the purged *in vivo* assay (Dean and Harper 1986). NO_x is produced from the molybdenum part and not the diaphorase moiety of NO_3^--reductase (Dean and Harper 1988). Animal tissues evolve small but significant amounts of NO from the terminal guanidino nitrogen atom of L-arginine (Hibbs *et al.* 1988; Rees *et al.* 1989). Bacteria of the Enterobacteriaceae family such as *E. coli*, *K. pneumoniae* and *Citrobacter* also evolve N_2O. Experiments with mutants defective in NO_3^--reductase indicate that N_2O formation from NO_2^- is also catalysed by this enzyme in these bacteria (see Stouthamer

1988). In all these organisms, the NO_x production does not seem to have a physiological function, but it may have impact on global NO_x production.

Denitrification and also N_2 fixation can easily be demonstrated in bacteria–plant associations. In a model system, germinated wheat plants and *Azospirillum* were grown in association for a week and then assayed for both activities under microaerobic conditions (Neuer *et al.*, 1985). The association evolved either N_2O by denitrification or reduced C_2H_2 by N_2 fixation, depending on the availability of NO_3^- in the medium. Both activities were strictly in associations of plants and bacteria. The *Azospirillum* strains varied in activity. The model system can be easily used to monitor the N_2 fixation and denitrification capabilities of associative bacteria.

Bacteria can excrete large amounts of NO_2^- in NO_3^--respiration. Media in which *A. brasilense* Sp 7 have been grown may contain up to 10 mM NO_2^- (Zimmer *et al.* 1984). *Azospirillum* lives in association with roots of plants, particularly of grasses, and was described to exert positive effects on the growth of cereals (Döbereiner and Pedrosa 1987). We recently tried to identify the biochemical components of this interaction. *Azospirillum* performs N_2 fixation, but the number of this bacterium in soils, though high, might not be sufficient for an effective transfer of N_2 fixation products to the plants under most environmental conditions. *Azospirillum* produces indole-3-acetic acid (IAA), if tryptophan is present in the medium. IAA excretion by bacteria such as *Azospirillum* may indeed, be a plant growth-promoting factor (Zimmer *et al.* 1988). Surprisingly, NO_2^- formed in NO_3^--respiration was found partly to replace IAA in several phytohormone assays, as in the straight growth test of *Avena* coleoptiles, in the formation of C_2H_2 by pea epicotyl segments, or in a test with wheat root segments in which the increase of wet weight is determined (Zimmer and Bothe 1988; Zimmer *et al.* 1988). Nitrite alone cannot exert such phytohormonal effects in the assays. Exogenously supplied ascorbate was found to enhance the effects of NO_2^- in several assays. Therefore NO_2^- was postulated to interact with a component such as ascorbate in the cells, and a reaction or degradation product may function as a phytohormone. The concentration of ascorbate in plant cells is about 1–2 mM, and thus sufficient to account for these effects. It remains to be shown whether bacteria like *Azospirillum* excrete about 0.1 mM NO_2^- into the rhizosphere of soils. The effects of IAA and NO_2^- on the root morphology of grasses are clearly different from each other. More recent, unpublished data from this laboratory indicate that NO_2^- particularly stimulates the formation of additional lateral roots in wheat plants.

The ecological factors which determine the denitrification activities in nature are poorly understood. The populations of bacteria which have the capability for denitrification may vary from soil to soil. We have recently started to quantify denitrifying bacteria in the population of microorganisms in soil samples by the DNA–DNA hybridization technique. For this approach the cloned genes for the chromosomal N_2O-reductase from *Pseudomonas stutzeri* (from W. G. Zumft, Karlsruhe) and for the plasmid-encoded enzyme from *Alcaligenes eutrophus* (from B. Friedrich, Berlin) was kindly made available to us. The experimental protocol consisted of the extraction of DNA from the organism chosen, quantification of the DNA to approximately the same amount, binding of the isolated DNA on to nitrocellulose or nylon filters, hybridization with the ^{32}P-labelled DNA probe from *Pseudomonas* or *Alcaligenes* (labelled by nick-translation) and detection of the positive strains which contain homologous sequences on the dot blots by autoradiography (see footnotes to Table 4.2).

Experiments were performed first with control organisms which are known to be, or not to be, denitrifiers (Table 4.2). The non-denitrifiers spinach, *Spermatozopsis* (a green alga), *Euglena*, and the cyanobacteria *Anacystis*, and *Anabaena* did not give a hybridization signal even at low stringencies with the probe from *Ps. stutzeri*. The experiments with the known denitrifiers *Alcaligenes*, *A. brasilense*, and the isolates E 2/2 and E 8/2 (probably *Pseudomonas* spp.) were positive. Unexpectedly, positive signals were obtained with *E. coli* and *Klebsiella* of the Enterobacteriaceae. As mentioned previously, *E. coli* reduces nitrate via nitrite to ammonia with the formation of small amounts of N_2O and NO. The occurrence of

Table 4.2. DNA-hybridization with control organisms using the cloned DNA probe for N_2O-reductase from *Alcaligenes eutrophus*

Agaricus (mushroom) (1)	*E. coli* (2)
spinach (0)	*Pseudomonas stutzeri* (3)
Spermatozopsis (0)	own isolate E 2/2 (1)
Euglena (0)	own isolate E 8/2 (1)
Klebsiella (1)	*Azospirillum brasilense* Sp 7 (1)
cyanobacteria	*Alcaligenes* (1)
(*Anacystis, Anabaena*) (0)	

Hybridization signal strength: 3 = very strong, 2 = strong, 1 = weak, 0 = none.
The DNA-preparation and hybridization employed the following procedures (Maniatis *et al.* 1982):

* incubation of the harvested cells with lysozyme or breakage in the French press/Waring blendor,
* lysis of the cells using 5 per cent sodium dodecylsulfate;
* proteinase incubation overnight at 50 °C;
* phenol/chloroform extraction of the proteins;
* ethanol precipitation of the DNA;
* estimation of the DNA concentration by ethidium bromide fluorescense;
* denaturating of the DNA by heating (95 °C);
* binding of equal amounts of the single stranded DNA on to nitrocellulose or nylon membranes;
* preparation of the cloned DNA-probes by the alkaline lysis method;
* electrophoretic separation of the plasmid fragments coding for denitrification genes;
* ^{32}P-labelling of the gene probe by nick-translation;
* denaturation of the labelled gene probe and incubation with the membrane bound DNA of the organisms to be investigated (16 h);
* detection of DNA sequences homologous to the gene probe by autoradiography.

Table 4.3. Correlation between the activity of N_2O-reductase and the DNA–DNA hybridization signal using the gene probe for N_2O-reductase from *Alcaligenes eutrophus*

Hybridization strength	Number of isolates	Number of isolates with N_2O-reductase activity	Correlation between the hybridization signal and activity
3	2	2	100%
2	5	4	80%
1	14	7	50%
0	16	1	94%

Hybridization signal strength: 3 = very strong; 2 = strong; 1 = weak; 0 = none.

Total DNA of 37 isolates from a soil of the botanical garden of our institute was analysed. The bacteria were associated with roots of the grass *Poa annua*. Hybridization was performed at high stringency using 50 per cent formamide, 42 °C and $1 \times SSC$ (see Maniatis *et al.* 1982). N_2O-reductase activity was measured by the formation of N_2O from NO_3^- in the presence of C_2H_2. The latter gas specifically blocks N_2O-utilization if N_2O-reductase is present.

a N_2O-reductase has not been described to our knowledge. This subject needs to be investigated further. The positive signal in the fungi is likely to be due to bacterial contaminants in this sample collected from the field.

The DNA–DNA hybridization technique was applied for determining the genetic capabilities of bacteria from a soil sample taken from the roots of the grass *Poa annua* (Table 4.3). The bacteria were isolated by the conventional dilution technique and by growing them on agar plates containing Difco yeast extract/tryptone medium. Thirty seven single colonies were isolated from the plates. The isolates were then grown in suspension cultures in the same medium. DNA was isolated as described in the footnotes to Table 2 and the hybridization was performed at high stringency (50 per cent formamide, 42 °C, $1 \times SSC$) which only allows hybridization to DNA regions with more than 90 per cent homology. Twenty one isolates (57 per cent) gave positive hybridizations. After growth under anaerobic conditions, all isolates were tested for the expression of N_2O-reductase which was determined by assaying the N_2O formation in the presence of C_2H_2. The latter gas specifically blocks N_2O utilization by N_2O-reductase in all denitrifying bacteria investigated so far (Payne 1985). Fourteen isolates formed N_2O which was in all cases either enhanced by or absolutely dependent on C_2H_2, indicating that the bacteria had expressed N_2O-reductase. There was a one to one correspondence between a very strong or strong DNA–DNA hybridization signal and the presence of the N_2O-reductase and also between a negative hybridization signal and the absence of N_2O-reductase, with the exception of one isolate in each case. Fourteen isolates which gave a weak positive signal in the Southern blots could not really be assigned because only seven of them had expressed N_2O-reductase (Table 4.3).

These investigations are only a beginning. Experiments with many samples have to be assayed to ascertain the reliability of the DNA–DNA hybridization technique as a diagnostic tool for the occurrence of denitrifying bacteria in soils. Other cloned DNA (e.g. of NO_2^--reductase) may be more useful for this approach. The use of antibodies against the enzymes involved in denitrification may also be helpful. Clearly, the DNA–DNA hybridization technique provides information about the genetic information in a given bacterium, whereas antibodies could detect the enzyme only if expressed in a bacterium under the special growth conditions. Perhaps a combination of both methods will give insights on the distributions of denitrifying bacteria in soils.

Acknowledgement

This work was kindly supported by grants from the Bundesminister für Forschung und Technologie.

References

Bazylinski, D. A. and Blakemoore, R. P. (1983). Denitrification and assimilatory nitrate reduction in *Aquaspirillum magnetotacticum*. *Applied Environmental Microbiology*, **46**, 1118–24.

Carr, G. J., Page, M. D., and Ferguson, S. J. (1989). The energy-conserving nitric-oxide-reductase system in *Paracoccus denitrificans*. *European Journal of Biochemistry*, **179**, 683–92.

Coyle, C. L., Zumft, W. G., Kroneck, P. M. H., and Jakob, W. (1985). Nitrous oxide reductase from denitrifying *Pseudomonas perfectomarina*. Purification and properties of a novel multi-copper enzyme. *European Journal of Biochemistry*, **153**, 459–67.

Danneberg, G., Zimmer, W., and Bothe, H. (1989). Energy transduction efficiencies in nitrogenous oxide respirations of *Azospirillum brasilense* Sp 7. *Archives of Microbiology*, **151**, 445–53.

Dean, J. V. and Harper, J. E. (1986). Nitric oxide and nitrous oxide production by soybean and winged bean during the *in vitro* nitrate reductase assay. *Plant Physiology*, **82**, 718–23.

Dean, J. V. and Harper, J. E. (1988). The conversion of nitrite to nitrogen oxide(s) by the constitutive NAD(P)H-nitrate reductase enzyme from soybean. *Plant Physiology*, **88**, 389–95.

Döbereiner, J.; Pedrosa, F. O. (1987) Nitrogen-fixing bacteria in nonleguminous crop plants. pp. 1–155, Science Tech. Publishers, Madison, Wisconsin/Springer, Berlin

Eiglmeier, K., Honoré, N., Iuchi, S., Lin, E. C. C., and Cole, S. T. (1989). Molecular genetic analysis of the FNR-dependent promoters. *Molecular Microbiology*, **3**, 869–78.

Ferguson, S. J. (1987). Denitrification: a question of the control and organization of electron and ion transport. *Trends in Biochemical Sciences*, **12**, 354–7.

Garber, E. A. E., Castignetti, D., and Hollocher, T. C. (1982). Proton translocation and proline uptake associated with reduction of nitric oxide by denitrifying. *Paracoccus denitrificans*. *Biochemistry and Biophysics Research Communications*, **107**, 1504–7.

Heiss, B., Frunzke, K., and Zumft, W. G. (1989). Formation of the N–N bond from nitric oxide by a membrane bound cytochrome bc complex of nitrate-respiring (denitrifying) *Pseudomonas stutzeri*. *Journal of Bateriology*, **171**, 3288–97.

Hibbs, J. B., Taintor, R. R., Vavrin, Z., and Rachlin, E. M. (1988). Nitric oxide: a cytotoxic activated macrophage effector molecule. *Biochemistry and Biophysics Research Communications*, **157**, 87–94.

Hochstein, L. I. and Tomlinson, G. A. (1988). The enzymes associated with denitrification. *Annual Review of Microbiology*, **42**, 231–61.

Hoglen, J. and Hollocher, T. C. (1989). Purification and some characteristics of nitric oxide reductase-containing vesicles from *Paracoccus denitrificans*. *Journal of Biological Chemistry*, **264**, 7556–63.

Iuchi, S. and Lin, E. C. C. (1987). Molybdenum effector of fumarate reductase expression and nitrate reductase induction in *E. coli* K-12. *J. Bacteriology*, **169**, 3720–5.

Iuchi, S. and Lin, E. C. C. (1988). ArcA (dye), a global regulatory gene in *Escherichia coli* mediating repression of enzymes in aerobic pathways. *Proceedings of the National Academy of Science, USA*, **85**, 1888–92.

Ji, X. B. and Hollocher, T. C. (1988). Reduction of nitrite to nitric oxide by enterobacteria. *Biochemistry and Biophysics Research Communications*, **157**, 106–8.

Kalman, L. V. and Gunsalus, R. P. (1988). The frd R gene of *Escherichia coli* globally regulates several operons involved in anaerobic growth in response to nitrate. *Journal of Bacteriology*, **170**, 623–9.

Lambden, P. R. and Guest, J. R. (1976). Mutants of *Escherichia coli* unable to use fumarate as an anaerobic electron acceptor. *Journal of Genetic Microbiology*, **97**, 145–60.

Li, S. F. and deMoss, J. A. (1988). Localisation of sequences in the nar promoter of *E. coli* required for the regulation by Fnr and Nar L. *Journal of Biological Chemistry*, **263**, 13700–5.

Maniatis, T., Fritsch E. F., and Sambrook, J. (1982). *Molecular cloning*, pp. 1–545. Cold Spring Harbor Laboratory.

McEwan, A. G., Jackson, J. B., and Ferguson, S. J. (1984). Rationalization of properties of nitrate reductases in *Rhodopseudomonas capsulata*. *Archives of Microbiology*, **137**, 344–9.

Neuer, G., Kronenberg, A., and Bothe H. (1985). Denitrification and nitrogen fixation by *Azospirillum*. III. Properties of a wheat-*Azospirillum* association. *Archives of Microbiology*, **141**, 364–70.

Payne, W. J. (1985). Diversity of denitrifiers and their enzymes. In *Denitrification in the nitrogen cycle* (ed. H. L. Golterman), pp. 47–65. Plenum Press, London.

Rees, D. D., Palmer, R. M. J., Hodson, H. F., and Moncada, S. (1989). A specific inhibitor of nitric oxide formation from L-arginine attenuates endothelium-dependent relaxation. *British Journal of Pharmacology*, **96**, 418–24.

Riester, J., Zumft, W. G., and Kroneck, P. M. H. (1989). Nitrous oxide reductase from *Pseudomonas stutzeri*. *European Journal of Biochemistry*, **178**, 751–62.

Robertson, L. A. and Kuenen, J. G. (1984). Aerobic denitrification: a controversy revived. *Archives of Microbiology*, **139**, 351–4.

Satoh, T. (1981). Soluble dissimilatory nitrate reductase containing cytochrome c from a photodenitrifier *Rhodopseudomonas sphaeroides* forma *denitrificans*. *Plant and Cell Physiology*, **22**, 443–52.

Sawers, R., Zehelein, E., and Böck, A. (1988). Two-dimensional gel electrophoretic analysis of *Escherichia coli* proteins: Influence of various anaerobic growth conditions and the fnr gene product on cellular protein composition. *Archives of Microbiology*, **149**, 240–4.

Shapleigh, J. P. and Payne W. J. (1985). Nitric oxide-dependent proton translocation in various denitrifiers. *Journal of Bacteriology*, **163**, 837–40.

Shaw, D. J. and Guest, J. R. (1982). Amplification and product identification of the fnr gene of *E. coli*. *Journal of Genetic Microbiology*, **128**, 2221–8.

Shaw, D. J., Rice, D. W. and Guest, J. R. (1983). Homology between C A P and Fnr, a regulator of anaerobic respiration in *Escherichia coli*. *Journal of Molecular Biology*, **166**, 241–7.

Soerensen, J. (1987). Nitrate reduction in marine sediment: Pathways and inter actions with iron and sulfur cycling. *Geomicrobiology Journal*, **5**, 401–21.

Spiro, S. and Guest, J. R. (1987). Regulation and over-expression of the fnr gene of *E. coli*. *Journal of Genetic Microbiology*, **133**, 3279–88.

Spiro, S. and Guest, J. R. (1988). Inactivation of the F N R protein of *Escherichia coli* by targeted mutagenesis in the N-terminal region. *Molecular Microbiology*, **2**, 701–7.

Stewart, V. (1982). Requirement of Fnr and Nar L functions for nitrate reductase expression in *Escherichia coli* K-12. *Journal of Bacteriology*, **151**, 1320–5.

Stewart, V. (1988). Nitrate respiration in relation to facultative metabolism in Enterobacteria. *Microbiological Review*, **52**, 190–232.

Stewart, V. and Berg, B. L. (1988). Influence of nar (nitrate reductase) gene expression in *E. coli* K-12. *Journal of Bacteriology*, **170**, 4437–44.

Stewart, V., Pardes, J., and Merkel, S. M. (1989). Structure of genes *nar* L and nar X of the *nar* (nitrate reductase) locus in *E. coli* K-12. *Journal of Bacteriology*, **171**, 2229–34.

Stouthamer, A. H. (1988). Dissimilatory reduction of oxidized nitrogen compounds. In: *Biology of anaerobic microorganisms* (ed. A. J. B. Zehnder), pp. 245–303. John Wiley and Sons, New York.

Trageser, M. and Unden, G. (1989). Role of cysteine residues and metal ions in the regulatory functioning of F N R, the transcriptional regulator of anaerobic respiration in *E. coli*. *Molecular Microbiology*, **5**, 93–9.

Unden, G. and Guest, J. R. (1985). Isolation and characterization of the F N R protein, the transcriptional regulator of anaerobic electron transport in *E. coli.* *European Journal of Biochemistry*, **146**, 193–9.

Unden, G. and Duchêne, A. (1987). On the role of cyclic A M P and the F N R protein in *E. coli* growing anaerobically. *Archives of Microbiology*, **147**, 195–200.

Unden, G. and Trageser, M. (1990). Regulation der Synthese des anaeroben Elektronentransports in *E. coli* durch Sauerstoff. *Forum Mikrobiologie*, **13**, 211–19.

Unden, G., Trageser, M., and Duchêne, A. (1990). Effect of positive redox potentials (> +400 mV) on the expression of anaerobic respiratory enzymes in *E. coli.* *Molecular Microbiology*, **4**, (in press).

Vosswinkel, R. and Bothe, H. (1990). Production of nitrous oxide and nitric oxide by some nitrate-respiring bacteria. In *Advances of the 2nd course on inorganic nitrogen metabolism* (ed. W. R. Ullrich). Springer, Berlin. (In press.)

Zimmer, W. and Bothe, H. (1988). The phytohormonal interactions between *Azospirillum* and wheat. *Plant and Soil*, **110**, 239–47.

Zimmer, W., Penteado Stephan, M., and Bothe, H. (1984). Denitrification by *Azospirillum brasilense* Sp 7 I. Growth with nitrite as respiratory electron acceptor. *Archives of Microbiology*, **138**, 206–11.

Zimmer, W., Roeben, K., and Bothe, H. (1988). An alternative explanation for plant growth promotion by bacteria of the genus *Azospirillum*. *Planta*, **176**, 333–42.

Zumft, W. G., Döhler, K., Körner, H., Löchelt, S., Viebrock, A., and Frunzke, K. (1988*a*). Defects in cytochrome cd$_1$-dependent nitrite respiration of transposon Tn5-induced mutants from *Pseudomonas stutzeri*. *Archives of Microbiology*, **149**, 492–8.

Zumft, W. G., Viebrock, A., and Körner, H. (1988*b*) . Biochemical and physiological aspects of denitrification. In *The nitrogen and sulfur cycles* (ed. J. A. Cole and S. J. Ferguson), pp. 245–79. Cambridge University Press, Cambridge.

5. Some aspects of the utilization of nitrate and ammonium by plants

D.J. PILBEAM and E.A. KIRKBY

Department of Pure and Applied Biology,
The University of Leeds,
Leeds LS2 9JT, UK

The two major forms of nitrogen taken up by plants are nitrate (NO_3^-) and ammonium (NH_4^+) ions. In most soils nitrate is the more abundant of these ions, although in acid soils ammonium predominates because nitrification is largely inhibited. Most plants, except those adapted to acid conditions, grow better when supplied with nitrate.

Although nitrate is usually the preferred nitrogen source for plant growth, the uptake and assimilation of this ion appear to be metabolically more expensive than the uptake and assimilation of ammonium, not least because the nitrate ion is reduced to ammonium after uptake. Raven (1985) has estimated that the metabolic cost to a plant assimilating ammonium is 296 photons per mol nitrogen assimilated, compared with the cheapest way in which nitrate can be assimilated, which is 305 photons per mol nitrogen.

Despite this extra cost of assimilation, plants have adapted to nitrate as the most common form of readily available nitrogen in the soil. Even for plants that grow better with nitrate, though, there is evidence that the growth rate can be increased when both forms of nitrogen are supplied simultaneously. Cox and Reisenauer (1973) reported that the yield of young wheat (*Triticum aestivum* L.) plants increased by up to 50 per cent dry weight when some of the nitrate in the nutrient medium was substituted with ammonium. Lewis and Chadwick (1983) also demonstrated that for barley (*Hordeum vulgare* L.) both the dry matter production and nitrogen concentration were higher when the plants were grown in 1 mol m^{-3} nitrate plus 1 mol m^{-3} ammoniun than in either 2 mol m^{-3} nitrate- or ammonium-containing nutrient solutions alone. The reason for the stimulating effect of the NH_4-N on the growth of plants supplied with NO_3 is not clear.

Where plants are supplied with both NO_3-N or NH_4-N the uptake of nitrate is depressed by the presence of ammonium, but the reverse is not true (Mengel and Viro 1978). It seems likely that ammonium may depress the rate of activation or synthesis of the carrier responsible for nitrate uptake (Doddema *et al.* 1978; MacKown *et al.* 1982 *a, b*) or inhibit the assimilation of nitrate in the plant (Radin 1975; MacKown *et al.* 1982 *b*).

There has been some debate as to whether ammonium inhibits nitrate influx or stimulates nitrate efflux from roots. MacKown *et al.* (1982a) showed that transferring N-starved maize (*Zea mays* L.) plants into solutions containing either $^{15}NO_3^-$ alone or $^{15}NO_3^-$ plus NH_4^+ led to a lower influx of $^{15}NO_3^-$ when NH_4^+ was present, whereas efflux of $^{14}NO_3^-$ was unaffected. In recent experiments Lee and Drew (1989) have shown that in barley plants supplied with different concentrations of ammonium but the same concentration of nitrate the influx of nitrate was inhibited to the same extent as total nitrate uptake, and over the same time scale.

The detailed mechanisms of uptake of nitrate and ammonium by plants are discussed elsewhere in this volume. In general the uptake of an NO_3^- ion is accompanied either by the uptake of a monovalent cation or the release of an OH^- equivalent into the growth medium. The uptake of an NH_4^+ ion is accompanied by the release of an H^+ equivalent or the uptake of a monovalent anion. Since nitrogen-containing ions are taken up in very large amounts this difference in uptake of the two nitrogen forms is responsible for the frequently reported alkalization of the rhizosphere during nitrate nutrition and acidification during ammonium nutrition.

The charge balance of plants under nitrate nutrition is shown in Fig. 5.1 as being maintained by OH^- efflux, although it could of course be brought about by H^+ co-transport (see Glass 1988). Similarly, although it is usually the uptake of ammonium that is referred to there is some suggestion that the NH_4^+ ion is de-protonated prior to transport across the plasmalemma.

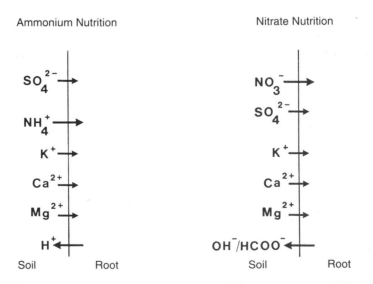

Fig. 5.1. Uptake of anions and cations during ammonium and nitrate nutrition.

However, the pK_a value of the $NH_3 + H_2O \rightleftharpoons NH_4^+ + OH^-$ equilibrium is 9.25, and so at normal soil or cytoplasmic pH any NH_4-N is present predominantly as the NH_4^+ ion, and it is this form to which plants are normally exposed.

Although the pH of the soil is influenced by the form of nitrogen taken up by the plants growing in it, this in turn influences the proportions of NO_3-N and NH_4-N taken up. At acid pH the uptake of nitrate is stimulated, whereas at basic pH values the uptake of ammonium is favoured. Michael *et al.* (1965) showed that at pH 4.0 the uptake of nitrate into barley seedlings was much higher than the uptake of ammonium, whereas at pH 6.8 the uptake of both ions was approximately the same. In sunflower (*Helianthus annuus* L.) plants grown in flowing culture at pH 6.5–6.8 the amount of nitrogen taken up by the plants over 51 days was only approximately 10 per cent lower for plants supplied with NH_4-N than for plants supplied with NO_3-N (Kurvits and Kirkby 1980). The growth of the plants supplied NH_4-N was depressed, but not to the extent seen in the experiments when the plants were grown in static water culture. This was presumably because the flowing nutrient solution did not allow the protons exchanged for ammonium to accumulate around the roots. The poor growth of many plants supplied with NH_4-N may be the result of effects of low pH rather than the ammonium ion *per se*.

These pH effects and the requirement of plants to maintain electro neutrality during uptake of cationic and anionic forms of nitrogen are obviously very important in any consideration of the utilization of nitrate and ammonium by plants. In this chapter aspects of the pH and charge balance are considered in relation to the distribution of both inorganic and organic ions. For more general coverage of the assimilation of nitrogen by plants the reader is referred to Kumar and Abrol (1990), Pilbeam and Kirkby (1990), and Stulen (1990).

The distribution of inorganic ions in plants during assimilation of ammonium and nitrate

A variety of factors are involved in differences in the uptake and distribution of inorganic anions and cations by plants supplied with the two forms of nitrogen. One of the most important of these factors is the alkalinization or acidification of the rhizosphere, as under acid conditions phosphate becomes less available in the soil, the concentration of aluminium and manganese increases in the soil solution and the uptake of calcium, magnesium, and potassium is depressed by H^+ ions in the rhizosphere. The effects of the pH of the growth medium on the uptake of potassium, calcium, and other ions is shown in sugar beet (*Beta vulgaris* L.) grown in NH_4-N from the work of Findenegg *et al.* (1989) (Table 5.1). Both the

Table 5.1. Concentrations of inorganic ions in shoots of *Beta vulgaris* grown at different pH values of the nutrient medium, NH_4-N supplied. From Findenegg *et al.* 1989, with permission.

pH of growth medium	4.0	5.0	6.0	7.0
Dry matter yield (mg per plant)	58	232	351	171
Concentration of ion in shoot ($mmol\ kg^{-1}$ dry matter)				
K^+	671	1677	2141	1656
Ca^{2+}	39	118	129	65
Mg^{2+}	363	305	287	189
$H_2PO_4^-$	776	945	914	537
Cl^-	279	940	1375	2557
NH_4^+	nd	31	135	656

concentrations and amounts of potassium and calcium in the shoots were depressed by low pH of the medium, as was also the uptake of ammonium.

For plants grown at the same pH, but with either NO_3-N or NH_4-N the uptake of calcium, magnesium, and potassium is usually higher for the NO_3-grown plants. In castor oil (*Ricinus communis* L.), plants grown in NH_4-N the uptake of those three ions was shown to be less than by plants grown in NO_3-N (Table 5.2, van Beusichem *et al.* 1988). The practical implications of this depression of potassium and calcium uptake are considerable. For example, tomato (*Lycopersicon esculentum* Mill.) plants supplied NH_4-N show a greater tendency to suffer from the physiological disorder blossom end rot than when NO_3-N is supplied (Mengel and Kirkby 1987). This disorder is caused by a shortage of calcium in the distal end of the fruit, a shortage that may be due either to lowered uptake of calcium or lowered mobility of the ion within the plant.

Table 5.2. Uptake of inorganic ions by *Ricinus communis* grown in either NO_3-N or NH_4-N. From van Beusichem *et al.* 1988, with permission.

Form of nitrogen nutrition	NO_3-N	NH_4-N
Uptake of ion ($mEq\ 100\ g^{-1}$ dry matter)		
K^+	116.8	63.5
Ca^{2+}	71.3	36.4
Mg^{2+}	32.6	22.1
PO_4^{2-}	22.4	25.5
Cl^-	2.5	4.9

The depression of uptake of potassium and calcium at low pH shown in Table 5.1 may be the cause of the lower yields of the plants grown at these low pH values, but it should be noted that shoot yield was also depressed at pH 7.0 relative to pH 6.0. This could have been caused by the accumulation of ammonium and chloride ions in the shoots. Mehrer and Mohr (1989) demonstrated inhibition of ribulose bisphosphate carboxylase and glyceraldehyde phosphate dehydrogenase activities in leaves of *Sinapis alba* L. plants with accumulation of NH_4^+ ions, inhibition that was not caused by pH effects. Ota and Yamamoto (1989) showed that radish (*Raphanus sativus* L.) supplied NH_4-N grew poorly and accumulated a high concentration of NH_4^+ ions in the leaves, whereas the addition of only a small amount of nitrate to the growth medium (5:1 mixture of NH_4^+: NO_3^-) gave rise to plants with considerably enhanced growth and very little NH_4^+ in the leaves.

Accumulation of NH_4^+ ions in the leaves is presumably not a common occurrence in soil-grown plants since it is observed at a pH at which nitrate would be the predominant nitrogen form in the soil. The assimilation of nitrate usually occurs mostly in the leaves of plants, but the stages of nitrate assimilation seem to be linked closely enough to preclude accumulation of ammonium. The reduction of nitrite to ammonium occurs in chloroplasts, but as the reaction requires reduced ferredoxin as reductant it occurs mainly in the light. This means that ammonium is only synthesized at a time when there is adequate reducing power and carbon skeletons for its incorporation into amino acids. Another source of NH_4^+ ions in leaves is the release which occurs from mitochondria during photorespiration, but this source too is light-dependent.

Allen *et al.* (1988) showed that all of the NH_4^+ taken up by non-nodulated *Phaseolus vulgaris* L. seedlings was assimilated in the roots. It is obvious however from the data already shown, that NH_4^+ ions can be translocated to the leaves. In Table 5.3 the concentrations of low molecular weight nitrogenous compounds in the xylem sap of castor oil, sugar beet, and sorghum [*Sorghum bicolor* L. (Moench)] grown in either NO_3-N or NH_4-N are shown (from van Beusichem *et al.* 1988; Arnozis and Findenegg 1986). It can be seen that with ammonium nutrition some NH_4^+ is translocated from roots to shoots, but this is usually a small proportion of the total nitrogen moving in the xylem. Some of the reduced nitrogen compounds may have originally been formed in the leaves, transferred to the roots, and then recycled in the xylem, and in studies on wheat fed $^{15}NO_3^-$ Cooper *et al.* (1986) showed that 50–70 per cent of the nitrogen in the xylem may be recycled in this way. However, even allowing for such recycling it appears from Table 5.3 that a considerable amount of the ammonium taken up by plants is assimilated in the roots.

When nitrate was the nitrogen source for the castor oil, sorghum, and sugar beet plants, the proportions of nitrogenous compounds in the xylem

Table 5.3. Concentrations of nitrogenous compounds in xylem sap of *Ricinus communis, Beta vulgaris*, and *Sorghum bicolor* grown in either NO_3-N or NH_4-N. From van Beusichem *et al.* 1988 (*R. communis*) and Arnozis and Findenegg 1986 (*B. vulgaris* and *S. bicolor*), with permission

Plant species	*R. communis*		*B. vulgaris*		*S. bicolor*	
Form of nitrogen nutrition	NO_3-N	NH_4-N	NO_3-N	NH_4-N	NO_3-N	NH_4-N
Concentration of compound (mmol N dm^{-3})						
Asparagine	2.4	11.6	0.96	8.16	7.62	9.34
Glutamine	13.9	42.3	12.84	65.70	10.20	20.76
Aspartate	0.3	0.2	0.67	0.29	0.16	0.14
Glutamate	0.6	0.6	1.71	2.22	0.31	0.00
Other amino acids	2.7	1.9	1.45	4.64	3.89	2.90
NO_3^-	19.4	0.0	52.43	15.00	41.62	3.35
NH_4^+	0.0	2.3	0.94	8.01	1.69	5.46
SUM	39.3	58.9	71.00	104.02	65.49	41.95
Total nitrogen determined	40.7	59.9	71.22	116.19	61.31	43.31

represented by the NO_3^- ion were much higher, indicating that nitrate is predominantly assimilated in the shoots of plants. In fact, it has been known for many years that the proportions of nitrate that are assimilated in the different parts of plants differs considerably between plant species (Pate 1973), and in the work of Arnozis and Findenegg (1986) it appears that the sorghum plants had more capacity to assimilate nitrate in the roots than the beet. From this observation it would be expected that sorghum should also have been capable of assimilating more ammonium in the roots than the beet, since the enzymes for ammonium assimilation are also required for the later stages of nitrate assimilation. Even in this species, though, some NH_4^+ was detected in the xylem sap. It may be concluded therefore that there is a potential for ammonium to accumulate in the leaves whenever ammonium is the sole nitrogen source for plants, but is more likely to occur when the pH of the growth medium is high. The detrimental effects of NH_4^+ in depressing the uptake of potassium and calcium are likely to be more common when the pH of the growth medium is very acid.

The form of nitrogen supplied also has effects on the uptake of other ions in plants. For example, chickpea (*Cicer arietinum* L.) plants have been shown to be much less affected by iron deficiency when supplied NH_4-N

Table 5.4. Uptake of inorganic ions by *Cicer arietinum* grown in either NO_3-N or NH_4-N, with iron supplied or withheld. From Alloush *et al.* 1990

Form of nitrogen nutrition	NO_3-N		NH_4-N	
Iron supplied/withheld	+ Fe	− Fe	+ Fe	− Fe
Dry matter yield				
(gram per plant)	3.31	0.92	3.14	3.55
Uptake of ion				
(mEq plant^{-1})				
K^+	3.10	0.85	1.38	2.12
Ca^{2+}	2.00	0.81	0.63	0.92
Mg^{2+}	0.92	0.32	0.66	0.96
Na^+	0.02	0.01	0.02	0.03
NH_4^+	–	–	7.82	6.76
$H_2PO_4^-$	0.71	0.17	0.63	0.81
SO_4^{2-}	0.79	0.26	0.82	1.10
Cl^-	0.05	0.03	0.12	0.15
NO_3^-	7.39	1.35	–	–

rather than NO_3-N (Alloush *et al.* 1990). The partial withdrawal of iron from chickpea plants supplied NH_4-N was shown to lead to increased uptake of potassium, calcium, and magnesium, and slightly decreased uptake of ammonium, whereas with the nitrate-grown plants there was a big decrease in the amount of potassium, calcium, and magnesium taken up (Table 5.4). Although deficiencies of these three cations may have been partially responsible for the restricted growth of the nitrate-grown plants their increased uptake by the iron-stressed plants supplied NH_4-N was probably a consequence of enhanced growth, not a cause. The cause of increased growth of these plants compared with the iron-sufficient, ammonium-grown plants is uncertain, but it seems likely that the ammonium-grown plants were less affected by partial iron withdrawal than the nitrate-grown plants because acidification of the rhizosphere with ammonium nutrition increased the availability of the small amounts of iron supplied.

Formation and distribution of organic ions during assimilation of ammonium and nitrate

During the reduction of nitrate to ammonium in plants

$$NO_3^- + 8H^+ + 8e^- \rightarrow NH_3 + 2H_2O + OH^-$$

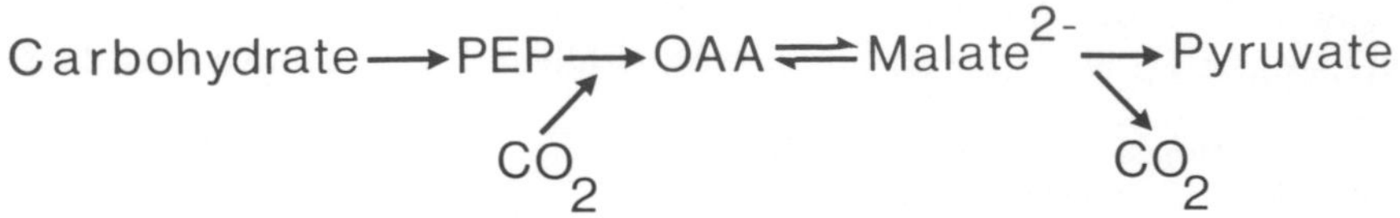

$$\text{Carbohydrate} \longrightarrow \text{PEP} \longrightarrow \text{OAA} \rightleftharpoons \text{Malate}^{2-} \longrightarrow \text{Pyruvate}$$

with CO_2 feeding into PEP and CO_2 released near Pyruvate.

Fig. 5.2. The mechanism of the cellular pH stat. An increase in cytoplasmic pH (from e.g. reduction of nitrate) stimulates the enzyme PEP carboxylase and results in the formation of oxalacetate and malate. A fall in cytoplasmic pH stimulates malic enzyme, which catalyses the decarboxylation of malate to pyruvate. (PEP = phosphoenolpyruvate, OAA = oxalacetate).

OH^- equivalents are produced. The presence of these OH^- equivalents causes carboxyl groups on organic acids to dissociate to the COO^-H^+ form, with the COO^- ions becoming the site of the negative charges released during nitrate assimilation. The formation of such organic anions is itself theoretically favoured by high pH through the postulated mechanism of the 'pH stat' (Davies 1973) (Fig. 5.2).

For plants supplied with ammonium the activity of phosphoenolpyruvate carboxylase (PEP carboxylase) should be less than in plants supplied with nitrate. This was confirmed by Schweizer and Erismann (1985) who showed that in non-nodulated plants of *P. vulgaris*, PEP carboxylase activity (expressed per amount of soluble protein) was high in leaves of plants supplied nitrate, but low in plants supplied ammonium or grown in nitrogen-free solutions. The difference was less obvious at pH 6.0 than pH 4.0, both on a soluble protein and a fresh weight basis, presumably because there was less uptake of nitrate at the higher pH. In the roots, PEP carboxylase activity became greater over the 13 days of the experiment in plants supplied ammonium at pH 4.0 than in plants supplied nitrate at the same pH, but the enzyme activity was at the higher level after 12 days growth in either ammonium or nitrate at pH 6.0 (based on soluble protein; but lower in nitrate-grown plants based on fresh weight).

The working of the pH stat mechanism is complicated by the fact that PEP carboxylase fulfils an anaplerotic function, allowing for the synthesis of carbon skeletons that are required for amino acid synthesis during the assimilation of both nitrate and ammonium. The higher activity of PEP carboxylase in the roots of ammonium-grown plants presumably reflects the need for more carbon skeletons to be available in the roots for amino acid synthesis when ammonium is the nitrogen source. When the nitrogen source is nitrate, the activity of the enzyme in the leaves should be higher as the assimilation of nitrate occurs predominantly in the shoots of plants.

An anaplerotic role for PEP carboxylase has been demonstrated in work on green algae. Smith *et al.* (1989) showed that when ammonium was supplied to nitrogen-limited cultures of *Selenastrum minutum* (Naeg.) Collins, there was an increase in PEP carboxylase activity. Furthermore, Guy *et al.* (1989) showed that whereas in nitrogen-starved cultures of *S. minutum* the pattern of discrimination against the ^{13}C isotope points to carbon assimilation occurring through the normal C3 pathway (catalysed initially by ribulose bisphosphate carboxylase), when ammonium was supplied to the culture up to 87 per cent of the carbon fixation occurred via carboxylation of PEP.

In recent studies on *Brassica nigra* (L.) Koch suspension cells, suspensions low in inorganic phosphate showed a fivefold higher activity of PEP carboxylase than suspensions adequately supplied with phosphate (Duff *et al.* 1989). Phosphorus-deficient plants may acidify the rhizosphere when grown with either nitrate or ammonium. The atypical acidification of the rhizosphere with nitrate nutrition may arise from protons from the dissociation of organic acids synthesized by enhanced PEP carboxylase activity under phosphorus deficiency. Le Bot *et al.* (1990) have observed an accumulation of organic acids in the roots of phosphorus-deficient chickpea plants grown with nitrate nutrition (Table 5.5), along with a slight drop in rhizosphere pH after a long period of withdrawal of phosphorus from the nutrient medium. As usual for ammonium-grown plants the concentrations of organic anions (shown in the table as alkalinity of the ash) were much lower than for nitrate-grown plants regardless of whether or not phosphorus was supplied, but the value for alkalinity of the ash was still nearly double in the roots of the phosphorus-deficient plants as compared with the phosphorus-sufficient plants.

Table 5.5. Accumulation of organic anions (shown as alkalinity of ash) in *Cicer arietinum* grown in either NO_3-N or NH_4-N, with phosphorus supplied or withheld. From Le Bot *et al.* 1990

Form of nitrogen nutrition Phosphorus supplied/withheld		NO_3-N		NH_4-N	
		+P	−P	+P	−P
Alkalinity of ash (mEq 100 g^{-1} dry weight)	Shoot	117	102	40	37
	Root	99	148	12	23

Accumulation of organic acids would occur if their synthesis was enhanced, if their conversion into amino acids was depressed, as their excretion from the plants was depressed, or any combination of these processes. The ability to be able to distinguish between these contributory processes would be essential to determine if an increase in PEP carboxylase activity could be the driving force behind accumulation of organic acids. What is required is measurement of the flux through the pathways of biosynthesis of organic acids.

Regardless of the exact role of PEP carboxylase the enzyme was a key component of earlier models to describe the assimilation of nitrate on a whole plant basis (Fig. 5.3). In the Ben Zioni/Dijkshoorn model, thought to hold for castor oil and other plants, organic acids produced during the assimilation of nitrate in the leaves are translocated as potassium salts to the root, where they break down to release OH^- equivalents to balance the NO_3^- uptake. In other plants, such as tomato, the organic acids produced in the leaves during the assimilation of nitrate mainly accumulate *in situ*, again as potassium salts, and potassium recirculation is relatively less important. In this scheme OH^- equivalents are not produced to exchange across the roots surface, but because potassium is required as a counter ion to the accumulating organic anions there is no charge imbalance arising from nitrate uptake. In this model OH^- exchange is only required to balance any assimilation of nitrate that occurs in the roots.

It now appears likely that the recirculation model does not hold for castor oil plants, one of the species to which it was originally ascribed. Van Beusichem *et al.* (1985) estimated that only 19–24 per cent of the HCO_3^- excreted from the roots of castor oil plants as the OH^- equivalents exchanged for incoming NO_3^- ions originated from reduction of nitrate and sulphate in the shoots (Fig. 5.4).

The excretion of 39.8 mEq HCO_3^- per plant represented the imbalance between cations and anions taken up by the plants, and approximately 53 per cent of the negative charges arising from the reduction of nitrate and sulphate were accumulated as organic anions in root and leaf vacuoles. When similar castor oil plants were supplied with NH_4-N instead of NO_3-N, the assimilation of nitrogen occurred almost exclusively in the roots (Table 5.2), and so the roots were the sites of the protons that were exchanged for the NH_4^+ ions (van Beusichem *et al.* 1988). The synthesis of organic anions was facilitated by a much greater amount of assimilation of sulphate in the ammonium-grown plants.

The implication here is that this amount of synthesis of organic acids must be essential in these plants, and the extra sulphate assimilation enables it to occur. That organic acid synthesis is essential is obvious, if only because of the demand for carbon skeletons for amino acid synthesis, but organic acids are also probably required for their role as osmotica during cell expansion.

Fig. 5.3. Ben Zioni – Dijkshoorn scheme (A) describing the movement of K^+, NO_3^- and malate in plants. From Ben Zioni *et al.* 1970, 1971; Dijkshoorn 1958. Modified Ben Zioni – Dijkshoorn scheme (B). From Kirkby 1974.

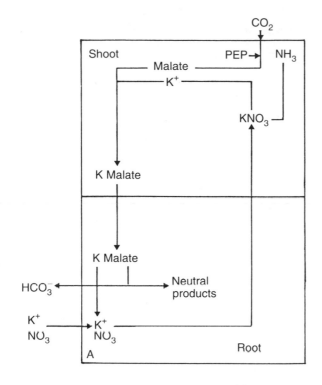

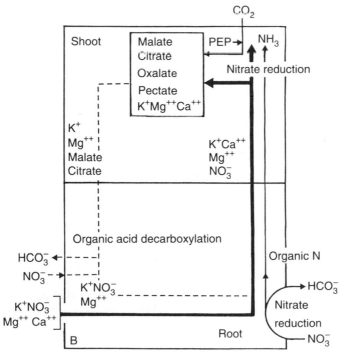

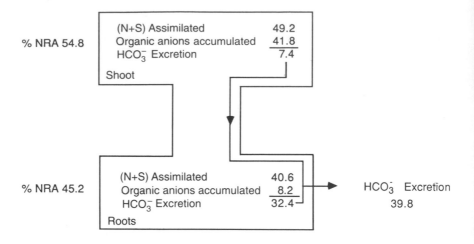

Fig. 5.4. Assimilation of N + S, accumulation of organic anions, and excretion of HCO_3^- from *Ricinus communis* plants grown for 40 days in NO_3-N. % NRA = percentage of total nitrate reductase activity. Other values expressed in mEq plant^{-1}. From van Beusichem *et al.* 1985, with permission.

Rates of expansion of leaves of *Polygonum cuspidata* Sieb. et Zucc. have been shown to be higher with increased availability of nitrogen (Hirose 1984), and increased availability of nitrogen also gives rise to greater leaf area in *Vicia faba* L. (Andrews *et al.* 1985). It has been suggested that nitrate may act as an osmoticum in younger leaves (Steingröver *et al.* 1986), so that leaf expansion may be dependent upon the flux of nitrate through the xylem; if the nitrate accumulates osmotic potential increases, but if the nitrate is reduced to ammonium organic acid synthesis is provided by the pH stat mechanism and the resulting organic anions maintain osmotic potential. In tomato, a plant where organic anions are known to accumulate in the leaves, the activity of nitrate reductase in each leaf increases until just before full expansion, and then declines (Bellaloui and Pilbeam 1990). The plant maintains an increasing capacity to reduce nitrate throughout vegetative growth as the growth of new leaves more than compensates for the decline in the assimilatory capacity of old leaves. It is probable that nitrate flux controls nitrate reductase activity, and these two factors together drive leaf expansion rather than vice versa.

This raises the question as to what happens under ammonium nutrition. Here, presumably, the increased uptake of anions, such as chloride, provides some increase in osmotic potential, but it is usually the case that plants supplied ammonium have a higher dry weight: fresh weight ratio than plants supplied nitrate. For this reason it is important that in experiments where comparisons are made between ammonium and nitrate nutrition dry weight values should be used.

Conclusions

It is obvious from this discussion that the relationship between organic acid synthesis and the assimilation of nitrogen is not well understood, and this will be a fruitful area for future research. Whereas Findenegg *et al.* (1989) have shown that in roots of *Beta vulgaris* supplied with ammonium there is higher PEP carboxylase activity in plants grown at higher pH than at lower pH, we have observed the reverse in experiments on tomato in our own laboratory. Here PEP carboxylase activity in roots of ammonium-grown plants maintained at pH 5.0 was higher than in plants at pH 5.5, which in turn had higher activity than plants at pH 6.0 (McKeever and Pilbeam unpublished). If ammonium uptake is greater at higher pH, PEP carboxylase activity would also be expected to be higher. However, it can be seen from Table 5.1 that more NH_4^+ ions are translocated to the leaves at higher pH, which implies that less assimilation occurs in the roots and so PEP carboxylase activity should be lower.

The lower yields of plants supplied with ammonium – nitrogen as compared with plants supplied nitrate have been shown to be due to a variety of factors. Plants may take up less ammonium than nitrate, they may take up smaller amounts of other cations when ammonium is supplied and they may accumulate NH_4^+ ions in the shoot. Recent work by Henry and Raper (1989) on *Nicotiana tabaccum* L. grown at pH 4.0 or 6.0 in either NO_3-N or NH_4-N illustrates this well. At the lower pH the plants grew poorly on NH_4-N because the uptake of ammonium was 50 per cent lower than those grown at pH 6.0. At the higher pH, the ammonium-grown plants performed as well as the nitrate-grown plants, and so presumably there was no excessive accumulation of NH_4^+ ions in the leaves. This pH is certainly the optimum seen for other plants supplied ammonium (Table 5.1). Different plant species, however, may be differently affected by ammonium nutrition. Plants that normally assimilate substantial proportions of nitrate in their roots would presumably have higher activities of all the enzymes of nitrogen assimilation in the roots than plants that assimilate nitrate almost exclusively in the leaves, and so they may be less likely to export toxic NH_4^+ ions to the leaves under ammonium nutrition.

The topic of ammonium and nitrate nutrition will continue to receive much attention because:

(1) both ammonium and nitrate fertilizers are used, and the use of fertilizers can be made more efficient;

(2) many soils are acidic, and uptake of ammonium from these soils will further depress pH and exacerbate low yields of crops;
(3) a better understanding of why plants supplied mixtures of ammonium and nitrate have higher yields than plants supplied either nitrogen source alone would help us guarantee these high yields.

Acknowledgements

Thanks are due to Jacques Le Bot, Kieran McKeever, and Francis Sanders for invaluable discussion during the preparation of this manuscript and to Mrs Sharon Hunter for its production.

References

Allen, S., Raven, J.A., and Sprent, J.I. (1988). The role of long-distance transport in *Phaseolus vulgaris* grown with ammonium or nitrate as nitrogen source, or nodulated. *Journal of Experimental Botany*, **39**, 513–28.

Alloush, G.A., Le Bot, J., Sanders, F.E., and Kirkby, E.A. (1990). Mineral nutrition of chickpea plants supplied with NO_3 or NH_4-N. I. Ionic balance in relation to iron stress. *Journal of Plant Nutrition*, **13**, 1575–90.

Andrews, M., MacFarlane, J.J., and Sprent, J.I. (1985). Carbon and nitrogen assimilation by *Vicia faba* L. at low temperature: the importance of concentration and form of applied N. *Annals of Botany*, **56**, 651–8.

Arnozis, P.A. and Findenegg, G.R. (1986). Electrical charge balance in the xylem sap of beet and sorghum plants grown with either NO_3 or NH_4 nitrogen. *Journal of Plant Physiology*, **125**, 441–9.

Bellaloui, N and Pilbeam, D.J. (1990). Reduction of nitrate in leaves of tomato during vegetative growth. *Journal of Plant Nutrition*, **13**, 39–55.

Ben-Zioni, A., Vaadia, Y., and Lips, S.H. (1970). Correlations between nitrate reduction, protein synthesis and malate accumulation. *Physiologia Plantarum*, **23**, 1039–47.

Ben-Zioni, A, Vaadia, Y., and Lips, S.H. (1971). Nitrate uptake by roots as regulated by nitrate products of the shoot. *Physiologia Plantarum*, **24**, 288–90.

Cooper, H.D., Clarkson, D.T., Johnston, M.G., Whiteway, J.N., and Loughman, B.C. (1986). Cycling of amino-nitrogen between shoots and roots in wheat seedlings. *Plant and Soil*, **91**, 319–22.

Cox, W.J. and Reisenauer, H.M. (1973). Growth and ion uptake by wheat supplied nitrogen as nitrate, or ammonium, or both. *Plant and Soil*, **38**, 363–80.

Davies, D.D. (1973) Metabolic Control in Higher Plants. In *Biosynthesis and its control in plants* (ed. B.V. Milborrow), pp. 1–20. Academic Press, London.

Dijkshoorn, W. (1958). Nitrate accumulation: nitrogen balance and cation–anion ratio during the regrowth of perennial rye grass. *Netherlands Journal of Agricultural Science*, **6**, 211–21.

Doddema, H., Hofstra, J. J., and Feenstra, W. J. (1978). Uptake of nitrate by mutants of *Arabidopsis thaliana*, disturbed in uptake of nitrate and chlorate. *Physiologia Plantarum*, **43**, 343–50.

Duff, S. M. G., Moorhead, G. B. G., Lefebvre, D. D., and Plaxton, W. C. (1989). Phosphate starvation inducible 'bypasses' of adenylate and phosphate dependent glycolytic enzymes in *Brassica nigra* suspension cells. *Plant Physiology*, **90**, 1275–8.

Findenegg, G. R., Nelemans, J. A., and Arnozis, P. A. (1989). Effect of external pH and Cl on the accumulation of NH_4 ions in the leaves of sugar beet. *Journal of Plant Nutrition*, **12**, 593–602.

Glass, A. D. M. (1988). Nitrogen uptake by plant roots. *ISI Atlas of Science: Animal and Plant Sciences*, **1**, 151–6.

Guy, R. D., Vanlerberghe, G. C., and Turpin, D. H. (1989). Significance of phosphoenolpyruvate carboxylase during ammonium assimilation. Carbon isotope discrimination in photosynthesis and respiration by the N-limited green alga *Selenastrum minutum*. *Plant Physiology*, **89**, 1150–7.

Henry, L. T. and Raper, C. D. Jr. (1989). Effects of root-zone acidity on utilization of nitrate and ammonium in tobacco plants. *Journal of Plant Nutrition*, **12**, 811–26.

Hirose, T. (1984). Nitrogen use efficiency in growth of *Polygonum cuspidatum* Sieb. et Zucc. *Annals of Botany*, **54**, 695–704.

Kirkby, E. A. (1974). Recycling of potassium in plants considered in relation to ion uptake and organic acid accumulation. *Proceedings of the 7th International Colloqium on Plant Analysis and Fertilizer Problems*, 557–68.

Kumar, P. A. and Abrol, Y. P. (1990). Ammonia assimilation in higher plants. In *Nitrogen in higher plants* (ed. Y. P. Abrol), pp. 159–79. Research Studies Press Ltd., Taunton.

Kurvits, A. and Kirkby, E. A. (1980). The uptake of nutrients by sunflower plants (*Helianthus annuus*) growing in a continuous flowing culture system supplied with nitrate or ammonium as nitrogen source. *Zeitschrift für Pflanzenernährung und Bodenkunde*, **143**, 140–9.

Le Bot, J., Alloush, G. A., Kirkby, E. A., and Sanders, F. E. (1990). Mineral nutrition of chickpea plants supplied with NO_3 or NH_4-N. II. Ionic balance in relation to phosphorus stress. *Journal of Plant Nutrition*, **13**, 1591–606.

Lee, R. B. and Drew, M. C. (1989). Rapid, reversible inhibition of nitrate influx in barley by ammonium. *Journal of Experimental Botany*, **40**, 741–52.

Lewis, O. A. M. and Chadwick, S. (1983). A ^{15}N investigation into nitrogen asssimilation in hydroponically grown barley (*Hordeum vulgare* L. cv. Clipper) in response to nitrate, ammonium and mixed nitrate and ammonium nutrition. *New Phytologist*, **95**, 635–46.

MacKown, C. T., Jackson, W. A., and Volk, R. J. (1982a). Restricted nitrate influx and reduction in corn seedlings exposed to ammonium. *Plant Physiology*, **69**, 353–9.

MacKown, C. T., Volk, R. J., and Jackson, W. A. (1982b). Nitrate assimilation by decapitated corn root systems: effects of ammonium during induction. *Plant Science Letters*, **24**, 295–302.

Mehrer, J. and Mohr, H. (1989). Ammonium toxicity: description of the syndrome in *Sinapis alba* and the search for its causation. *Physiologia Plantarum*, 77, 545-54.

Mengel, K. and Kirkby, E.A. (1987). *Principles of plant nutrition* (4th edn). International Potash Institute, Bern.

Mengel, K. and Viro, M. (1978). The significance of plant energy status for the uptake and incorporation of NH_4^- nitrogen by young rice plants. *Soil Science and Plant Nutrition*, 24, 407-16.

Michael, G., Schumacher, H., and Marschner, H. (1965). Uptake of ammonium and nitrate nitrogen from labelled ammonium nitrate and their distribution in the plant. *Zeitschrift für Pflanzenernährung und Bodenkunde*, 110, 225-38.

Ota, K. and Yamamoto, Y. (1989). Promotion of assimilation of ammonium ions by simultaneous application of nitrate and ammonium ions in radish plants. *Plant and Cell Physiology*, 30, 365-71.

Pate. J.S. (1973). Uptake, assimilation and transport of nitrogen compounds by plants. *Soil Biology and Biochemistry*, 5, 109-19.

Pilbeam, D.J. and Kirkby, E.A. (1990). The physiology of nitrate uptake. In *Nitrogen in higher plants* (ed. Y.P. Abrol), pp. 39-64. Research Studies Press Ltd, Taunton.

Radin, J.W. (1975). Differential regulation of nitrate reductase induction in roots and shoots of cotton plants. *Plant Physiology*, 55 178-82.

Raven, J.A. (1985). Regulation of pH and generation of osmolarity in vascular plants:a cost benefit analysis in relation to efficiency of use of energy, nitrogen and water. *New Phytologist*, 101, 25-77.

Raven, J.A. and Smith, F.A. (1976). Nitrogen assimilation and transport in vascular land plants in relation to intracellular pH regulation. *New Phytologist*, 76, 415-31.

Schweizer, P. and Erismann, K.H. (1985). Effect of nitrate and ammonium nutrition of nonnodulated *Phaseolus vulgaris* L. on phosphoenolpyruvate carboxylase and pyruvate kinase activity. *Plant Physiology*, 78, 455-8.

Smith, R.G. Vanlerberghe, G.C., Stitt, M., and Turpin, D.H. (1989). Short-term metabolite changes during transient ammonium assimilation by the N-limited green alga *Selenastrum minutum*. *Plant Physiology*, 91, 749-55.

Steingröver, E., Woldendorp, J., and Sijtsma, L. (1986). Nitrate accumulation and its relation to leaf elongation in spinach leaves. *Journal of Experimental Botany*, 37, 1093-102.

Stulen, I. (1990). Interactions between carbon and nitrogen metabolism in relation to plant growth and productivity. In *Nitrogen in higher plants* (ed. Y.P. Abrol), pp. 297-312. Research Studies Press Ltd., Taunton.

van Beusichem, M.L., Baas, R., Kirkby, E.A., and Nelemans, J.A. (1985). Intracellular pH regulation during NO_3^- assimilation in shoot and roots of *Ricinus communis*. *Plant Physiology*, 78, 768-73.

van Beusichem, M.L., Kirkby, E.A., and Baas, R. (1988). Influence of nitrate and ammonium nutrition on the uptake, assimilation and distribution of nutrients in *Ricinus communis*. *Plant Physiology*, 86, 914-21.

6. Uptake and assimilation of nitrate under nitrogen limitation

C.M. LARSSON, M. MATTSSON,* P. DUARTE,
M. SAMUELSON, E. ÖHLÉN, P. OSCARSON,*
B. INGEMARSSON, M. LARSSON, and
T. LUNDBORG*

*Department of Botany, Stockholm University, S-106 91 Stockholm, Sweden, and *Department of Crop Genetics and Plant Breeding, Swedish Agricultural University, S-268 00 Svalöv, Sweden.*

Introduction

Nitrogen concentrations in plant tissue normally range from 0.5–5 per cent of total plant dry matter, depending on plant species, tissue type, tissue age, and nitrogen availability. Nitrogen reservoirs in soils are, with few exceptions, massive compared to the amount of nitrogen contained in living biomass (Post *et al.* 1985; Haynes 1986). However, release of nitrogen in mineralization may be slow, and its availability to individual plants further limited by factors affecting its mobility, by intra- and interspecific competition, or by denitrification. Thus, nitrogen availability may be a major factor limiting plant productivity in natural stands. Likewise, nitrogen input is required in agriculture to compensate for output (harvest, gaseous losses, runoff, and leaching) and may often be yield-limiting (detailed reviews of these subjects are available in Haynes, 1986).

While not neglecting the importance of abiotic factors other than nitrogen nor the role of interactions with other organisms in the rhizosphere, it is clear that plants would benefit greatly from inherent abilities to acclimatize to restrictions in nitrogen availability. Regulation of nitrogen assimilation and physiological responses to nitrogen limitation have consequently been central areas in plant physiology for a considerable time (reviewed by Clarkson 1986; Jackson *et al.* 1986; Haynes 1986). For two general reasons, however, many of the data currently available are difficult to interpret within the context of long-term nitrogen-limited growth. First, experimental procedures often induce a transient imbalance between nitrogen availability and nitrogen demand, whereas experiments on balanced nitrogen limitation over long periods (thus allowing for acclimatization) are scarce. Second, with regulation there is often understood to be limitation; that is, certain rate-limiting steps adjust the overall rate of nitrogen assimilation to the demand set by growth. This is not a relevant approach in relation to nitrogen-limited growth, since the ultimate limitation is the rate by which nitrogen is made available to the plant, not by any of the steps devoted to processing of nitrogen in the plant. Strategies for investigating stable nitrogen-limited growth in water-culture and basic characteristics of mainly root function under these conditions will be dealt with in this chapter.

Nutrient application in water-culture: relations to plant–soil systems

The traditional, and for many purposes justified, method of nutrient application in water-culture is to supply a balanced mixture of elements to the plant as a single dose, and then to renew completely the solution at regular time intervals. In studies of nutrient limitation this batch procedure may lead to severe fluctuations in nutrient concentrations, but this can be avoided by using flowing nutrient solution techniques (Asher and Edwards 1983). Here, nutrient solution is continuously pumped over the root system, thus decreasing nutrient depletion while in contact with the roots to practically zero. Modern microcomputer-controlled versions of such systems which allow for simultaneous control of concentration and uptake have also been described (Blom-Zandstra and Jupijn 1987; Glass *et al.* 1987).

Any nutrient added at any concentration will, however, not be limiting in a strict sense, since the source of nutrients is inexhaustible. This means that each ion taken up is instantaneously replaced by another ion of the same species. Above the lowest concentration required to support maximum growth, uptake will be controlled by the sink strength of the plant which, in turn, depends mainly on the growth rate and developmental stage (Allen *et al.* 1986). Below this concentration, the sink strength depends on the properties of the acquisition system. Put in a plant–soil context, this would represent the perhaps less common situation where water moves freely through the root system. In reality, water movement is normally directed to the root system, ultimately driven by the water potential gradient between the soil and the ambient air. Nutrients move to roots both with the water, and by diffusion (detailed analysis carried out by Nye and Tinker 1977). The ultimate limitation to the absolute amount of nutrients received by the roots is, under these circumstances, set by the rate of nutrient release at the soil–water interface. The concentration of nutrients at the root–water interface becomes primarily a function of the uptake properties of the roots. Viewed in this way, the concentration at the root surface should be regarded as a dependent variable rather than a decisive variable when put in relation to uptake.

It thus seems that experimental studies of nutrient-limited growth in water-culture require methodology for adequate control of nutrient fluxes.

A major contribution in this regard is the programmed nutrient addition technique, in which nutrients are added in accordance to, or in fixed relation to, a previously established growth curve (Asher and Cowie 1970, reviewed by Asher and Edwards 1983). Ingestad and co-workers (Ingestad and Lund 1979, reviewed by Ingestad 1982; Ingestad and Lund 1986) developed the relative addition rate (RA) concept in which nutrients are added at a fixed rate (i.e. relative to the amounts of nutrients already in biomass), set to produce a desired relative rate of plant nutrient increment (RN). Assuming exponential growth with stable internal nutrient levels after acclimatization, relative growth rates (RGR) should equal RA, i.e.

$$N_t = N_0 e^{RAt} \tag{6.1}$$

where N_0 and N_t are the nutrient contents initially and after time t, respectively, and

$$RA = RN = RGR \tag{6.2}$$

at nutrient limitation. A further prerequisite is that shoot RGR equals root RGR. Thus, the exponential increase in root size will match the exponential increase in nitrogen additions. If these prerequisits are met, the RA principle resembles that of chemostats, i.e. the nutrient input is constant relative to the amount of biomass absorbing the nutrients. In cultures of higher plants, however, ontogeny-linked changes may occur in both growth rates, growth patterns, and tissue nitrogen levels. Nevertheless, in a number of woody perennials, close agreements between observed data and data predicted from (eqn 6.1) and (eqn 6.2) have been obtained (Ingestad and Lund 1979; Ingestad 1980; Ingestad and Kähr 1985), which probably can be attributed to a great extent to the long generation time and long life span of each individual organ in these species.

Dry matter production and allocation under nitrogen limitation

In the following experiments in which plants kept initially in a nitrogen-free basal medium and then fed a complete nutrient solution, with nitrate as the nitrogen source, will be reviewed. The addition rate has been calculated for nitrogen, i.e. RA will denote relative rates of nitrate-N additions, whereas other elements are present in surplus. The studies have centered on a variety of annual crop species (data on *Pisum sativum* L. cv. Marma, and *Hordeum vulgare* L. cv. Golf will be dealt with, unless otherwise indicated), and the aquatic monocot *Lemna gibba* L. (duckweed). The *Lemnaceae* are in some respects highly suitable for these studies. They propagate rapidly and purely vegetatively in laboratory culture, thus randomizing ontogenetic variations with time and between cultures, which justifies the analogy to chemostats. Furthermore, the slender, unbranched roots, consisting of three to four

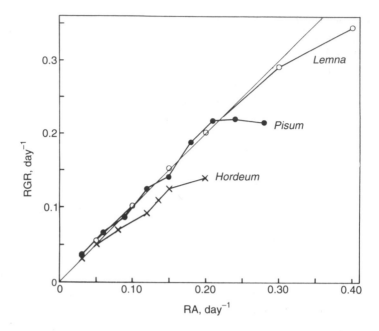

Fig. 6.1. Relative growth rate (RGR) of *Lemna*, *Pisum*, and *Hordeum* as a function of the relative addition rate (RA) of nitrate-N. Growth was measured as dry matter increments over the interval 25 to 30 days after sowing (*Pisum* and *Hordeum*), or fresh weight increments in cultures of indefinite age (*Lemna*). The continuous line indicates RGR = RA. Data for *Lemna* and *Pisum* from Oscarson *et al.* (1989a, with permission).

layers of cells surrounding the vascular tissue with little microscopically evident longitudinal heterogeneity, provides an apparently homogeneous material for studies of root activity. Interpreted cautiously, *Lemna* can thus serve as a 'reference' for material with changing distribution of matter and activity with age.

Figure 6.1 shows RGR during vegetative growth measured as total dry weight increments as a function of RA. For two species, *Lemna* and *Pisum*, good agreement is observed between RA and RGR, whereas there is some discrepancy in *Hordeum* (also observed with *Triticum* and *Helianthus*, data not shown). Root RGR is less affected by RA in the annuals, and the discrepancy between RA and root RGR increases progressively with higher RA (Fig. 6.2A). Thus, root contributions to total dry weight will change with time, but will invariably decline with increasing RA (Fig. 6.2B). Nevertheless, stability is sufficient to allow some calculations of utilization and translocation rates for other constituents, as affected by RA. Table 6.1 gives data for carbon in strongly nitrogen-limited *Pisum* cultured at RA 0.06

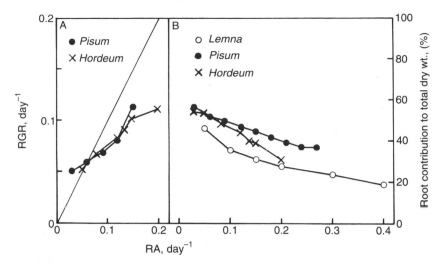

Fig. 6.2. A. Root RGR as a function of RA. B. Root contributions to total dry weights as a function of RA. Other details as in legend to Fig. 6.1. Data for *Lemna* and *Pisum* from Oscarson *et al.* (1989a, with permission).

day^{-1}, based on measurements of net CO_2 fixation, respiration in roots, carbon flux in the xylem, measured growth rates of roots and shoots, root and shoot elemental composition, and, by inference, rates of phloem translocation to roots. The overall relative rate of carbon utilization (RC) corresponds well to that preset by RN (0.067 vs 0.060). An unbalanced loss of carbon corresponding to approximately 15 per cent of net carbon fixation in shoots was recorded over the whole RA range studied. It is likely that this carbon fraction is lost as root exudates.

Table 6.1. Rates of carbon acquisition, translocation, and utilization in *Pisum* acclimatized to RA 0.06 day^{-1}. The higher figure for translocation to roots via phloem is calculated assuming that the unbalanced carbon loss occurs in roots via, e.g. root exudation

Process	Rate (mg C g^{-1} plant dry weight day^{-1})
Net carbon in shoots	40.8
Carbon increment in shoots	14.4
Carbon increment in roots	11.5
Translocation to shoot (xylem)	4.6
Translocation to root (phloem)	25.4–31.0
Root carbon loss in respiration	9.3
Unbalanced carbon loss	5.6

Nitrate acquisition at limited nitrogen supply

Net uptake kinetics

A wealth of information in the literature shows that if nitrogen or any other macronutrient is witheld from a plant, the uptake rates after readdition are substantially higher than before starvation (although in the case of nitrate a proper induction might be required; see, for example, Lee and Rudge 1986; Clarkson 1986; Oscarson *et al.* 1989*b*, and references therein). Prolonged starvation may cause uptake rates to decrease again. Clarkson (1986) reconciled these observations as representing initial de-repression of carrier synthesis, followed by retardation of growth and eventual deterioration of tissue. The transient increase in uptake rates can be viewed as a response to nitrogen demand, i.e. approximately RGR, which in terms of dry matter increments is affected very little by a short period of nitrogen starvation (Lee and Rudge 1986; Mattsson *et al.* 1988). The increase in nitrogen uptake thus often represents recovery rather than acclimatization to the restricted nitrogen supply. It is also probably inappropriate to use nitrogen sufficient growth as a 'control' for the behaviour under nitrogen limitation, since the two growth conditions may require fundamentally different regulatory modes. It would be more appropriate to study a range of nitrogen-limited growth rates, as can be done with the RA technique.

Data in Fig. 6.3 show that V_{max} for net nitrate uptake in RA-limited cultures of several species and cultivars increases on a dry weight basis with RA up to a transition point, which in the cases of *Lemna* and *Pisum* coincides with the transition from growth-limiting to non-limiting RA (cf. Fig. 6.1). The transition occurs at lower values of RA in the three barley cultivars. Data for *Lemna* fronds are included, showing some uptake in this tissue also, but with much lower V_{max} values. Less variation was obtained for the apparent K_m values (Oscarson *et al.* 1989*b*). These data thus show that when cultures maintained at a range of values of RA are compared, uptake rates actually decrease with decreased nitrogen availability, contrary to what might have been expected from traditional experiments on short-term nitrogen limitation.

Another experimental approach is that of spatial nutrient limitation, in which nitrogen supply is restricted to only one part of the root system. The response is usually a decline in growth of the depleted root part, whereas increased growth and proliferation (Drew and Saker 1975; Lambers *et al.* 1982) and increased uptake rate per unit root weight (Drew and Saker 1975) is recorded in the part of the root receiving nitrogen. However, problems arise again with the definition of limitation. Although there is a spatial limitation, this does not imply that the absolute amount of nitrogen present is limiting growth. We may thus again be confronted with a recovery situa-

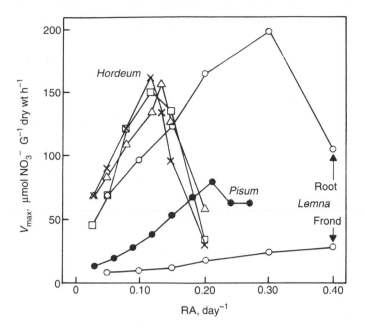

Fig. 6.3. V_{max} for net nitrate uptake expressed on a root dry weight basis or frond dry weight basis as a function of RA. Net nitrate uptake was determined in depletion experiments from the rate of decline in UV-absorbtion (202 or 220 nm). Initial nitrate concentrations ranged from 10 to 150 μM. Data are shown for three genotypes of *Hordeum vulgare*; cv. Golf (crosses), cv. Mette (squares), and cv. *laevigatum* (triangles). Data for *Lemna* and *Pisum* from Oscarson *et al.* (1989*b*, with permission).

tion rather than with a situation where the plants acclimatize to limitation in nitrogen availability.

In our approach to this problem, we cultured barley at RA 0.09 day^{-1} (i.e. under overall nitrogen limitation) where the nitrate was fed unevenly (in 100:0, 80:20, 70:30, and 60:40 ratios) or evenly (control) to root halves separated in a split root system. An initial phase of uneven growth rates in the unevenly treated root parts led to contributions to total root dry matter ranging from approximately 25 per cent (0 per cent nitrogen root) to 75 per cent (100 per cent nitrogen root). During the subsequent 15 days of culturing, however, root weight proportions remained constant, indicating that RGR of the root parts at this stage was unaffected by the nitrogen input ratios. V_{max} for net nitrate uptake was at this stage only slightly affected by the different input ratios, with the notable exception of the 0 per cent nitrogen root (Fig. 6.4). These data conform to those shown in Fig. 6.3

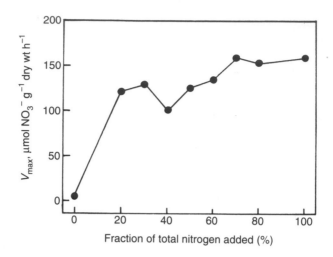

Fig. 6.4. V_{max} for net nitrate uptake in *Hordeum* on a root dry weight basis, determined as in legend to Fig. 6.3., as a function of the fraction of the total nitrogen addition fed to individual root parts in split root systems. Overall RA was 0.09 day^{-1} in all cases. Data for acclimatized plants (see text for explanations).

in the sense that there is a certain relation between V_{max} and RGR in the acclimatized stage. They also reinforce the observation that external nitrate is required for full induction of uptake machinery (review by e.g. Larsson and Ingemarsson 1989). Induction apparently cannot be mediated by any sort of signal transferred from the nitrate-treated root to the root receiving no external nitrogen.

Influx/efflux

Net uptake data are sometimes difficult to interpret because of unknown contributions to net uptake rates of the unidirectional flux components, influx and efflux. Nitrate efflux is often substantial, particularly at high levels of nitrate nutrition (Morgan *et al.* 1973; Breteler and Nissen 1982). In case of RA-limited cultures, however, short-term measurements of nitrate influx using the short-lived isotope ^{13}N showed that at external concentrations approaching saturation, efflux is a minor (*Pisum*; Oscarson *et al.* 1987) or insignificant (*Lemna*; Ingemarsson *et al.* 1987a) component of net uptake. Significant efflux was, however, observed at lower concentrations in both species. An experiment with *Hordeum* (Fig. 6.5), where influx and net uptake rates were compared at different external nitrate concentrations, point in the same direction; that is, no discrepancy was observed close to saturation, whereas discrepancy increased at lower concentrations. It is thus

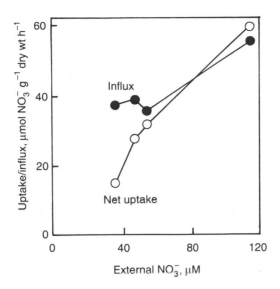

Fig. 6.5. Comparison of net nitrate uptake rates and nitrate influx rates on a plant dry weight basis, as a function of external nitrate concentration in *Hordeum vulgare* cv. Flare. Net uptake and influx were determined simultaneously, using ^{13}N-nitrate as an influx tracer (methodology given by Oscarson *et al.* 1987).

concluded that values of V_{max} represent valid estimates of maximum influx rates (I_{max}), whereas K_m can only be used as a 'practical' measure of affinity, unrelated as it is to the affinity of the influx mechanism.

The physiological significance of nitrate efflux is currently not clear. Regulatory roles of efflux were proposed by Deane-Drummond and Glass (1983a) and subsequently in the nitrate/nitrate exchange and substrate cycling models put forward by Deane-Drummond (1984, 1986). The common denominator is that uptake rates change (e.g. in relation to nitrogen limitation) as a consequence of changed efflux rates. Other experiments (using ^{13}N-labelled nitrate) indicate that change from nitrate-sufficient to nitrogen-limited growth causes, when measured at near-saturating nitrate concentrations, increases in influx whereas efflux is less affected (Lee and Drew 1986; Oscarson *et al.* 1987; Ingemarsson *et al.* 1987a). Likewise, as discussed above, differences in V_{max} obtained at a range of nitrogen limitations are most probably related to changed I_{max}. Under conditions of nitrogen limitation, however, concentrations at the root surface must be far from saturation, and probably considerably below K_m (Oscarson *et al.* 1989b). Efflux appears here to be an increasingly important parameter (cf. Fig. 6.5, Oscarson *et al.* 1987; Ingemarsson *et al.* 1987a). The significance of efflux should be assessed in experiments employing low external concentrations comparable to those likely to be encountered during nitrogen limitation. No such experiments have, to the authors' knowledge, yet been carried out.

Uptake in relation to internal nitrate

A well-documented aspect of nitrate nutrition is that of inducibility of the putative plasma membrane nitrate transporter (Larsson and Ingemarsson 1989). In a recent examination of this problem, using ^{13}N-nitrate as an influx tracer, Siddiqi *et al.* (1989) concluded that influx rates in barley roots during induction were proportional to the resulting tissue nitrate concentration up to approximately 50 μmol g^{-1} fresh weight. A decline in influx was observed at higher tissue nitrate concentrations, possibly indicating negative nitrate feedback.

In RA-limited cultures of *Pisum* and *Hordeum*, large increases in V_{max} for nitrate uptake were observed within a very narrow range of root nitrate concentrations. Declines in V_{max} at the highest values of RA coincided with drastic increases in root nitrate concentrations (Fig. 6.6 cf. Fig. 6.3). These data do not readily support the contention of a quantitative relationship between root nitrate concentrations and the extent of expression of the nitrate uptake system at low RA, but, in agreement with the data of Siddiqi *et al.* (1989), they indicate that nitrate feedback becomes operative at high tissue nitrate concentrations. Data obtained for *Lemna* are possible to reconcile with the latter hypothesis. RA-limited *Lemna* grown on ammonium and

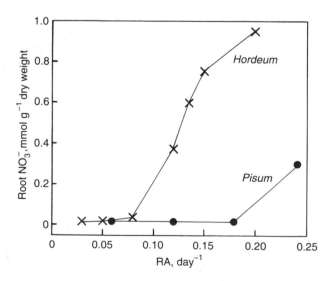

Fig. 6.6. Root nitrate levels in *Pisum* and *Hordeum* as a function of RA.

in the presence of tungstate (which leads to synthesis of an inactive nitrate reductase) take up nitrate following a proper induction with 2.5 μmol nitrate g^{-1} fresh weight for 24 h. If induction is made with 250 μmol nitrate, very high internal nitrate concentrations are obtained, and uptake rates decline to zero. More interestingly, this cessation of uptake could be specifically attributed to inhibition of influx (measured as uptake of ^{13}N) and to nitrate itself, since accumulation of any product of nitrate assimilation was blocked by the tungstate treatment (Ingemarsson *et al.* 1987*a*).

Roles of C,N-intermediates and the ammonium effect

Metabolic regulation of nitrate uptake via C,N-intermediates would, in contrast to regulation by nitrate itself, have the advantage that the regulator is mobile in both xylem and phloem; information on nitrogen status could thus be transmitted from roots to shoots and vice versa. Its disadvantage would be that not only primary assimilation, but also protein turnover and (in shoots) photorespiration would affect C,N-intermediates in the transport pool both quantitatively and qualitatively (Joy 1988).

Indeed, cycling of nitrogen from shoots to roots is often substantial, even under nitrogen limitation (Lambers *et al.* 1982). In *Hordeum* growing at RA 0.08 day^{-1}, phloem-delivered nitrogen makes up at least 50 per cent of the nitrogen finally integrated in root tissue (M. Mattsson, unpublished data). It also appears that, at least during nitrogen sufficient growth, a substantial portion of the nitrogen (notably recently assimilated nitrogen) delivered to roots rapidly recycles — i.e. returns to the shoot without significant mixing with the bulk nitrogen of the root (Cooper and Clarkson 1989). Through cycling, increased shoot demand for amino acids, for example, would also affect the root amino acid pool, thereby possibly offsetting feedback inhibition of nitrate uptake (Cooper and Clarkson 1989). In barley plants deprived of sulphur, influx rates of nitrate decreased, but was rapidly restored by sulphur or methionine replenishment (Clarkson *et al.* 1989). It was suggested that sulphur deficiency caused a build-up of non-sulphur amino acids, thereby decreasing nitrate influx. Likewise, there are data obtained in experiments on microalgae and cyanobacteria that are compatible with the view that nitrate uptake is regulated by some, as yet unidentified, C,N-intermediate (Larsson and Larsson 1987).

Ammonium uptake is often, in the short-term, considerably more rapid than nitrate uptake, as is also its assimilation (Ingemarsson *et al.* 1984). Thus, in the presence of ammonium, amino acid pools including, perhaps, a regulatory C,N-derivative, increase. It is also an almost general phenomenon that addition of ammonium leads to suppression of nitrate uptake, although there are cases where the pattern is more complex (Breteler and Siegerist 1984). Experiments using ^{13}N-labelled nitrate by Lee and Clarkson (1986), Ingemarsson *et al.* (1987*b*), Oscarson *et al.* (1987) and recently by

Lee and Drew (1989) have all pointed to nitrate influx being effected by the presence of ammonium. This, however, contrasts with data on ^{36}Cl-chlorate (a nitrate analogue) fluxes that indicate stimulation of efflux by ammonium (Deane-Drummond and Glass 1983*b*; Deane-Drummond 1986).

Regardless of this contradiction, it appears logical that if ammonium inhibits nitrate uptake via build-up of regulatory C,N-intermediates, then its effects should be abolished by inhibitors of ammonium assimilation. Relief from ammonium inhibition by the glutamine synthetase inhibitor methionine sulphoximine (MSO) is a strong argument in favour of the hypothetic regulatory role of C,N-intermediates in cyanobacteria and eukaryotic micro-algae (Larsson and Larsson 1987). A similar effect of MSO was recorded under some conditions in *Phaseolus* roots (Breteler and Siegerist 1984). By contrast, MSO could not relieve nitrate uptake from inhibition by ammonium in RA-limited *Lemna* (Ingemarsson *et al.* 1987*b*). During long-term growth of *Lemna* under RA limitation with a mixed nitrogen source, (2 ammonium:3 nitrate, molar ratio), the expression of the nitrate uptake system is actually enhanced (Fig. 6.7). Ammonium inhibition is possibly limited to phases of excess and non-steady state uptake rates causing

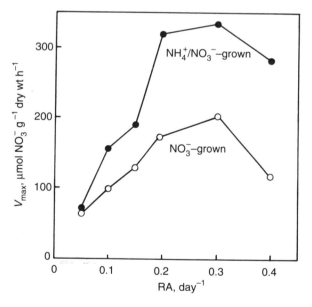

Fig. 6.7. V_{max} for net nitrate uptake in *Lemna* roots as a function of RA. Comparison between nitrate-grown (cf. Fig. 6.3.) and ammonium/nitrate-grown (molar ratio 2:3) cultures. Measurements of nitrate uptake were, in both cases, performed in the absence of ammonium.

depolarization of the plasma membranes, thus reducing driving forces for co-transport mechanisms (Ullrich *et al.* 1984). Thus, ammonium appears to be a non-specific inhibitor of anion transport in *Lemna*, as judged from experiments where phosphate uptake was also inhibited (Ullrich *et al.* 1984). Similar effects were, however, not observed in barley (Lee and Drew 1989). It does not appear possible to couple the ammonium effect to accumulation of a regulatory C,N-intermediate (cf. also Lee and Drew 1989), and any involvement of such intermediates in regulation of nitrate uptake in higher plants is still hypothetical.

Relations of root activity to root nitrogen concentrations

Plots of root RGR, V_{max} for nitrate uptake, and (in *Pisum*) root respiration vs. root nitrogen concentrations yield fairly linear relationships in both *Lemna* and *Pisum* (Fig. 6.8). Extrapolated intercepts to zero activity show at least moderate agreement within each species. Since non-protein nitrogen in these cases makes up 5 per cent or less of total nitrogen, these data, in a strictly formal way, link tissue activity to tissue protein concentration. They also identify a minimum nitrogen concentration at zero activity which formally identifies the threshold nitrogen concentration required to sustain growth, and which may provide physiological indicator of senescent material, and material bound in structures which do not actively participate in plant work. 'Activity' is thus given by a proportionality factor (i.e. the

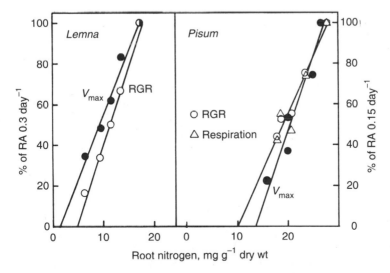

Fig. 6.8. Relationship between root activities and root nitrogen concentrations. Based partly on data from Oscarson *et al.* (1989*a,b*, with permission).

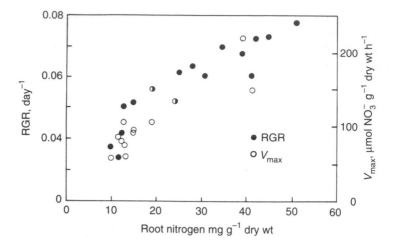

Fig. 6.9. As Fig. 6.8., but data for *Hordeum*.

slopes of the lines in Fig. 6.8) and a constant (i.e. the minimum nitrogen concentration). In *Hordeum* there is a curvilinear relationship (Fig. 6.9). This difference from the other two species can be attributed to the large build-up of soluble nitrogen pools, also when RA is still growth-limiting (cf. Figs 6.1 and 6.6), whereas at lower RAs the relationship may still very well be linear.

The simplest interpretation of these data is that under steady state nutrition and stable growth, synthesis of the putative nitrate transporter occurs in a fixed proportion to other proteins engaged in plant growth and maintenance. This would mean that the expression of the nitrate uptake system is not specifically regulated under nitrogen limitation, provided that nitrate is actually present in amounts sufficient to induce the transporter, but under general regulation in the same way as overall protein synthesis. Final elucidation of this point requires identification and quantification of the nitrate transport protein (Larsson and Ingemarsson 1989).

Concluding remarks: The growth–response nature of acclimatization to nitrogen limitation

Collectively, the data described in this chapter indicate two major responses to nitrogen limitation; changed protein concentration, and changed protein allocation with respect to its partitioning to roots and shoots. As a result, the plants will have a slower growth rate and, in relative terms, a larger root system at low values of RA. The capacity to acquire nitrate is a consequence of this growth response. This can be illustrated in plots where the relative V_{max}, i.e. nitrogen uptake per unit nitrogen (whole plant) and unit time, is plotted against whole plant RGR. The theoretical minimum uptake activity is set by relative $V_{max} = RGR$. At RA that is saturating or super-saturating for growth, relative V_{max} values come close to the minimum line, indicating that uptake rates under these conditions are effectively controlled by growth. The lower the RA employed, the higher the relative V_{max} becomes in both *Lemna* and *Hordeum*, whereas the opposite trend occurs in *Pisum* (Fig. 6.10). Since tissue nitrogen concentrations and V_{max} both decrease (cf. Figs 6.8 and 6.9), the increased relative V_{max} observed at low RA in *Lemna* and *Hordeum* depends almost exclusively on the increased relative sizes of the root systems (in the case of *Hordeum* only up to approximately RA 0.12 day^{-1}). The range of species investigated is presently too narrow to allow for generalizations. It can, however, be suggested on the basis of these observations that changed root:shoot partitioning is the main acclimatization response of the nitrate acquisition system to limiting RA. The poorer performance of *Pisum* can be explained by the normal habit of a legume to acclimatize to nitrogen deficiency by nodulation. This also implies

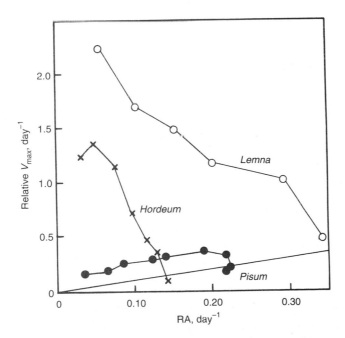

Fig. 6.10. Relationship between relative V_{max} (i.e. nitrogen taken up per unit nitrogen in whole plant and unit time) and whole plant RGR. The continuous line indicates relative $V_{max} = RGR$. Data for *Lemna* and *Pisum* from Oscarson *et al.* (1989*b*; with permission).

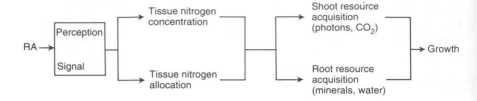

Fig. 6.11. Schematic representation of plant responses to RA, and their implications for plant resource acquisition.

that the uptake system is not under specific regulation during nitrogen-limited growth. The uptake system is, rather, under general regulation, and coordinated with other growth-related activities. Specific regulation in the form of feedback rate control comes into play only when RA approaches maximum RGR. Data discussed here indicate that nitrate itself may be an important influx regulator under these circumstances.

Acclimatization to RA is schematically illustrated in Fig. 6.11, which incorporates some of the elements of Thornley's model (Thornley 1972; Reynolds and Thornley 1982) for control of shoot and root growth rates, and shoot:root ratios in plants. Basically, the Thornley model considers shoot and root growth to be dependent on substrate pools in shoots and roots (notably C and N), and on the transport resistances involved in communicating substrates between organs. The applicability of the model to changes in shoot:root ratios induced by a variety of environmental perturbations was reviewed by Wilson (1988). There are, however, certain differences in reasoning between Fig. 6.11 and the Thornley model. Tissue nitrogen in Fig. 6.11 does not represent substrate pools but rather sizes of the assimilatory/utilizing machinery in shoots and roots. Data in Fig. 6.8 indicate that the relationship between growth and tissue nitrogen concentrations in *Lemna* and *Pisum* is linear, not that of a saturation curve as in the Thornley model. In *Hordeum*, a saturation curve is obtained, but this relationship is most likely explained by storage of substrates to be used at later stages of growth (Fig. 6.9).

The principal problem is, however, understanding the events leading to the growth response and thus to acclimatization to nitrogen limitation (Fig. 6.11, box). In the classical view (see, for example, Wilson 1988), nitrogen limitation increases the proportion of nitrogen taken up that is retained by roots, making less nitrogen available for translocation to shoots. As discussed previously, however, roots import considerable amounts of nitrogen from shoots, and it is equally feasible that nitrogen limitation increases the relative sink strength of the root for phloem-delivered nitrogen. Hormonal effects may be important here. Integration of current knowledge of hormonal regulation of root growth and development with data on acclimatization of the nitrate acquisition system would seem a fruitful approach in future studies.

Acknowledgements

Financial support from the Swedish Natural Science Research Council and the Swedish Council for Forestry and Agricultural Research is acknowledged.

References

Allen, S., Thomas, G. E., and Raven J. A. (1986). Relative uptake rates of inorganic nutrients by NO_3^-- and NH_4^+-grown *Ricinus communis* and by two *Plantago* species. *Journal of Experimental Botany*, **37**, 419–28.

Asher, C. J. and Cowie, A. M. (1970). Programmed nutrient addition — a simple method for controlling the nutrient status of plants. *Proceedings of the Australian Plant Nutrition Conference*, Section 1(b), pp. 28–32. Mt Gambier, South Australia.

Asher, C. J. and Edwards, D. G. (1983). Modern solution culture techniques. In *Encyclopedia of Plant Physiology*, Vol. 15A, Inorganic plant nutrition (ed. A. Läuchli and R. L. Bieleski), pp. 94–119. Springer Verlag, Berlin.

Blom-Zandstra, M. and Jupijn, G. L. (1987). A computer-controlled multi-titration system to study transpiration, OH^- efflux and nitrate uptake by intact lettuce plants (*Lactuca sativa* L.) under different environmental conditions. *Plant, Cell and Environment*, **10**, 545–50.

Breteler, H. and Nissen, P (1982). Effects of exogenous and endogenous nitrate concentration on nitrate utilization by dwarf bean. *Plant Physiology*, **70**, 754–9.

Breteler, H. and Siegerist, M. (1984). Effect of ammonium on nitrate utilization by roots of dwarf bean. *Plant Physiology*, **75**, 1099–103

Clarkson, D. T. (1986). Regulation of the absorption and release of nitrate by plant cells. A review of current ideas and methodology. In *Fundamental, ecological and agricultural aspects of nitrogen metabolism in higher plants* (ed. H. Lambers, J. J. Neeteson, and I. Stulen), pp. 3–27. Martinus Nijhoff Publishers, Dordrecht.

Clarkson, D. T., Saker, L. R., and Purves, J. V. (1989). Depression of nitrate and ammonium transport in barley plants with diminished sulphate status. Evidence of co-regulation of nitrogen and sulphate intake. *Journal of Experimental Botany*, **40**, 953–63.

Cooper, H. D. and Clarkson, D. T. (1989). Cycling of amino–nitrogen and other nutrients between shoots and roots in cereals — a possible mechanism integrating shoot and root in the regulation of nutrient uptake. *Journal of Experimental Botany*, **40**, 753–62.

Deane-Drummond, C. E. (1984). Mechanism of nitrate uptake in *Chara corallina* cells: lack of evidence for obligatory coupling to proton pump and a new NO_3^-/NO_3^- exchange model. *Plant, Cell and Environment*, **7**, 317–23.

Deane-Drummond, C.E. (1986). Nitrate uptake into *Pisum sativum* cv. Feltham First seedlings: commonality with nitrate uptake into *Chara corallina* and *Hordeum vulgare* through a substrate cycling model. *Plant, Cell and Environment,* **9**, 41–48.

Deane-Drummond, C.E. and Glass, A.D.M. (1983*a*). Short-term studies of nitrate uptake into barley plants using ion-specific electrodes and $^{36}ClO_3^-$. I. Control of net uptake by NO_3^- efflux. *Plant Physiology,* **73** 100–4.

Deane-Drummond, C.E. and Glass, A.D.M. (1983*b*). Short-term studies of nitrate uptake into barley plants using ion-specific electrodes and $^{36}ClO_3^-$. II. Regulation of NO_3^- efflux by NH_4^+. *Plant Physiology,* **73**, 105–10.

Drew, M.C. and Saker, L.R. (1975). Nutrient supply and the growth of the seminal root system in barley. II. Localized, compensatory increases in lateral root growth and rates of nitrate uptake when nitrate supply is restricted to only part of the root system. *Journal of Experimental Botany,* **26**, 79–90.

Glass, A.D.M., Saccomani, M., Crookall, G., and Siddiqi, M. (1987). A microcomputer-controlled system for the automatic measurement and maintenance of ion activities in nutrient solutions during their absorption by intact plants in hydroponic facilities. *Plant, Cell and Environment,* **10**, 375–81.

Haynes, R.J. (1986). *Mineral nitrogen in the plant-soil system.* Academic Press Inc., Orlando, Florida.

Ingemarsson, B., Johansson, L., and Larsson, C.M. (1984). Photosynthesis and nitrogen utilization in exponentially growing nitrogen-limited cultures of *Lemna gibba. Physiologia Plantarum,* **62**, 363–9.

Ingemarsson, B., Oscarson, P., af Ugglas, M., and Larsson, C.M. (1987*a*). Nitrogen utilization in *Lemna.* II. Studies of nitrate uptake using $^{13}NO_3^-$. *Plant Physiology,* **85**, 860–4.

Ingemarsson, B., Oscarson, P., af Ugglas, M., and Larsson, C.M. (1987*b*). Nitrogen utilization in *Lemna.* III. Short-term effects of ammonium on nitrate uptake and nitrate reduction. *Plant Physiology,* **85**, 865–7.

Ingestad, T. (1980). Growth, nutrition, and nitrogen fixation in grey alder at varied rate of nitrogen addition. *Physiologia Plantarum,* **50**, 353–64.

Ingestad, T. (1982). Relative addition rate and external concentration. Driving variables used in plant nutrition research. *Plant, Cell and Environment,* **5**, 443–53.

Ingestad, T. and Kähr, M. (1985). Nutrition and growth of coniferous seedlings at varied relative nitrogen addition rate. *Physiologia Plantarum,* **65**, 109–16.

Ingestad, T. and Lund, A.B. (1979). Nitrogen stress in birch seedlings. I. Growth technique and growth. *Physiologia Plantarum,* **45**, 137–48.

Ingestad, T. and Lund, A.B. (1986). Theory and techniques for steady state mineral nutrition and growth of plants. *Scandinavian Journal of Forestry Research,* **1**, 439–53.

Jackson, W.A., Pan, W.L., Moll, R.H., and Kamprath, E.J. (1986). Uptake, translocation and reduction of nitrate. In *Biochemical basis of plant breeding,* Vol 2, Nitrogen metabolism (ed. C.A. Neyra), pp. 73–108. CRC Press, Boca Raton.

Joy, K.W. (1988). Ammonia, glutamine, and asparagine: a carbon–nitrogen interface. *Canadian Journal of Botany,* **66**, 2103–9.

Lambers, H, Simpson, R.J., Beilharz, V.C., and Dalling, M. (1982). Growth and translocation of C and N in wheat (*Triticum aestivum*) grown with a split root system. *Physiologia Plantarum*, **56**, 421–9.

Larsson, C.M. and Ingemarsson, B. (1989). Molecular aspects of nitrate uptake in higher plants. In *Molecular and genetic aspects of nitrate assimilation* (ed. J.L. Wray and J.R. Kinghorn), pp. 3–14. Oxford Science Publications, Oxford.

Larsson, C.M. and Larsson, M. (1987). Regulation of nitrate utilization in green algae. In *Inorganic nitrogen metabolism* (ed. W.R. Ullrich, P.J. Aparicio, P.J. Syrett, and F. Castillo), pp. 203–7. Springer-Verlag, Berlin.

Lee, R.B. and Clarkson, D.T. (1986). Nitrogen-13 studies of nitrate fluxes in barley roots. I. Compartmental analysis from measurements of ^{13}N efflux. *Journal of Experimental Botany, 37*, 1753–67.

Lee, R.B. and Drew, M.C. (1986). Nitrogen-13 studies of nitrate fluxes in barley roots. II. Effect of plant N-status on the kinetic parameters of nitrate influx. *Journal of Experimental Botany*, **37**, 1768–79.

Lee, R.B. and Drew, M.C. (1989). Rapid, reversible inhibition of nitrate influx in barley by ammonium. *Journal of Experimental Botany*, **40**, 741–52.

Lee, R. and Rudge, K.A. (1986). Effects of nitrogen deficiency on the absorption of nitrate and ammonium by barley plants. *Annals of Botany, 67*, 471–86.

Mattsson, M., Lundborg, T., and Larsson, C.M. (1988). Nitrate utilization in barley: relations to nitrate supply and light/dark cycles. *Physiologia Plantarum*, **73**, 380–6.

Morgan, M.A., Volk, R.J., and Jackson, W.A. (1973). Simultaneous influx and efflux of nitrate during uptake by perennial ryegrass. *Plant Physiology*, **51**, 267–72.

Nye, P.H. and Tinker, P.B. (1977) *Solute movement in the soil-root system*, Studies in ecology, Vol. 4. Blackwell Scientific Publications, Oxford.

Oscarson, P., Ingemarsson, B., af Ugglas, M., and Larsson, C.M. (1987). Short-term studies of NO_3^- uptake in *Pisum* using $^{13}NO_3^-$. *Planta, 170*, 550–5.

Oscarson, P., Ingemarsson, B., and Larsson, C.M. (1989*a*). Growth and nitrate uptake properties of plants grown at different relative rates of nitrogen supply. I. Growth of *Pisum* and *Lemna* in relation to nitrogen. *Plant, Cell and Environment, 12*, 779–85.

Oscarson, P., Ingemarsson, B., and Larsson, C.M. (1989*b*). Growth and nitrate uptake properties of plants grown at different relative rates of nitrogen supply. II. Activity and affinity of the nitrate uptake system in *Pisum* and *Lemna* in relation to nitrogen availability and nitrogen demand. *Plant, Cell and Environment, 12*, 787–94.

Post, W.M., Pastor, J., Zinke, P.J., and Stangenberger, A.G. (1985). Global patterns of soil nitrogen storage. *Nature, 317*, 613–16.

Reynolds, J.F. and Thornley, J.H.M. (1982). A shoot:root partitioning model. *Annals of Botany, 49*, 585–97.

Siddiqi, M.Y., Glass, A.D.M., Ruth, T.J., and Fernando, M. (1989). Studies of the regulation of nitrate influx by barley seedlings using $^{13}NO_3^-$. *Plant Physiology, 90*, 806–13.

Thornley, J.H.M. (1972). A balanced quantitative model for root:shoot ratios in vegetative plants. *Annals of Botany, 36*, 431–41.

Ullrich, W.R., Larsson, M., Larsson, C.M., Lesch, S., and Novacky, A. (1984). Ammonium uptake in *Lemna gibba* G1, related membrane potential changes, and inhibition of anion uptake. *Physiologia Plantarum,* **61**, 369–76.

Wilson, J.B. (1988). A review of evidence on the control of shoot:root ratio, in relation to models. *Annals of Botany,* **61**, 433–49.

7. NO_3^- assimilation in root systems: with special reference to *Zea mays* (cv. W64A × W182E)

ANN OAKS and DEBORAH M. LONG

Department of Botany, University of Guelph, Guelph, Ontario N1G 2W1, Canada

Nitrate reductase (NR, EC 1.6.6.1), the first enzyme in the sequence of reactions involved in the assimilation of NO_3^-, was described initially by Evans and Nason (1953). It was shown to require reduced pyridine nucleotides, NADH or NADPH (Evans and Nason 1953), and to be substrate inducible (Tang and Wu 1957). Early results from Hageman's laboratory showed that light was also required for the induction (Hageman and Flesher 1960). When NO_3^- was removed from the medium (Schrader *et al.* 1968), or when the lights were turned off (Travis *et al.* 1969), nitrate reductase activity disappeared. More recently this regulation with respect to NO_3^- or light has been demonstrated to occur at the protein and mRNA levels (Melzer *et al.* 1989; Somers *et al.* 1983, Remmler and Campbell 1986; Oaks *et al.* 1988; Bowsher *et al.* 1991; Deng *et al.* 1990). End products of NO_3 assimilation, NH_4^+ or amino acids, have been shown to inhibit the induction of NR in *Neurospora crassa* (Marzluf 1981). In higher plants, similar inhibitions with low levels of NH_4^+ or amino acids have not been so clearly defined (Oaks *et al.* 1977, 1979; Oaks 1979). Although in tissue culture systems, amino acids clearly inhibit the induction of NR (Filner 1966), a closer investigation indicated that it was the uptake of NO_3^- and not its conversion to NO_2^- that was the phase of metabolism most sensitive to amino acid additions (Heimer and Filner 1970). Of the amino acids tested, glutamine appears to hold the key in the *Neurospora* system (reviewed by Marzluf 1981; Hurlburt and Garrett 1988) and perhaps in plants as well (Oaks 1974).

Because of easy access to either shoot tissues or cells in culture, these tissues have been used most extensively in studies related to NR. However, in one early report Sanderson and Cocking (1964) described a pyridine nucleotide dependent reduction of NO_3^- in tomato roots. Subsequently Oaks and coworkers (Oaks *et al.* 1972; Aslam and Oaks 1976) demonstrated a synthesis of NR in maize roots in response to NO_3^- and a rapid loss when NO_3 was removed from the system. Comparisons of root tip and mature root sections indicated a faster *de novo* synthesis and re-induction in root tips, but a faster degradation in mature root sections. Redinbaugh and

Campbell (1981) and Oji *et al.* (1988) were successful in purifying NR from maize and barley roots respectively. Redinbaugh and Campbell (1981) also demonstrated an NAD(P)H bispecific NR in addition to the NADH-NR in maize roots. This observation of two enzymes was confirmed for barley roots by Warner *et al.* (1987). In their system, the two NRs are controlled by separate genes, *nar* 1 and *nar* 7. A mutation in *nar* 1 results in the loss of the NADH enzyme in both root and shoot, an observation which suggests that the same NADH enzyme is active in root and shoot tissues. Similarly *nar* 7 appears to code for the NAD(P)H -NR in barley roots and shoots. We have isolated a cDNA clone for the maize root NR (Long *et al.*, unpublished data). A preliminary comparison of the sequence shows an approximate 75 per cent homology in the translated region between the root clone and a corn shoot clone. To date, attempts to characterize and purify the root NRs at the protein level have been difficult because of both the low levels of NR in root tissues and the *in vitro* instability of the enzyme. It may well be that characterization at the level of the gene will solve problems of the structure and regulation of root as well as shoot NR (for a review see Daniel-Vedele *et al.* 1989).

Experiments to define the properties of the root NR are difficult in another aspect, which we think has not been fully appreciated by more biochemically/molecularly oriented researchers. Roots in field grown plants can be extensive, and in maize, at least as large in mass as the shoot tissues (Epstein 1973). Their morphology is dependent on soil conditions such as pH, nitrogen source and associated micro-organisms (Marschner *et al.* 1986; Vermeer and McCully 1982; Fyson and Oaks 1987). One may expect the physiology to be as plastic as the morphology. In attempts to measure NO_3^- reduction in intact roots or the assimilation of NH_4^+, Pate and his co-workers pioneered techniques designed to measure nitrogen contents in xylem exudates (Pate 1973). They found variability in the levels of reduced nitrogen between a variety of plant species (Wallace and Pate 1965; Pate 1973; Wallace 1986). Nitrogen constituents in legumes were also found to change in response to infection with *Rhizobium* (Herridge *et al.* 1978). For example, in cow pea, asparagine is the major nitrogen component in the xylem sap of uninfected plants. After infection allantoin is the major constituent. They postulated that the proportion of reduced nitrogen in uninfected control plants reflected the capacity of the root system to reduce nitrogen and that this depended on the activity of root nitrate reductase or at least on root metabolism (Atkins 1987). They were able to show, for example, that there was no NR activity in *Xanthium* roots which exported NO_3^-, and considerable NR activity in field pea roots which exported asparagine to the shoots (Wallace and Pate 1965; Pate 1973; Wallace 1986). They were also able to show that the allantoin production in lupins turned on by infection with *Rhizobium* could be inhibited by allopurinol, an inhibitor of xanthine oxidoreductase (Atkins *et al.* 1988). This was a clear demonstration that nitrogen constituents in the xylem sap reflected the capacity of the root to assimilate NO_3^- or NH_4^+.

When $^{15}NO_3^-$ was used as an external nitrogen source, Rufty *et al.* (1982) were able to show that in a soybean system $^{15}NO_3^-$ was a major ion species in the xylem exudate whereas reduced nitrogen had very little ^{15}N. This observation suggests that reduced nitrogen in the xylem exudate need not reflect root metabolism, or at least recent root metabolism. Rufty *et al.* (1982) postulated that NO_3^- reduced in the leaf tissue was recycled to the roots where it could be stored or exported to the shoot. Previous NO_3^- nutrition will also affect this balance (Andrews 1986). With a different experimental design and using an *in vivo* nitrate reductase assay, Radin (1978) was able to show that soybeans stored more NO_3^- in their roots and transferred less to the xylem exudate than did cotton plants, and that soybeans had relatively more NR in their root tissues. In addition, the NR activity in the soybean leaf responded dramatically to exogenous NO_3^- whereas the cotton leaf NR had adequate endogenous NO_3^- to support a maximum activity. This observation suggests that the soybean leaf NR is limited by a low endogenous supply of NO_3^- and not by the enzyme *per se*. Rufty *et al.* (1989, and Chapter 8 of this book) were also able to show that nitrogen assimilation in roots is dependent on recent photosynthate. Thus a number of factors, uptake of NO_3^-, its transfer to the xylem and the supply of carbohydrate from the shoot may limit the activity of the root nitrate reductase. It may be, therefore, that *in vivo* activity represents a true value for NO_3^- reduction and that it need not parallel NR values obtained *in vitro*.

In this chapter we illustrate two aspects of this overall problem:

1. The stabilization of nitrate reductase in maize roots with chymostatin, a protease inhibitor.
2. The influence of experiments designed to perturb nitrate reductase activity in the root system on the levels of reduced nitrogen in the xylem sap.

Nitrate reductase in maize roots

Oaks *et al.* (1972), Wallace (1975) and Aslam and Oaks (1976) demonstrated the presence of an active NR in roots of 2–3-day-old seedling maize plants. In more mature plants there was characteristically no detectable NR activity in root tissues (Wallace 1975). Wallace was able to stabilize the enzyme to some extent with additions of casein or PMSF. In our hands, similar treatments were not successful.

Recently we identified another protease inhibitor, chymostatin, which is effective in stabilizing the root NR in maize (Long and Oaks 1990). The principal features of NR when chymostatin is included in the extraction buffer are:

1. That in contrast to agar grown roots, roots of seedlings grown hydroponically for 24 h in the presence of KNO_3 (10 mM) and extracted with buffer containing chymostatin (10 μM), contained higher levels of NR in mature root sections (Table 7.1). In earlier experiments (Oaks *et al.* 1972; Wallace 1975), higher levels of NR were found in the root tips. With chymostatin, the enzyme was relatively stable for at least 2 h after extraction.

2. Redinbaugh and Campbell (1981), Warner *et al.* (1987) and Oji *et al.* (1988) demonstrated an NADH/NADPH bispecific NR in addition to the standard NADH-NR in maize and barley roots. With chymostatin in the extraction buffer it is apparent that NADH-NR may be prominent in the root tips but that the bispecific NR is the major enzyme in mature root sections (Table 7.1).

3. When NO_3^- is removed from the system, there is a more rapid loss of the bispecific enzyme relative to the NADH-NR in root tip sections (Fig. 7.1). In contrast to earlier results by Oaks *et al.* (1972), where the *in vivo* stability of the enzyme was examined, the NR in mature root sections is more stable that the root tip NR (Fig. 7.1).

Table 7.1. Effect of extraction buffer on the level of nitrate reductase in maize roots

Treatment	Nitrate reductase activity (NO_2^- produced h^{-1} GFW^{-1})	
Experiment 1	1 cm	2.3 cm
Agar grown plants	tip	from tip
NADH-NR	1.26	0.17
NADPH-NR	0.26	0.03
Experiment 2		
Hydroponically grown plants		
NADH-NR	1.96	3.08
NADPH-NR	1.17	2.62

Maize kernels (*Zea mays* cv. W64A × W182E) were germinated for 48 h on Petri plates containing 1 per cent (w/v) agar made up in 1/10 Hoagland's solution either with (Experiment 1) or without (Experiment 2) 10 mM KNO_3. In experiment 2 the seedlings were transferred to an aerated hydroponic system which contained 1/10 Hoagland's solution and 10 mM KNO_3 for an additional 24 h. The 1 cm root tips or sections 2–3 cm from the tips were excised, blotted dry, frozen in liquid nitrogen, ground to a fine powder, and stored for a maximum of 3 days at −70 °C. Root powders from Experiment 1 were extracted with a Tris-HCl Buffer (25 mM; pH 8.2) which contained EDTA (1 mM), FAD (20 μM), BSA (1 per cent w/v), DTT (1 mM), and cysteine (10 mM). The extraction buffer in Experiment 2 contained chymostatin (10 μM) dissolved in DMSO in addition to the other ingredients.

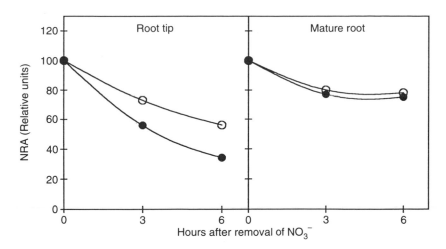

Fig. 7.1. *In vivo* stability of nitrate reductase in maize root tip and mature root sections. Seedlings were grown on agar for 48 h before transfer to a hydroponic system which contained 10 mM KNO_3 as described in Table 7.1 (Experiment 2). Roots were induced for 24 h before transfer to a minus NO_3^- hydroponic system. Samples were collected at 0, 3 and 6 h after the removal of NO_3^-. Initial values (in μmol NO_2^- produced h^{-1} g FWt^{-1}) were for the root tip NADH-NR 1.65, NAD)(P)H-NR 0.93 and for the mature root NADH-NR 1.30, NAD(P)H-NR 1.28.

In another series of experiments we extracted NR from whole roots or shoots of seedlings grown from 2–10 days in a hydroponic system (Fig. 7.2). Both the NADH and the NAD(P)H-bispecific NR in root tissues showed maximum activities at 4 days. The relatively high levels of NADH-NR probably relate to the high level of lateral roots that develop under these conditions. The shoot-NR peaked at about 6 days. In subsequent experiments we have used whole roots from seedlings grown for 5–6 days. With this system we can now examine the effect of external factors on the regulation of root NR. For example, Bowsher *et al.* (1991) have shown that NR activity and mRNA fluctuate in a diurnal fashion in maize shoot tissues and that these effects are much less pronounced in root tissues.

Because of ambiguous results and interpretations obtained with additions of NH_4Cl (Oaks 1979) and common beliefs stated in the literature, we tested the effect of NH_4Cl additions in our hydroponic system. In *Aspergillus* or *Neurospora*, additions of NH_4Cl lead to a disappearance of NR activity (reviewed by Marzluf 1981). Typically in higher plants such additions lead to a mild enhancement of NR activity (Oaks *et al.* 1977, 1979; Oaks, 1979; Mohanty and Fletcher 1976). The results in Table 7.2 show that

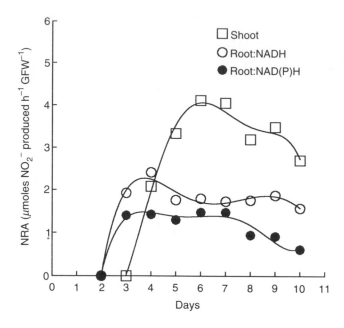

Fig. 7.2. Change in root and shoot NR with age: maize kernels were planted on agar and after 48 h were transferred to a hydroponic system as described in Table 7.1 (Experiment 2). Plants were grown in continuous NO_3^- (10 mM).

Table 7.2. The effect of NO_3^- and NH_4^+ on the induction of nitrate reductase in maize shoots and roots

Treatment	Shoot (NADH)	Root (NADH) $\mu mol\ NO_2$-produced	Root (NADPH) $h^{-1}\ GFWt^{-1}$
No nitrogen	0.11	0	0
KNO_3 (5 mM)	1.45	0.83	0.61
NH_4Cl (5 mM)	0.19	0.11	0.06
$KNO_3 + NH_4Cl$	1.51	0.95	0.84

Seedlings were grown for 48 h on 1 per cent (w/v) agar containing 1/10 Hoagland's solution before transfer to a hydroponic system as described in Table 7.1, Experiment 2. The seedlings were grown for an additional 4 days. KNO_3 (10 mM) was added 6 h before extraction. The whole root system was excised, dry blotted, frozen in liquid N_2 and extracted with our standard buffer which contained chymostatin.

there is a mild enhancement of the NADH-NR in either root or shoot tissue when NH$_4$Cl is added to the system, but a much greater enhancement in the activity of the NAD(P)H-bispecific NR.

From these experiments we can conclude that under certain conditions, there are significant levels of NR in maize roots, levels that reach up to 60–80 per cent of the levels found in the shoot tissue, and that these levels of NR should be sufficient to account for the reduction of ^{15}NO$_3^-$ found *in vivo* by Gojon *et al*. (1986).

Nitrate reduction in maize roots

We know from the early work of Pate (1973) and from more recent work by Andrews (1986), Reed and Hageman (1980), and Wallace (1986) that the proportion of exogenous nitrate accumulated or reduced in shoot or root tissue depends to a large extent on the external NO$_3^-$ concentrations. Rufty *et al*. (1982, 1989) have also shown that the reduction of NO$_3^-$ in root or shoot tissue depends on the availability of photosynthate and that reduced nitrogen in the xylem exudate need not represent NO$_3^-$ assimilation in the root. Nevertheless we decided to examine environmental parameters such as

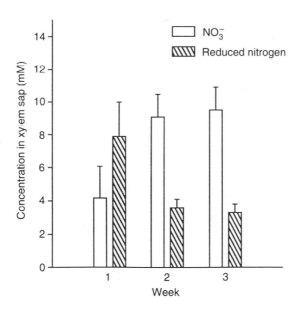

Fig. 7.3. Effect of age on the capacity of maize roots to reduce nitrogen. Plants were grown for the appropriate time in greenhouse loam. They were watered with 1/10 Hoagland's solution which contained 1 mM KNO$_3$. Plants were watered 30 min before de-topping and the stumps were rinsed with distilled water before the exudate was collected. Collection times did not exceed 1 h after de-topping.

age of the seedling, external NO_3^- concentration, and the removal of the endosperm on levels of NO_3^- and reduced nitrogen in the xylem sap.

When seedlings were grown in greenhouse loam (Fig. 7.3) or in turface (data not shown), and were watered with 1 mM KNO_3, NO_3^- was concentrated in the xylem sap, reaching concentrations of about 10 mM in 2 or 3 week-old plants. The concentration of NO_3^- in the xylem sap was lower at Week 1. At this age, endosperm reserves are still an important source of nitrogen (Srivastava *et al*. 1976) and this may be important in regulating either the initial uptake of NO_3^- or its transfer to the xylem. Levels of reduced nitrogen were also higher in Week 1 than in Weeks 2 or 3. We think this reflects the reduced nitrogen supplied to the seedling by the endosperm. Ratios of NO_3^--N to reduced nitrogen at Weeks 2 and 3 were similar to those reported earlier by Pate (1973) and Gojon *et al*. (1986).

When the plants were grown for 20 days in greenhouse loam and were watered with 1/10 strength Hoagland's salts containing 1, 5, or 10 mM KNO_3, the concentration of either NO_3^--N or reduced nitrogen was only slightly higher at the two higher levels of NO_3^- (Table 7.3, and Oaks 1986). Increases in shoot growth were also minor with higher levels of added KNO_3. Thus there are constraints on the capacity of the root system to reduce or accumulate external NO_3^-.

In another type of experiment, seeds were allowed to germinate on nutrient agar for 40 h. They were then transferred to 'turface', either with (control) or without (treatment) their endosperms. They were watered with 1/10 strength Hoagland's salts which contained 1 mM KNO_3 for an additional 4 days. Xylem exudates from the minus endosperm seedlings contained much higher levels of both NO_3^--N and reduced nitrogen than did the control seedlings. However, the ratio of NO_3^--N to total nitrogen was not altered by the treatment (Table 7.4). Levels of K^+ in the xylem exudate

Table 7.3. Effect of the concentration of NO_3^- on growth and nitrogen concentration of xylem exudate

KNO_3 (mM)	1	5	10
Xylem sap			
Total reduced nitrogen (mM)	10.0	13.5	16.5
NO_3^--N (mM)	10.5	14.4	15.6
Asparagine (% reduced nitrogen)	1.3	3.8	8.9
Glutamine (% reduced nitrogen)	35.8	48.6	42.4
Shoot dry wt (mg plant^{-1})	472	565	549

Plants (*Zea mays* cv. W64A × W182E) were grown for 19 days in greenhouse loam. They were watered with 1/10 Hoagland's salts and the appropriate concentration of KNO_3 every other day.

Table 7.4. Constituents of xylem exudate in response to endosperm removal

Treatment	NO$_3^-$	α-NH$_2$N	Total nitrogen	K$^+$	NO$_3^-$/Total nitrogen (%)
			(mM)		
+ Endosperm (control)	2.5	4.2	8.0	20.6	31.2
- Endosperm	6.0	6.6	16.6	19.8	36.1

Seedlings were grown on 1 per cent agar made up in 1/10 strength Hoagland's solution for 48 h. At this time the endosperm was removed and the seedlings (+ or − endosperm) were transferred to turface and watered daily with a 1/10 strength Hoagland's solution modified to contain 1.0 mM KNO$_3$. The shoots were removed 4 days after transfer to turface and the xylem exudate collected for periods of 1 h after de-topping.

were not altered by the removal of the endosperm. Thus at least part of the constraint on NO$_3^-$ accumulation or assimilation appears to reside in metabolites (reduced nitrogen perhaps) supplied by the endosperm.

Conclusion

There is an active nitrate reductase in maize roots which is measurable in *in vitro* assays, and which is present in amounts sufficient to account for the *in vivo* reduction of NO$_3^-$ reported by Pate (1973), Reed and Hageman (1980), and Gojon *et al.* (1986). However, the *in vitro* assay alone is not sufficient to indicate the importance of NR in the assimilation of NO$_3^-$. It is quite probable that the seedling NR which is measured *in vitro* has no real function *in vivo* while the endosperm reserves are being supplied to the seedling (Table 7.4, and Srivastava *et al.* 1972).

The capacity of NR to reduce NO$_3^-$ appears to be suppressed by metabolites in the endosperm. However, in older plants (20 days after sowing), NO$_3^-$ is still the major nitrogen constituent in the xylem exudate. This indicates that other factors limit the *in vivo* levels of NR activity.

References

Andrews, M. (1986). The partitioning of nitrate assimilation between root and shoot of higher plants. *Plant, Cell and Environment*, **9**, 511–19.

Aslam, M. and Oaks, A. (1976). Comparative studies on the induction and inactivation of nitrate reductase in corn roots and leaves. *Plant Physiology*, **53**, 572–6.

Atkins, C.A. (1987). Metabolism and translocation of fixed nitrogen in the nodulated legume. *Plant and Soil*, **100**, 157–69.

Atkins, C.A., Sanford, P.J., Stoner, P.J., and Pate, J.S. (1988). Inhibition of nodule functioning in cow pea by a xanthine oxidoreductase inhibitor, allopurinol. *Plant Physiology*, **88**, 1229–34.

Bowsher, C. G., Long, D. M., Oaks, A., and Rothstein, S. J. (1991). The effect of light/dark cycles on expression of nitrate assimilatory genes in maize shoots and roots. *Plant Physiology*, **95**, 281–5.

Daniel-Vedele, F., Dorbe, M. F., Caboche, M., and Rouzé, P. (1989). Cloning and analysis of the tomato nitrate reductase-encoding gene: protein domain structure and amino acid homologies in higher plants. *Gene*, **85**, 371–80.

Deng, M. D., Moureaux, T., Leydecker, M. T., and Caboche. (1990). Nitrate reductase expression is under the control of a circadian rhythm and is light inducible in *Nicotiana tabacum* leaves. *Planta*, **180**, 257–61.

Epstein, E. (1973). Roots. *Scientific American*, **228**, 48–58.

Evans, H. J. and Nason, A. (1953). Pyridine nucleotide nitrate reductase from extracts of higher plants. *Plant Physiology*, **28**, 233–54.

Filner, P. (1966). Regulation of nitrate reductase in cultured tobacco cells. *Biochimica Biophysica Acta*, **118**, 299–310.

Fyson, A. and Oaks, A. (1987). Physical factors involved in the formation of soil sheaths in corn seedling roots. *Canadian Journal of Soil Science*, **67**, 591–600.

Gojon, A., Soussana, J-F., Passama, L., and Robin, P. (1986). Nitrate reduction in roots and shoots of barley (*Hordeum vulgare* L.) and corn (*Zea mays* L.) seedlings. I. ^{15}N study. *Plant Physiology*, **82**, 254–60.

Hageman, R. H. and Flesher, D. (1960). Nitrate reductase activity in corn seedlings as affected by light and nitrate contents of nutrient media. *Plant Physiology*, **35**, 700–8.

Heimer, Y. M. and Filner, P. (1970). Regulation of the nitrate assimilation pathway of cultured tobacco cells. II. Properties of a variant cell line. *Biochimica Biophysica Acta*, **215**, 152–65.

Herridge, D. F., Atkins, C. A., Pate, J. S., and Rainbird, R. M. (1978). Allantoin and allantoic acid in the nitrogen economy of the cow pea (*Vigna unguiculata*, L. J. Walp). *Plant Physiology*, **62**, 495–8.

Hurlburt, B. K. and Garrett, R. H. (1988). Nitrate assimilation in *Neurospora crassa*: Enzymatic and immunoblot analysis of wild-type and nit mutant protein products in nitrate-induced and glutamine-repressed cultures. *Molecular and General Genetics*, **211**, 35–40.

Long, D. M. and Oaks, A. (1990). Stabilization of nitrate reductase in maize roots by chymostatin. *Plant Physiology*, **93**, 846–50.

Long, D. M., Oaks, A., and Rothstein, S. J. (1991). Comparison of nitrate reductase clones obtained from maize roots and shoots. *Plant Physiology* (submitted).

Marschner, H., Romheld, V., Horst, W. J. and Martin, P. (1986). Root-induced changes in the rhizosphere: Importance for the mineral nutrition of plants. *Zeitschrift für Pflanzenernährung und Bodenkunde*, **149**, 441–56.

Marzluf, G. A. (1981). Regulation of nitrogen metabolism and gene expression in fungi. *Microbiological Reviews*, **45**, 437–61.

Melzer, J. M., Kleinhofs, A., and Warner, R. L. (1989). Nitrate reductase regulation: effects of nitrate and light on nitrate reductase mRNA accumulation. *Molecular and General Genetics*, **217**, 341–6.

Mohanty, B. and Fletcher, J. S. (1976). Ammonium influence on the growth and nitrate reductase activity of Paul's Scarlet Rose Suspension Cultures. *Plant Physiology*, **58**, 152–5.

Oaks, A. (1974). The regulation of nitrate reductase in suspension cultures of soybean cells. *Biochimica Biophysica Acta*, **372**, 122-6.

Oaks, A. (1979). Nitrate reductase in roots and its regulation. In *Nitrogen assimilation in plants* (ed. E. J. Hewitt and C. V. Cutting), pp. 217-24. Academic Press, London.

Oaks, A. (1986). Biochemical aspects of nitrogen metabolism in a whole plant context. In *Fundamental, ecological, and agricultural aspects of nitrogen metabolism in higher plants* (ed. H. Lambers, J. J. Neeteson, and I. Stulen). Publ. Martinus Nijhoff, Dordrecht.

Oaks, A., Aslam, M., and Boesel, I. (1977). Ammonium and amino acids as regulators of nitrate reductase in corn roots. *Plant Physiology*, **59**, 391-4.

Oaks, A., Stulen, I., and Boesel, D. (1979). The effect of amino acids and ammonium in the assimilation of $K^{15}NO_3$. *Canadian Journal of Botany*, **57**, 1824-9.

Oaks, A., Wallace, W., and Stevens, D. (1972). Synthesis and turnover of nitrate reductase in corn roots. *Plant Physiology*, **50**, 649-54.

Oaks, A., Poulle, M., Goodfellow, V. J., Cass, L. A., and Deising, H. (1988). Role of light and NO_3^- and NH_4^+ ions in the regulation of nitrate reductase in corn (*Zea mays* W64A × W182E). *Plant Physiology*, **88**, 1067-72.

Oji, Y., Takahashi, M., Wagai, Y., and Wakiuchi, W. (1988). NADH-dependent nitrate reductase from two-row barley roots: purification, characteristics and comparison with leaf enzyme. *Physiology of Plants*, **72**, 311-15.

Pate, J. S. (1973). Uptake, assimilation and transport of nitrogen compounds by plants. *Soil Biology and Biochemistry* **5**, 109-19.

Radin, J. W. (1978). A physiological basis for the division of nitrate assimilation between roots and leaves. *Plant Science Letters*, **13**, 21-25.

Redinbaugh, M. G. and Campbell, W. H. (1981). Purification and characterization of NAD(P)H: nitrate reductase and NADH: nitrate reductase from corn roots. *Plant Physiology*, **68**, 115-20.

Reed, A. J. and Hageman, R. H. (1980). Relationship between nitrate uptake, flux, and reduction and the accumulation of reduced nitrogen in maize (*Zea mays* L.). II Effect of nutrient nitrate concentration. *Plant Physiology*, **66**, 1184-9.

Remmler, J. L. and Campbell, W. H. (1986). Regulation of corn leaf nitrate reductase. II Synthesis and turnover of the enzyme's activity and protein. *Plant Physiology*, **80**, 442-7.

Rufty, T. W., Volk, R. J., McClure, P. R., Israel, D. W., and Raper, C. D. (1982). Relative content of NO_3^- and reduced N in xylem exudate as an indicator of root reduction of concurrently absorbed. $^{15}NO_3^-$. *Plant Physiology*, **69**, 166-70.

Rufty, T. W., MacKown, C. T., and Volk, R. J. (1989). Effects of altered carbohydrate availability on whole plant assimilation of $^{15}NO_3^-$. *Plant Physiology*, **89**, 457-63.

Sanderson, G. W. and Cocking, E. C. (1964). Enzymatic assimilation of nitrate in tomato plants. I. Reduction of nitrate to nitrite. *Plant Physiology*, **39**, 416-22.

Schrader, L. E., Ritenour, G. L., Eilrich, G. L., and Hageman, R. H. (1968). Some characteristics of nitrate reductase in higher plants. *Plant Physiology*, **43**, 930-40.

Somers, D. A., Kuo, T. K., Kleinhofs, A., Warner, R. L., and Oaks, A. (1983). Synthesis and degradation of barley nitrate reductase. *Plant Physiology*, **72**, 949-52.

Srivastava, H.S., Oaks, A., and Bakyta, I. (1976). The effect of nitrate on early seedling growth in *Zea mays*. *Canadian Journal of Botany*, **54**, 923–9.

Tang, P.S. and Wu, H.Y. (1957). Adaptive formation of nitrate reductase in rice seedlings. *Nature*, **179**, 1355–6.

Travis, R.L., Jordan, W.R., and Huffaker, R.C. (1969). Evidence for an inactivating system of nitrate reductase in *Hordeum vulgare* during darkness that required protein synthesis. *Plant Physiology*, **44**, 1150–6.

Vermeer, J. and McCully, M.E. (1982). The rhizosphere in *Zea*: New insights into its structure and development. *Planta*, **156**, 45–61.

Wallace, W. (1975). A re-evaluation of the nitrate reductase content of maize root. *Plant Physiology*, **55**, 774–7.

Wallace, W. (1986). Distribution of nitrate assimilation between root and shoot of legumes and a comparison with wheat. *Physiology of Plants*, **66**, 630–6.

Wallace, W. and Pate, J.S. (1965). Nitrate reductase in the field pea (*Pisum arvense* L.). *Annals of Botany*, **29**, 655–71.

Warner, R.L., Narayanan, K.R., and A. Kleinhofs. (1987). Inheritance and expression of NAD(P)H nitrate reductase in barley. *Theoretical and Applied Genetics*, **74**, 714–17.

8. Relationship between carbohydrate availability and assimilation of nitrate

THOMAS W. RUFTY, Jr.,*
RICHARD J. VOLK,[†] and ANTHONY D.M. GLASS[‡]

*USDA-ARS, Dept. of Crop Science, North Carolina State University, Raleigh, N.C. 27695–7620; [†]Dept. of Soil Science, N.C. State University, Raleigh, N.C. 27695–7619; [‡]Dept. of Botany, University of British Columbia, Vancouver, B.C., Canada V6T 2B1.

Introduction

Nitrogen and carbon assimilatory processes in crop plants are interdependent. If the activity of either assimilation system is disrupted, adjustments occur in the other. When the exogenous supply of nitrogen is limited, for example, distinct changes occur in photosynthetic rate (Natr 1975) and capacity, i.e. leaf area expansion (Watson 1947; Tolley-Henry and Raper, 1986). Nitrogen stress also alters partitioning of fixed carbon between starch and sucrose in leaves (Robinson and Baysdorfer 1985; Radin and Eidenbock 1986), utilization of carbohydrate in sink tissues (Rufty et al. 1988), and distribution of carbon between the shoot and root (Brouwer 1962; Ingestad 1979). Reciprocally, when plants are utilizing NO_3^-, maximal assimilation depends on sustained photosynthetic activity and provision of carbohydrate (Hageman and Flesher 1960; Beevers and Hageman 1972; Mengel and Viro 1978; Aslam et al. 1979). Moreover, rates of NO_3^- assimilation in leaves in darkness are closely related to carbohydrate status (Aslam and Huffaker 1984), and NO_3^- uptake and reduction in roots decline rapidly when carbohydrate availability is restricted by decreasing the aerial CO_2 concentration, darkness, shoot decapitation, and stem ringing (Breteler and Hanisch ten Cate 1980; Jackson et al. 1980; Hanisch ten Cate and Breteler 1981; Aslam and Huffaker 1982; Pace et al. 1990). While there is little doubt that carbon and nitrogen assimilation activities are closely coupled, numerous questions remain concerning the mechanistic basis for responses when imbalances occur.

In this chapter we further examine aspects of the relationship between NO_3^- assimilation and carbohydrate availability. As just indicated, there is compelling empirical evidence linking activity of the NO_3^- assimilation pathway in vivo and plant carbohydrate status. Theoretically, the relationship is understandable. The enzymatic reactions involved in the assimilation of NO_3^- to protein and nucleic acids, the primary metabolically active end-products, require continuous availability of carbon skeletons, reductant and ATP. Furthermore, transport processes in roots governing NO_3^- uptake and release to the xylem, and thus the provision of NO_3^- substrate for endogenous assimilation, are energy dependent (Pitman 1977; Hanson 1978; Glass 1988). As a result, virtually all of the components of the NO_3^- assimilation pathway are subject to disruption when carbohydrate availability declines.

There have been few attempts to characterize alterations in NO_3^- assimilation in response to carbohydrate limitation on a whole-plant basis. In the first section below, we describe the results of such an experiment. Plants, progressively depleted of carbohydrate reserves in darkness, were exposed to $^{15}NO_3^-$ to define relative sensitivities and, possibly, specific points of control. Some of these results were published previously (Rufty *et al.* 1989). In the second section, we examine some anomalous results from recent experimentation investigating carbohydrate/energy relations and NO_3^- transport and reduction in roots.

Nitrate assimilation in the whole plant

An experiment was initiated to evaluate changes in NO_3^- assimilation into protein as plants become deficient in carbohydrate. Young tobacco plants (*Nicotiana tabacum* L. cv. 2326) growing in solution culture were exposed to 99A per cent $^{15}NO_3^-$ for 6 h intervals during a light/dark cycle and an additional 30 h of extended darkness.

Carbohydrate availability during the course of the experiment could be separated into four distinct phases (Fig. 8.1). Non-structural carbohydrate (presumably available for metabolism), was highest in the photosynthetic period (Phase 1). There was rapid utilization of the internal reserves in the shoot in the first 12 h of darkness (Phase 2). In the next 12 h, the rate of utilization of carbohydrate remaining in the shoot slowed noticeably (Phase 3); plants evidently were becoming carbohydrate deficient. Little further decline occurred in the minimal reserves present during the last 18 h of darkness, as plants experienced severe carbohydrate deficiency.

The carbohydrate content of the root was relatively low. The root contained less than 15 per cent of the plant's carbohydrate reserve at the beginning of the dark period. Also, root carbohydrate was relatively unavailable, declining at a slow, steady rate throughout (Fig. 8.1). The carbohydrate utilization pattern is consistent with the notion that root metabolic activities are largely dependent on carbohydrate being supplied from the shoot (Wann and Raper 1979; Jackson *et al.* 1980). Accordingly, root respiration changed little in the light and initial 12 h of darkness, but then decreased approx-

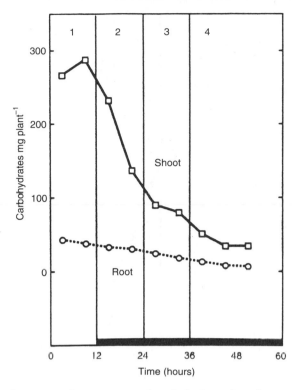

Fig. 8.1. Total content of non-structural carbohydrate (starch, sucrose, hexose) in the shoot and root. For details of carbohydrate analyses, refer to Rufty *et al.* 1988.

imately 50 per cent in Phase 3 (data not shown) when shoot carbohydrate was considerably depleted. Respiration stabilized at the lower rate in Phase 4.

$^{15}NO_3^-$ uptake and whole-plant $^{15}NO_3^-$ reduction

Uptake of $^{15}NO_3^-$ was severely restricted as internal carbohydrate was depleted over the course of the experiment (Fig. 8.2). Uptake in the initial 12 h of darkness was approximately 80 per cent of that in the light, but then decreased sharply in Phase 3, coincident with the decline in root respiration. Even though the uptake process was sensitive to decreases in carbohydrate status, reduction of absorbed $^{15}NO_3^-$ was even more so. Whole-plant $^{15}NO_3^-$ reduction decreased sharply immediately after the light period and was restricted to a greater extent than uptake throughout. Clearly, endogenous factors were limiting assimilation of absorbed NO_3^-.

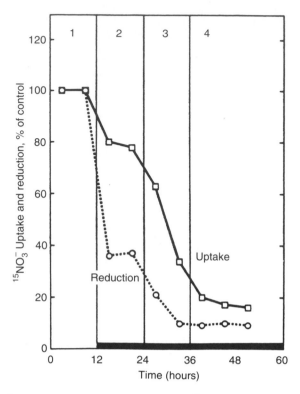

Fig. 8.2. Uptake and whole plant reduction of $^{15}NO_3^-$ during each 6 h exposure interval as a percentage of control values in the light. Mean values for uptake and reduction in the light were 520 and 309 μmol plant^{-1}. Reduction represents total ^{15}N in SRN and IRN fractions. Refer to Rufty *et al*. 1984 for methodological details.

^{15}N assimilation in the root

Analysis of different nitrogen fractions in root tissues indicated that, although assimilation of absorbed NO_3^- was markedly impaired in darkness on a whole plant basis, the proportion of absorbed ^{15}N incorporated into root protein (insoluble reduced nitrogen, IRN), was unaffected (Fig 8.3). Partitioning into reduced ^{15}N fractions (IRN + soluble reduced nitrogen, SRN) decreased by approximately 35 per cent in Phase 3, suggesting decreased $^{15}NO_3^-$ reduction. While the methodology used here does not allow precise estimation of changes in root $^{15}NO_3^-$ reduction, other studies have consistently indicated that reduction is more sensitive than uptake when roots are carbohydrate deficient (Jackson *et al*. 1980; Hanisch ten Cate and Breteler 1981).

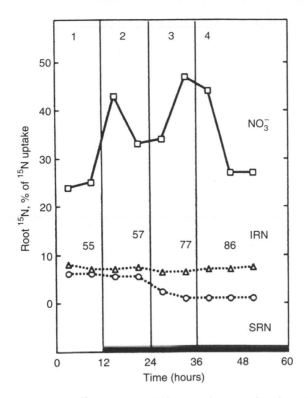

Fig. 8.3. Partitioning of ^{15}N among different nitrogen fractions in the root, expressed as a percentage of total ^{15}NO$_3^-$ uptake. Inset values represent proportion of total reduced ^{15}N in IRN fraction.

The tendency for sustaining incorporation of absorbed ^{15}N into protein under low energy conditions, even as ^{15}NO$_3^-$ reduction decreases, is not limited to tobacco. It was also observed in experiments with corn (Morgan *et al.* 1985, 1986) and soybean (Rufty and Israel, unpublished). The phenomenon is difficult to reconcile, mechanistically. Protein synthesis is an energy-intensive process, much more so than NO$_3^-$ reduction, yet it is affected less than reduction. A hierarchy in energy utilization is implied.

In recent experimentation with soybean, we have determined that incorporation of absorbed ^{15}N into root protein occurs along the entire length of the primary root (Table 8.1), and thus is not confined to meristematic centres even though it is most intense there. Evidently the priority for protein synthesis is a general characteristic of root cells.

With little evidence in hand, we speculate that the relative decreases in NO$_3^-$ reduction in carbohydrate deficient roots reflect a deficiency of

Table 8.1 Assimilation of ^{15}N into IRN (protein) along the primary root of soybean (*Glycine max* L. Merrill, 'Ransom')

Segment	Fresh weight	^{15}N
cm from apex	mg	ng mg^{-1} FWt
0.0 – 0.5	36	1.224
0.5 – 2.0	133	0.354
2.0 – 6.0	427	0.279
6.0 – 10.0 lateral zone	990	0.498
> 10.0 lateral zone	1614	0.403

Six-day-old plants growing in solution culture in a growth room were exposed to 0.5 mM 99A per cent ^{15}NO$_3^-$ for 10 min. Root segments were separated and analysed. For details on analytical methods refer to Rufty *et al.* 1984.

reductant. Although decreases in active nitrate reductase (NR) protein cannot be ruled out, excess NR activity was present in corn roots low in carbohydrate when NO$_3^-$ reduction decreased (Pace *et al.* 1990). It seems unlikely that the decline in NO$_3^-$ reduction was due to limited NO$_3^-$ substrate, because root NO$_3^-$ was elevated in the same time interval that reduction decreased (Fig. 8.3, Phase 3). This evidence is not definitive in itself, since NO$_3^-$ can accumulate in cellular compartments separate from NR (MacKown *et al.* 1983; Rufty *et al.* 1986). The observation, however, that reduction of ^{15}NO$_3^-$ entering the root was restricted, as shown in other studies (Jackson *et al.* 1980; Pace *et al.* 1990), and implied here, clearly argues against a substrate limitation.

^{15}N translocation to the shoot

It has been reported previously in experiments with soybean that translocation of absorbed NO$_3^-$ from the root to the shoot was inhibited in the dark phase of the diurnal cycle (Rufty *et al.* 1984). Similar decreases in translocation to the xylem in darkness have been observed with other species (Ngambi *et al.* 1980; Pearson and Steer 1977; Pearson *et al.* 1981). This pool of NO$_3^-$ is efficiently mobilized in the following light period, providing a major portion of the NO$_3^-$ reduced in photosynthetic tissues (Rufty *et al.* 1987). Root retention of NO$_3^-$ absorbed in the dark thus functions to coordinate substrate delivery with conditions most favourable energetically for assimilation in the shoot.

In the present experiment with tobacco, translocation of absorbed ^{15}N was decreased in the initial 12 h of darkness, as expected (Fig. 8.4). After that

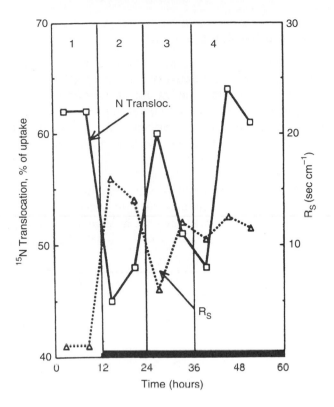

Fig. 8.4. Translocation of ^{15}N, estimated from shoot ^{15}N accumulation and expressed as a percentage of total $^{15}NO_3^-$ uptake, and R_s of leaves.

time, however, translocation fluctuated rhythmically. The mechanism(s) controlling translocation apparently were unrelated to the plant carbohydrate status.

The fluctuations in ^{15}N translocation were reciprocal to changes in stomatal resistance (R_s) through Phase 3. The changes in R_s reflect a circadian rhythm in stomatal opening and closure (Heath 1984). It is unlikely, however, that translocation was controlled completely by transpiration. Other studies have demonstrated that transport of ions and water into the xylem varies rhythmically, and the rhythms persist in detached roots (Grossenbacher 1938; Hagan 1949; Vaadia 1960; Fiscus 1986). It is reasonable that the rhythm in ^{15}N translocation also is controlled by an oscillator(s) within the root. If so, translocation and R_s may have been entrained through Phase 3 by changes in root turgor in response to the R_s rhythm. The lack of synchrony in Phase 4 would support the contention that the rhythms were not necessarily coupled.

^{15}N assimilation in the shoot

The pattern of changes in ^{15}N accumulation in the shoot with extended darkness (data not shown) resembled that for uptake (Fig. 8.2). Some differences occurred, as ^{15}N translocation out of the root fluctuated independently, but the declines over time were quantitatively similar. Since the aim of this chapter is to define relative responses, shoot ^{15}N data are expressed as percentages of the total ^{15}N accumulated there during each 6 h interval (Fig. 8.5). This allows assessment of the efficiency of assimilation of available ^{15}NO$_3^-$ substrate.

Past studies have demonstrated that NO$_3^-$ reduction occurs in leaf tissues in darkness, but at rates lower than in the light (Aslam and Huffaker 1982, 1984; Reed *et al.* 1983). That was the case here also, as the proportion of shoot ^{15}N accumulating as reduced ^{15}N (SRN + IRN) decreased markedly

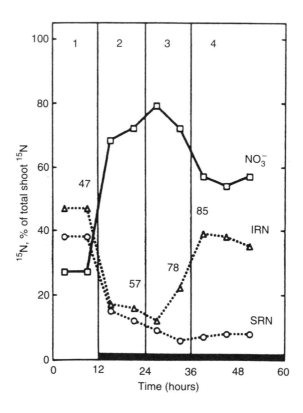

Fig. 8.5. Partitioning of ^{15}N among different nitrogen fractions in the shoot, expressed as a percentage of the total shoot ^{15}N. Inset values represent proportion of total reduced ^{15}N in IRN fraction.

in Phase 2 (Fig 8.5). Of the total reduced ^{15}N accumulated, a much larger proportion was I R N as the dark period progressed (Fig. 8.5, numbers beside I R N values). Thus, once NO_3^- reduction occurred, ^{15}N was incorporated into protein relatively efficiently.

Shoot NO_3^- reduction (as reflected in the accumulation of SRN + IRN) was more sensitive to darkness than any other process examined. It was chiefly responsible for the sharp decline in whole plant reduction shown in Fig. 8.2. A decrease was evident in the initial 6 h of darkness, when the shoot contained a large amount of carbohydrate and when NO_3^- uptake and apparent reduction in the root (SRN + IRN accumulation; Phase 2, Fig. 8.3) were showing little indication of carbohydrate stress. Reduction evidently was extremely sensitive to the lower energy state following the photosynthetic period and/or N R activity was restricted severely. There is evidence that N R can be rapidly inactivated in darkness (Remmler and Campbell 1986; Campbell 1988), and *de novo* synthesis of active N R may be light-dependent (Oaks *et al.* 1988). The relationship between light and synthesis of active N R is complicated by the possible regulatory involvement of phytochrome (Jones and Sheard 1975; Rajasekhar *et al.* 1988; Melzer *et al.* 1989) and endogenous rhythms at the transcriptional level (Galangau *et al.* 1988). If decreases in active N R are responsible for decreased NO_3^- reduction *in situ*, then the regulatory effects must be substantial because N R activity can be present in leaf tissues in large excess (refer to Harper 1974, Fig. 3; Huffaker and Rains 1978).

It is unlikely that NO_3^- substrate limited reduction in darkness (Fig. 8.5). It has been suggested that transport of NO_3^- out of storage, possibly vacuolar, compartments into metabolic pools is inhibited in darkness (Aslam *et al.* 1976) In this experiment, however, $^{15}NO_3^-$ was supplied exogenously to roots, so it would have entered the metabolic (cytoplasmic) pool of leaf cells (Shaner and Boyer 1976) before being accumulated in vacuoles. The methodology did not permit an assessment of $^{15}NO_3^-$ retention in vascular tissues in darkness.

Summary

The general picture emerging from this ^{15}N study is that there are multiple responses in the NO_3^- assimilation pathway when availability of carbohydrate decreases, and sensitivities of the various pathway components are different. Nitrate reduction in leaves apparently is the most sensitive process. Although protein synthesis is energy intensive, there is no indication of it being acutely sensitive to carbohydrate status either in the shoot or root.

The amount of ^{15}N reaching the shoot can change in darkness with decreasing carbohydrate availability. It is a function of uptake into the root, which decreases as the carbohydrate supply from the shoot declines, and of rhythmic fluctuations in NO_3^- translocation into the xylem, which may be controlled by an oscillator in the root, and not by carbohydrate availability. The data indicate that the root transport responses are coordinative in nature and not rate-limiting. This contrasts with the situation in the light, where the activity of root transport processes and the associated delivery of NO_3^- to the shoot regulate the rate of shoot NO_3^- reduction and protein synthesis (Shaner and Boyer 1976; Kannangara and Woolhouse 1967; Rufty *et al.* 1987).

Nitrate transport and reduction in roots

In this section, results of recent experiments are presented which have implications for the relationship between the carbohydrate/energy status and uptake and reduction of NO_3^- by roots. The results appear inconsistent with concepts currently accepted or being considered.

$^{13}NO_3^-$ influx kinetics

Within the past decade, the radioisotope ^{13}N has been used in a number of research programs examining NO_3^- uptake characteristics of roots (e.g. Presland and McNaughton 1984; Lee and Drew 1986; Lee and Clarkson 1986; Oscarson *et al.* 1987; Siddiqi *et al.* 1989). Due to increased detection sensitivity, ^{13}N has advantages over ^{14}N and the stable isotope ^{15}N in measuring undirectional influx into root cells. Experiments can be limited to relatively short time intervals (5–10 min), minimizing the confounding effects of efflux and the regulatory involvement of endogenous compartments other than the cell cytosol.

In recent experiments with barley (*Hordeum vulgare* L. cv. Klondike), $^{13}NO_3^-$ was used to characterize influx over a range of external concentrations (Fig. 8.6A, B). The plants had been exposed to 100 μM $^{14}NO_3^-$ for 24 h prior to the experiment, so the influx system was fully induced (Siddiqi *et al.* 1989). Two distinct transport systems were evident; one saturable at approximately 0.5 mM $[NO_3^-]_0$ and the other linear beyond 1.0 mM. The saturable system was sensitive to temperature changes ($Q_{10} \sim 3$–4) and the presence of metabolic inhibitors (Glass *et al.* 1990). The characteristics of the saturable system are therefore consistent with the contention, based on thermodynamic considerations, that NO_3^- influx across the plasmalemma of root cells is active (Higinbotham *et al.* 1967; Clarkson 1986; Glass 1988). On the other hand, the linear response to $[NO_3^-]_0$ indicates down-hill, channel-mediated influx. Accordingly, that system is relatively insensitive to temperature, Q_{10} values approach 1, and to metabolic inhibitors. A similar linear system was observed above 1.0 mM in NO_3^- depletion experiments with maize (Pace and McClure 1986).

The point of issue is that the linear phase occurs at such a low external concentration of NO_3^-. From compartmental analysis studies with barley, it has been estimated that the concentration of NO_3^- in the cytosol is approximate 26 mM (Lee and Clarkson 1986). Similar analysis of the plants used in these experiments (Fig. 8.6) indicates a cytoplasmic concentration of approximately 18 mM. Assuming a concentration of 20 mM and an electrical potential difference at the plasmalemma of $-$ 150 mV, for example, passive influx of external NO_3^- should not occur until external concentrations are in the molar range (approximately 8 M). Viewed another way, if passive flux occurs at an external concentration of 1.0 mM, then the cytosolic NO_3^- concentration must be approximately 2.0 μM, seemingly an unacceptably low value for physiological function.

We can propose two possible explanations for the apparent inconsistency. One is that estimates for the cytoplasmic NO_3^- connection are invalid. This could reflect inherent limitations in compartmental analysis (Cheeseman 1986) and its inability to distinguish the cytoplasmic NO_3^- concentration just inside the plasma membrane of the specific cells involved in the uptake process. Another possibility is that the linear system represents NO_3^- uptake

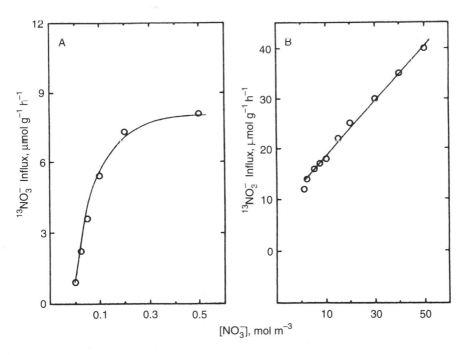

Fig. 8.6. $^{13}NO_3^-$ influx A. from 0.005 to 0.5 mol m^{-3} [NO_3^-]$_0$ and B. from 1 to 50 mol m^{-3} [NO_3^-]$_0$ into roots of induced plants pretreated with 0.1 mol m^{-3} $^{14}NO_3^-$ for 24 h.

into a compartment separate from the primary root symplasm. This could involve movement through meristematic and undifferentiated cell zones, where concentrations of NO_3^- may be very low and where secondary deposits are minimal, possibly allowing direct entry into the vascular system through intercellular pathways. Regardless of the exact mechanism, research to resolve the physiological basis for the linear response may well need to define influx into specific cells or cell regions of the root and not continue to treat the root as a homogeneous entity. It is noteworthy, however, that Serra *et al.* (1978), working with the diatom *Skeletonema costatum*, reported linear uptake kinetics at external NO_3^- concentrations beyond those at which the high affinity uptake system saturated (approximately $10\,\mu M$). Clearly, tissue heterogeniety cannot be invoked to explain such kinetics in single-celled organisms.

Response to glucose

Previous studies have indicated that addition of glucose to nutrient media can increase uptake and reduction of NO_3^- and NR activity in roots (e.g. Aslam and Oaks 1975; Jackson *et al.* 1980; Hanisch ten Cate and Breteler 1981). Additional examples of the response to glucose are presented in Table 8.2, which contains data from experiments with maize. In both experiments, plants had been supplied previously with NO_3^-, so uptake and reduction systems were induced. In Experiment 1, following shoot excision roots were exposed to 5 mM glucose for 2 h, with $^{15}NO_3^-$ uptake and reduction measured in the second hour. Uptake increased 21 per cent relative to the control and accumulation of reduced ^{15}N increased 115 per cent. In Experiment 2, roots of intact seedlings growing in the light were exposed to $^{15}NO_3^-$ and 5 mM glucose for 2 h. Uptake was increased by 20 per cent and reduced ^{15}N accumulation by 50 per cent.

Table 8.2 Alterations in $^{15}NO_3^-$ uptake and accumulation of reduced-^{15}N in roots of two maize genotypes exposed to 5 mM glucose in nutrient solution

Experiment	Glucose addition	$^{15}NO_3^-$ uptake	Reduced-^{15}N-accumulation
	5 mM	$\mu mol\ g^{-1} FWt\ h^{-1}$	
1	−	7.88	1.50
	+	9.57 (121)	3.22 (215)
2	−	11.64	2.10
	+	12.82 (110)	3.78 (180)

Plants were growing in solutions containing $^{14}NO_3^-$ prior to the $^{15}NO_3^-$ exposure periods. In Experiment 1, 10-day-old seedlings (DeKalb XL45) were de-topped and exposed to glucose for 2 h and to 5 mM $^{15}NO_3^-$ during the final 1 h. In Experiment 2, 7-day-old intact seedlings (Pioneer 3369A) were exposed to 5 mM glucose and 0.3 mM $^{15}NO_3^-$ for 2 h. Data are expressed as percentage of control in parenthesis.

The glucose stimulation suggests that uptake and reduction of NO_3^- in control root were carbohydrate limited. Examination of roots respiration, however, provides opposing evidence. Oxygen uptake by segments of roots used in the experiments shown in Table 8.2 revealed a marked sensitivity to salicylhydroxamic acid (SHAM) in the presence of 5 mM glucose and in its absence (data not shown). The degree of inhibition ranged from 20–40 per cent in the presence of 10 to 20 mM SHAM, depending on the segment of primary root examined. Moreover, in a separate experiment utilizing excised roots (tip to 5 cm section), 30 min exposure to 0.3 mM $^{15}NO_3^-$ resulted in a 30 per cent increase in reduced ^{15}N accumulation in the presence of 5 mM glucose, while a 40 per cent inhibition of O_2 uptake was measured in the presence of 15 mM SHAM in similar segments over the same time-frame. If it is accepted that the SHAM sensitivity represents engagement of the alternative respiratory pathway (Moller *et al.* 1988) and that the energy overflow hypothesis (Lambers 1982, 1983, 1985) is correct, then carbohydrate was present in control roots in excess.

Although the mechanistic basis for the apparent paradox is obscure, we can offer two possible explanations. One is that glucose directly affects transport into root cells and, to an even larger extent, increases its availability for reduction. This assumes that excess carbohydrate is available to provide the increased energy and carbon skeletons required for more rapid ^{15}N assimilation. That glucose might increase NO_3^- availability for reduction was suggested previously in a study with leaves (Aslam *et al.* 1976). Another explanation is that glucose stimulates metabolism, active uptake NO_3, and NO_3 reduction at a cellular site which is carbohydrate deficient. Previous experiments with maize indicated that NO_3^- reduction occurred in cells at the root periphery (Rufty *et al.* 1986). Those epidermal and outer cortical cells are the farthest symplastic region from vascular tissues and the associated supply of carbohydrate. Thus, a carbohydrate deficiency could exist in peripheral cells, explaining the glucose stimulation there, while the great majority of root cells possess carbohydrate in excess and are responsible for the characteristic SHAM response.

References

Aslam, M. and Huffaker, R.C. (1982). *In vivo* nitrate reduction in roots and shoots of barley seedlings in light and darkness. *Plant Physiology*, **70**, 1009–13.

Aslam, M. and Huffaker, R.C. (1984). Dependency of nitrate reduction on soluble carbohydrates in primary leaves of barley under aerobic conditions. *Plant Physiology*, **75**, 623–8.

Aslam, M. and Oaks, A. (1975). Effect of glucose on the induction of nitrate reductase in corn roots. *Plant Physiology*, **56**, 634–39.

Aslam, M. Huffaker, R.C., Rains, D.W., and Rao, K.P. (1979). Influence of light and ambient carbon dioxide concentration on nitrate assimilation by intact barley seedlings. *Plant Physiology*, **63**, 1205–9.

Aslam, M., Oaks, A., and Huffaker, R. C. (1976). Effect of light and glucose on the induction of nitrate reductase and on the distribution of nitrate in etiolated barley leaves. *Plant Physiology*, **58**, 588–91.

Beevers, L. and Hageman, R. H. (1972). The role of light in nitrate metabolism in higher plants. *Photophysiology*, **7**, 85–113.

Breteler, H. and Hanisch ten Cate, C. H. (1980). Fate of nitrate during initial nitrate utilization by nitrogen-depleted dwarf bean. *Physiologia Plantarum*, **48**, 292–6.

Brouwer, R. (1962). Nutritive influences on the distribution of dry matter in the plant. *Netherlands Journal of Agricultural Science*, **10**, 399–408.

Campbell, W. H. (1988). Nitrate reductase and its role in nitrate assimilation in plants. *Physiologia Plantarum*, **74**, 214–19.

Cheeseman, J. M. (1986). Compartmental efflux analysis: an evaluation of the technique and its limitations. *Plant Physiology*, **80**, 1006–11.

Clarkson, D. T. (1986). Regulation of the absorption and release of nitrate by plant cells: A review of current ideas and methodology. In *Fundamental, ecological, and agricultural aspects of nitrogen metabolism in higher plants* (ed. H. Lambers), pp. 3–27. Martinus Nijhoff, Dordrecht.

Fiscus, E. L. (1986). Diurnal changes in volume and solute transport coefficients of *Phaseolus* roots. *Plant Physiology*, **80**, 752–9.

Galangau, F., Daniel-Vedele, F., Moureaux, T., Dorbe, M. F., Leydecker, M. T., and Caboche, M. (1988). Expression of leaf nitrate reductase genes from tomato and tobacco in relation to light-dark regimes and nitrate supply. *Plant Physiology*, **88**, 383–8.

Glass, A. D. M. (1988). Nitrogen uptake by plant roots. *ISI Atlas of Science: Animal and Plant Sciences*, **1**, 151–56.

Glass, A. D. M., Siddiqi, M. Y., Ruth, T. J., and Rufty, T. W. Jr. (1990). Studies of the uptake of nitrate in barley. II. Energetics of $^{13}NO_3^-$ influx. *Plant Physiology*, **92**, 1585–9.

Grossenbacher, K. A. (1938). Diurnal fluctuations in root pressure. *Plant Physiology*, **4**, 669–76.

Hagan, R. M. (1949). Autonomic diurnal cycles in the water relations of nonexuding detopped root systems. *Plant Physiology*, **24**, 441–54.

Hageman, R. H. and Flesher, D. (1960). Nitrate reductase activity in corn seedlings as affected by light and nitrate content of nutrient media. *Plant Physiology*, **35**, 700–8.

Hanisch ten Cate, C. H. and Breteler, H. (1981). Role of sugars in nitrate utilization by roots of dwarf bean. *Physiologia Plantarum*, **52**, 129–35.

Hanson, J. B. (1978). Application of chemiosmotic hypothesis to ion transport across the root. *Plant Physiology*, **62**, 402–5.

Harper, J. E. (1974). Soil and symbiotic nitrogen requirements for optimum soybean production. *Crop Science*, **14**, 255–60.

Heath, O. V. S. (1984) Stomatal opening in darkness in the leaves of Commelina communis, attributed to an endogenous circadian rhythm: control of phase. *Proceedings of the Royal Society London B*, **220**, 399–414.

Higinbotham, N., Etherton, B., and Foster, R. J. (1967). Mineral ion contents and cell transmembrane electropotentials of pea and oat seedling tissue. *Plant Physiology*, **42**, 37–46.

Huffaker, R. C. and Rains, D. W. (1978). Factors influencing nitrate acquisition by plants; assimilation and fate of reduced nitrogen. In *Nitrogen in the environment* (ed. D. R. Nielsen and J. G. MacDonald), Vol. 2, pp. 1–43. Academic Press, New York.

Ingestad, T. (1979). Nitrogen stress in birch seedlings. *Physiologia Plantarum*, **45**, 149–57.

Jackson, W. A., Volk, R. J., and Israel, D. W. (1980). Energy supply and nitrate assimilation in root systems. In *Carbon–nitrogen interaction in crop production* (ed. A. Tanaka), pp. 25–40. Japanese Society for the Promotion of Science, Tokyo.

Jones, R. W. and Sheard, R. W. (1975). Phytochrome, nitrate movement, and induction of nitrate reductase in etiolated pea terminal buds. *Plant Physiology*, **55**, 954–9.

Kannangara, C. G. and Woolhouse, H. W. (1967). The role of carbon dioxide, light, and nitrate in the synthesis and degradation of nitrate reductase in leaves of *Perilla frutescens*. *New Phytology*, **66**, 553–61.

Lambers, H. (1982). Cyanide-resistant respiration: a non-phosphorylating election transport pathway acting as an energy overflow. *Physiologia Plantarum*, **55**, 478–85.

Lambers, H. (1983). The functional equilibrium, nibbling on the edges of a paradigm. *Netherlands Journal Agricultural Science*, **31**, 305–11.

Lambers, II. (1985). Respiration in intact plants and tissues: Its regulation and dependence on environmental factors, metabolism, and invaded organisms. In *Higher plant cell respiration* (cd. R. Douce and D. A. Day), pp. 418–73. Springer-Verlag, New York.

Lee, R. B. and Clarkson, D. T. (1986). Nitrogen-13 studies of nitrate fluxes in barley roots. I. Compartmental analysis from measurements of ^{13}N efflux. *Journal of Experimental Botany*, **37**, 1753–67.

Lee, R. B and Drew, M. C. (1986). Nitrogen-13 studies of nitrate fluxes in barley roots. II. Effect of plant N-status on the kinetic parameters of nitrate influx. *Journal of Experimental Botany*, **37**, 1768–79.

MacKown, C. T., Jackson, W. A., and Volk, R. J. (1983). Partitioning of previously accumulated nitrate to translocation, reduction, and efflux in corn roots. *Planta*, **157**, 8–14.

Melzer, J. M., Kleinhofs, A., and Warner, R. L (1989). Nitrate reductase regulation: effects of nitrate and light on nitrate reductase mRNA accumulation. *Molecular and General Genetics*, **217**, 341–6.

Mengel, K. and Viro, M. (1978). The significance of plant energy status for the uptake and incorporation of NH_4^+-nitrogen by young rice plants. *Soil Science and Plant Nutrition*, **24**, 407–16.

Moller, I. M., Berczi, A., Linus, H. W. van der Plas, and Lambers, H. (1988). Measurement of the activity and capacity of the alternative pathway in intact plant tissues: Identification of problems and possible solutions. *Physiologia Plantarum*, **72**, 642–9.

Morgan, M. A., Jackson, W. A., and Volk, R. J. (1985). Uptake and assimilation of nitrate by corn roots during and after induction of the nitrate uptake system. *Journal of Experimental Botany*, **363**, 859–69.

Morgan, M. A., Jackson, W. A., Pan, W. L., and Volk, R. J. (1986). Partitioning of reduced nitrogen derived from exogenous nitrate in maize roots: initial priority for protein systhesis. *Plant and Soil*, **91**, 343–7.

Natr, L. (1975). Influence of mineral nutrition on photosynthesis and the use of assimilates. *International Biology Programme*, **5**, 537–55.

Ngambi, J. M., Champigny, M. L., Mariotti, A., and Moyse, A. (1980). Assimilation des nitrates et photosynthese d'un Mil. Pennisetum americanum 23D B, an cours d'un nyethemere. *Comptes Rendus des Séances de l'Académie des Sciences, Paris* **291D**, 109–12.

Oaks, A., Poulle, M., Goodfellow, V. J., Cass, L. A., and Deising, H. (1988). The role of nitrate and ammonium ions and light on the induction of nitrate reductase in maize leaves. *Plant Physiology*, **88**, 1067–72.

Oscarson, P., Ingemarsson, B., af Ugglas, M., and Larson, C. M. (1987). Short-term studies of NO_3^- uptake in *Pisum* using $^{13}NO_3^-$. *Planta*, **170**, 550–5.

Pace, G. M. and McClure, P. R. (1986). Comparison of nitrate uptake kinetic parameters across maize imbred lines. *Journal of Plant Nutrition*, **9**, 1095–111.

Pace, G. M., Volk, R. J., and Jackson, W. A. (1990). Nitrate reduction in response to CO_2-limited photosynthesis. *Plant Physiology*, **92**, 286–92.

Pearson, C. J. and Steer, B. T. (1977). Daily changes in nitrate uptake and metabolism in *Capsicum annuum*. *Planta*, **137**, 107–12.

Pearson, C. J., Volk, R. J., and Jackson, W. A. (1981). Daily changes in nitrate influx, efflux, and metabolism in maize and pearl millet. *Planta*, **152**, 319–24.

Pitman, M. G. (1977). Ion transport into the xylem. *Annual Review of Plant Physiology*, **28**, 71–88.

Presland, M. R. and McNaughton, G. S. (1984). Whole plant studies using radioactive 13-nitrogen. II. Compartmental model for the uptake and transport of nitrate ions by *Zea mays*. *Journal of Experimental Botany*, **35**, 1277–88.

Radin, J. W. and Eidenbock, M. P. (1986). Carbon accumulation during photosynthesis in leaves of nitrogen and phosphorus stressed carbon. *Plant Physiology*, **82**, 869–71.

Rajasekhar, V. K., Gowri, G., and Campbell, W. H. (1988). Phytochrome-mediated light regulation of nitrate reductase expression in squash cotyledons. *Plant Physiology*, **88**, 242–4.

Reed, A. J., Canvin, D. T., Sherrard, J. H., and Hageman, R. H. (1983). Assimilation of [^{15}N] nitrate and [^{15}N] nitrite in leaves of five plant species under light and dark conditions. *Plant Physiology*, **71**, 291–4.

Remmler, J. L. and Campbell, W. H. (1986). Regulation of corn leaf nitrate reductase. II. Synthesis and turnover of the enzyme's activity and protein. *Plant Physiology*, **80**, 442–7.

Robinson, J. M. and Baysdorfer, C. (1985). Inter-relationships between photosynthetic carbon and nitrogen metabolism in mature soybean leaves and isolated leaf mesophyll cells. In *Regulation of carbon partitioning in photosynthetic tissue* (ed. R. L. Heath and J. Preiss), pp. 333–57. American Society of Plant Physiologists, Rockville, MD.

Rufty, T. W., Jr., Israel, D. W., and Volk, R. J. (1984). Assimilation of $^{15}NO_3^-$ taken up by plants in the light and in the dark. *Plant Physiology*, **76**, 769–75.

Rufty, T. W., Jr., Thomas, J. F., Remmler, J. L., Campbell, W. H., and Volk, R. J. (1986). Intercellular localization of nitrate reductase in roots. *Plant Physiology*, **82**, 675–80.

Rufty, T. W., Jr., Volk, R. J., and MacKown, C. T. (1987). Endogenous NO_3^- in the root as a source of substrate for reduction in the light. *Plant Physiology*, **84**, 1421–6.

Rufty, T. W., Jr., Huber, S. C., and Volk, R. J. (1988). Alterations in leaf carbohydrate metabolism in response to nitrogen stress. *Plant Physiology*, **88**, 725–30.

Rufty, T. W., Jr., MacKown, C. T., and Volk, R. J. (1989). Effects of altered carbohydrate availability on whole-plant assimilation of $^{15}NO_3^-$. *Plant Physiology*, **89**, 457–63.

Serra, J. L., Llama, M. J., and Cadenas, E. (1978). Nitrate utilization by the diatom *Skeletonema costatum*. I. Kinetics of nitrate uptake. *Plant Physiology*, **62**, 987–90.

Shaner, D. L. and Boyer, J. S. (1976). Nitrate reductase activity in maize leaves. I. Regulation by nitrate flux. *Plant Physiology*, **58**, 499–504.

Siddiqi, M. Y., Glass, A. D. M., Ruth, T. J., and Fernando, M. (1989). Studies of NO_3^- influx by barley seedlings using $^{13}NO_3^-$. *Plant Physiology*, **90**, 806–13.

Tolley-Henry, L. and Raper, C. D., Jr. (1986). Expansion and photosynthetic rate of leaves of soybean plants during onset of and recovery from nitrogen stress. *Botanical Gazette*, **147**, 400–6.

Vaadia, Y. (1960). Autonomic diurnal fluctuations in rate of exudation and root pressure of decapitated sunflower plants. *Physiologia Plantarum*, **13**, 701–17.

Wann, M. and Raper, C. D., Jr. (1979). A dynamic model for plant growth: adaptation for vegetative growth of soybean. *Crop Science*, **19**, 461–7.

Watson, D. J. (1947). Comparative physiological studies on the growth of field crops. II. The effect of varying nutrient supply on net assimilation rate and leaf area. *Annals of Botany*, **11**, 375–407.

9. Transport of nitrate and ammonium through plant membranes

Institut für Botanik, Technische Hochschule, Schnittspahnstr. 3, D-6100 Darmstadt, Germany

Introduction

Among the many ions required for growth and development of plant cells inorganic nitrogen ions are of special importance, qualitatively because of the fundamental role of nitrogen compounds in living matter, and quantitatively in comparison with most other ions which are required in smaller amounts. Because the uptake of these ions is subjected to complicated patterns of regulation and because of their rapid metabolism once inside the plant, this uptake has only recently reached a certain degree of mechanistic clarification (Syrett 1981; Ullrich 1983; Glass 1988). Although transport of nitrate, nitrite, and ammonium as the respective ions very often proceeds towards the sites of consumption, i.e. is inwardly directed, it cannot simply be regarded as a net influx system.

In assimilation of inorganic nitrogen, uptake is necessarily the first step, and this step is self-regulated, and also extensively regulated by the subsequent steps of nitrate and nitrite reduction, and by the assimilation of ammonium to amino acids. This chapter is restricted to work concerning the uptake mechanisms.

Definitions

According to P. Mitchell's theoretical discussion of ion transport across membranes three possibilities have to be taken into account: uniport, antiport and symport.

1. Uniport of anions, as nitrate or nitrite, encounters an inside-negative membrane potential (E_m) usually of -100 to $-250\,mV$ at the plasmalemma. Thus, a much higher external than internal concentration of the respective ion would be required for an energetically balanced, 'passive', inward-directed transport. The electrical consequence would be hyperpolarization of the plasmalemma. In the absence of such a concentration gradient, energy would be required for the transport step itself and for the regeneration of the original E_m. The action of so-called permeases has been postulated, especially for micro-organisms.

This energy requirement does not apply to ammonium, which, as a cation, in most cases can follow its electrochemical gradient, but would also require energy for the permanent re-establishment of the original electrical gradient (E_m).

2. Antiport, as a counter-transport of ions with the same electrical charge in a mechanistically coupled mechanism, can be energetically equilibrated if the stoichiometry and the driving forces ($\Delta\tilde{\mu}$) are in the right order of magnitude. Thus, antiport can be self-energized, but it also requires a permanent re-establishment of the driving forces, in this case more on the chemical side, by pump activities. E_m can remain almost or entirely unchanged in such a case, depending on the antiport stoichiometries.

3. Symport or co-transport includes directly coupled transport of ions with opposite charges at a more or less fixed stoichiometry. This can again be electrically balanced, but in the case of transport against an electrochemical gradient for one of the ions, excess energy has to be provided from $\Delta\tilde{\mu}$ of the other ion. Hence, normally, a transient electrical disequilibrium (in the case of anions a depolarization) is to be expected. The energy for maintaining this symport is again that necessary to permanently re-establish the original conditions during the transport process. Plasmalemma H^+-ATPases (so-called proton pumps) have recently been isolated and claimed to be responsible for stable E_m and pH at the plasmalemma, and within the cytosol. Although final discrimination between H^+ co-transport and OH^- antiport is not possible at present, the action of plasmalemma H^+-ATPases indicates that for anion and for non-electrolyte transport there is an H^+ co-transport mechanism rather than OH^- antiport. As mentioned before, cations such as NH_4^+ do not need this energetic support across a wide concentration range. They have enough inward-directed driving force, since the plasmalemma is negatively charged inside.

Criteria for the mechanism

Two criteria have usually been used to discriminate between these three possibilities; namely, pH changes and transient E_m changes (ΔE_m). In the case of uncharged solutes or anions, titration stoichiometries between external alkalinization and anion uptake have been measured, and used as a direct indication for co-transport stoichiometries with protons. This would imply that the whole amount of protons or hydroxyl ions transported across the membrane remains dissociated in the new environment also. However, this is rarely the case (Stewart 1983; Good 1988) because H^+ and OH^- are very weak ions whose concentrations depend almost exclusively on that of stronger ions, more precisely on the difference between the concentrations of strong cations and strong anions, in any compartment. This basic chemical fact has often been disregarded. Hence, if a strong anion such as nitrate is taken up, its disappearance in the medium will, by itself, change SID (= strong ion concentration difference (Stewart 1983)) towards an excess of strong cations, and thus cause an increase in $[OH^-]$ independent of whether the charge equilibrium disturbed by nitrate transport has been re-established by OH^- or another weak ion. However, if nitrate was exchanged for another strong ion (e.g. Cl^-) no alkalinization would be observed. Co-transport of nitrate with K^+ would result in no pH change in the case of a 1:1 stoichiometry, whereas co-transport with H^+ would, although this would be independent of the actual stoichiometry.

Complexity of nitrate and ammonium uptake systems

In the study of inorganic nitrogen uptake three systems of different complexity have been used:

(1) unicellular microalgae or micro-organisms without storage vacuoles;

(2) vacuolate algae or small higher water plants;

(3) roots (attached or detached) of higher terrestrial plants with both storage vacuoles and long-distance transport of either inorganic nitrogen (nitrate) or organic nitrogen (amino acids) within the roots or to the stems and leaves. These different systems will be discussed in the following section.

Case 1

In unicellular microalgae such as *Chlorella, Monoraphidium,* and *Cyanidium* or *Chlamydomonas*, nitrogen uptake is usually limited by the rate of metabolic consumption as soon as the small cytosolic storage capacity is reached and hence there is an equilibrium between medium and cytosol. This equilibrium can, of course, be energetically supported and thus be far from the Nernst equilibrium. The kinetics of nitrate uptake has two components, of which one is saturable and shows a low half-saturation constant (K_s) in the micromolar range (for a review see Ullrich 1983). The other component has a low affinity or may even appear linear (Serra *et al.* 1978; Ullrich *et al.* 1990), but can also become saturated at higher concentrations, in *Monoraphidium* at between 3 and 10 mM NO_3^- (Ullrich 1974, 1983). This is similar to the kinetics of nitrate reductase, whose K_m is usually in the order of 0.1–0.2 mM, its activity being supported by nitrate accumulation in the cytosol due to the action of the saturable nitrate uptake system. In the absence of CO_2 only small amounts of inorganic nitrogen are converted to amino acids, also in the light, and hence nitrate uptake ceases in many cases. However, there are several reports of nitrate uptake in the absence of CO_2 through the low-affinity transport system (Ullrich 1983; Ullrich *et al.* 1990), when nitrate is reduced under special conditions as, for example, at alkaline pH in the medium (Eisele and Ullrich 1975, 1977; Vega and Menacho 1990). In this case, ammonia is released to the medium in amounts almost stoichiometric to nitrate disappearance. This situation can be induced by methionine sulfoximine (MSX), a potent inhibitor of glutamine synthetase (Rigano *et al.* 1979; Fuggi 1990, for *Cyanidium*). However, at low CO_2 concentration or in the presence of MSX, the ammonia from photorespiration also leaks out of the cells at high rates (Larsson *et al.* 1982). The regulation of this system is only partly understood, as is the very rapid and strong inhibition by ammonium of nitrate uptake in algae (see below). Membrane potential (E_m) measurements are scarce and less reliable for unicellular algae without vacuoles. A strong regulation of nitrate uptake rather than nitrate reduction, but also of the formation of active nitrite reductase, is exerted by light, made visible by using monochromatic light (Azuara and Aparicio 1984; Quiñones and Aparicio 1990). Nitrate uptake in *Chlorella* and *Monoraphidium* immediately after changing the light quality shows much lower rates in monochromatic blue or red light than in white light, although nitrite and ammonium uptake, or the rate of photosynthetic O_2 evolution, exhibit the same rates under these conditions (Calero *et al.* 1980).

Case 2

Plants or algae with vacuolate cells can take up nitrate, and also ammonium to some extent, even when the ions are not consumed by metabolism, but after replenishment of the cytoplasmic pool transport across the tonoplast and metabolism will determine the uptake rate. As a typical example of a higher plant *Lemna gibba* has been studied. Nitrate uptake in *Lemna* is highly inducible, as is the concomitant transient plasmalemma depolarization which is even quantitatively related to uptake kinetics (Novacky *et al.* 1978; Ullrich and Novacky 1981). Besides the active, inducible, system which seems to be mainly responsible for depolarization, a constitutive system with lower transport rates and probably also lower affinity for nitrate can be found in *Lemna*, either in non-induced plants or by inhibition of the inducible system. Such an inhibition is produced by ammonium, especially in the dark in nitrogen-starved plants, less in nitrogen-supplied *Lemna* (Ullrich *et al.* 1984). Nitrate uptake is completely restored to the former rate when ammonium in the medium is consumed or removed. Parallelism between nitrate uptake and membrane depolarization, and between the strong depolarization by ammonium and transport inhibition of nitrate and phosphate, led to an electrophysiological interpretation of ammonium inhibition (see p. 129) (Ullrich *et al.* 1984).

Case 3

Roots of terrestrial higher plants also show uptake characteristics with a constitutive and an inducible system (Behl *et al.* 1988; Ullrich *et at.* 1989), of

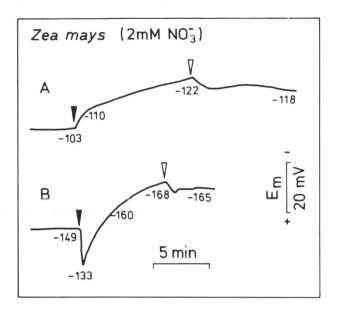

Fig. 9.1. Membrane potential (E_m) responses to addition (closed arrow) and removal (open arrow) of 1 mM $Ca(NO_3)_2$ in root segments of 7-day-old maize plants. E_m values (in mV) shown at the traces. Perfusion with 1 mM $CaSO_4$/1 mM $Ca(NO_3)_2$/1 mM $CaSO_4$. Plants grown without external nitrogen source. A: typical hyper-polarization response of non-induced root. B: combined primary depolarization and subsequent hyper-polarization in a nitrate-induced root (several 10 min treatments with 2 mM NO_3^- over several hours).

which the inducible one has a much higher nitrate affinity. Extracellular K^+ seems to be more essential for roots than for *Lemna* and, in contrast to *Lemna*, stimulates nitrate uptake to some extent, probably as a charge- and pH-balancing cation for the proportion of nitrate translocated to other plant parts (Blevins *et al.* 1974; Touraine and Grignon 1982; Tischner 1990). A characteristic difference to *Lemna* (and to the similar water plant *Limnobium stoloniferum*, Ullrich and Novacky 1990) seems to be that upon addition of nitrate usually only a smaller and slower hyper-polarization is observed instead of the rapid transient depolarization (Fig. 9.1 A) (Thibaud and Grignon 1981; Ullrich 1987; Ullrich *et al.* 1989; Mc Clure *et al.* 1989; Spanswick 1989), and an even more rapid hyper-polarization in suspension cultures of higher plant cells (Rona *et al.* 1990). To some extent the primary depolarization signal upon addition of nitrate can be induced in roots of soybean and maize, but so far not to the same order of magnitude (Ullrich 1987) (Fig. 9.1 B, Table 9.1). Ammonium inhibition of nitrate uptake, as far as currently investigated, is not very strong in roots (MacKown *et al.* 1982), and

Table 9.1 Maximum (or mean) E_m changes in plant cells upon addition of nitrate

Species	ΔE_m (depol- arization)	ΔE_m (hyper- polarization)	Reference
Root cortex			
Glycine max	+15mV	−30 to −45mV	Ullrich 1987, and unpublished
Zea mays	+26mV	−25 to −40mV	Ullrich, unpublished
"	+6mV*	−22mV*	Spanswick 1989
"	−	−20 to −30mV	Thibaud and Grignon 1981
Avena sativa	+2mV**	−14mV**	Ullrich 1987
Green fronds:			
Lemna gibba	+65mV*	mainly re-polarization	Ullrich and Novacky 1981
Root hairs:			
Limnobium stoloniferum	+55mV*	−	Ullrich and Novacky 1990
Suspension cells			
Catharanthus roseus	−	−17mV[†]	Rona *et al.* 1990

*Mean values; **value taken from few data only, [†]rapid hyper-polarization within approximately 20 s, single, representative experiment.
Depolarization occurs within 30s, hyper-polarization is slower and often incomplete. Examples for the time course are shown in Fig. 9.1.

neither is the corresponding plasmalemma depolarization by ammonium (Ullrich, unpublished results).

In all three cases discussed above, it can be shown by various experimental conditions, or merely from the time course, that at a steady state, net uptake of the nitrate and nitrite and, in the same way, of ammonium, is mainly controlled by the subsequent steps of metabolism and/or transport. Limitation by the uptake step was only found in the first phase of uptake upon addition of the nitrogen-containing ion. This could be confirmed by use of inhibitors of either metabolism or uptake (Ullrich *et al.* 1990). Tungstate did not yield clear-cut inhibition; it has been claimed to specifically inhibit only nitrate reductase but not transport, but doubt has been cast on this specificity (Maldonado *et al.* 1990; Ullrich *et al.* 1990).

Mechanisms of nitrate, nitrite, and ammonium transport

As uniport of anions into a negatively charged compartment may require specific, ATP-energized permeases which have not yet been shown to function in plant membranes, the discussion in recent years has focused mainly on the mechanisms of either anion/H^+ co-transport (symport) or anion/anion antiport. The criteria are not completely conclusive, as was shown in the introduction. As quantitative measures of co-transport or antiport stoichiometries, pH shifts are equivocal (see p. 122). Unless other strong ions are involved in the electrical balance, an observed stoichiometry of 1:1 will reflect the pH change by uptake of the respective ion only. There is little chance to find the excess protons or hydroxyl ions transported in either direction during the same period of time, except by current measurements in single cells (Sanders *et al.* 1983). Thus, in tissues depolarization and/or hyper-polarization are the only indicators of a temporarily unbalanced symport or antiport. Since active H^+-ATPases (proton pumps) have been widely investigated within the last few years, it is much more plausible, and in accordance with general convention, to assume that there is H^+ co-transport rather than OH^- antiport, whatever the stoichiometry may be. Hyper-polarization (Fig. 9.1 A) would thus indicate either uniport of an anion for a very short time, or symport of less protons than anions and depolarization (Fig. 9.1 B) symport with a stoichiometry of more than one H^+ per anion. Balancing fluxes of stronger ions, but also of H^+ or OH^-, will usually soon reduce the de- or hyper-polarization rates or even cause rapid recovery which is usually attributed to H^+-ATPases (Ullrich and Novacky 1981; Ullrich and Novacky 1990).

Nitrate transport mechanisms and pH changes

Among the two nitrate transport systems mentioned earlier the high-affinity system may be interpreted as a secondary-active nitrate/H^+ co-transport, driven by the protonmotive force $\Delta \bar{\mu} H^+$. As shown by Glass (1988), for a wide range of pH values and nitrate concentrations, at an assumed E_m of -150 mV, with 2 H^+, an inward-directed driving force for nitrate transport remains under most physiological conditions (Ullrich and Novacky 1981). Investigations on nitrate transport mutants and respective proteins strongly suggest that this transport mechanism is performed by a carrier protein (see below). The low-affinity system is likely to be a passive transport system, representing uniport by a carrier protein or, more likely, by anion channels (McClure *et al.* 1987; Glass 1988; Keller *et al.* 1989).

The transport systems allow nitrate efflux from the cells or tissue to the medium (Morgan *et al.* 1973). Because of this efflux, which mainly occurs with nitrogen-sufficient cells or plant organs, Deane-Drummond (1982, 1984) proposed a mechanism consisting of mere anion exchange in which both anions involved were nitrate. For this to occur passively at a membrane potential of -120 to -180 mV (an often-registered value for E_m in roots of oats, soybean, and maize) a considerable gradient between low internal and high external nitrate concentration would be necessary for equilibrium, and an even higher gradient for net nitrate influx. Nitrate efflux has also been found as a component that diminishes net nitrate uptake in $^{13}N/^{14}N$ experiments, especially at low external nitrate concentrations (Oscarson *et al.* 1987). In *Lemna* an efflux of nitrate, like that of other ions, occurs particularly upon addition of nitrate under N_2, when E_m drops and the energy supply within the tissue is low (Ullrich *et al.* 1990). This suggests the involvement of anion channels which can be opened at low E_m, as described for guard cells (Keller *et al.* 1989). The amounts released indicate that nitrate leaking out from the cytosol originates, to a great extent, from the vacuolar storage pool. Roos (1989) has reported experiments with vacuoles of *Penicillium* in which amino acids are released in low-ATP conditions (ATP concentration in the order of 0.1 mM) whereas in high-ATP conditions (≥ 1 mM ATP) amino acids are rapidly taken up. Nitrate fluxes at the tonoplast may have the same ATP dependence and may be mediated by anion channels or anion/anion antiport (Marigo *et al.* 1990) in a low-energy state. This would also explain the results of experiments with various isotopes (^{13}N-nitrate, ^{15}N-nitrate, or $^{36}ClO_3^-$) and pre-loaded cells. Energetically this system is much more likely than anion exchange at the plasmalemma. In any case, nitrate efflux will occur close to the saturation equilibrium, even in a co-transport system, but will never provide driving force for net nitrate influx. Other antiport models propose OH^- or HCO_3^- exchange for nitrate. Principally, OH^- antiport would show the same result as H^+ symport, whereas HCO_3^- antiport would be different, in its pH effect intermediate between OH^- and Cl^-, or another strong acid. From respiration in non-green tissue, or in green cells in the dark, there would be enough intracellular HCO_3^-, and thus driving force, available for nitrate influx. However, in green cells in the light, when the highest nitrate uptake rates are found, this is very unlikely because of the rapid consumption and, hence, an inward-directed flux of CO_2/HCO_3^-. Green *Lemna* fronds also show depolarization in the light, however to a lesser degree. This fact is also not very compatible with an efficient antiport under these circumstances. It would imply a stoichiometry of > 1 $HCO_3^-/1$ NO_3^-.

On condition that no other stronger ions are involved, nitrate/OH^- antiport, as well as nitrate/H^+ symport, would result in an extracellular alkalinization of 1:1, independent of the original stoichiometry in nitrate transport itself. The alkalinization stoichiometry depends only on the strength of the ion used for charge equilibration. Thus, for example, concomitant K^+ uptake will prevent all pH changes or even lead to external acidification, when an excess of K^+ is taken up (Dijkshoorn 1962; Kirkby and Mengel 1967; Blevins *et al.* 1974; Ullrich *et al.* 1989).

Another problem is that of the intracellular (cytosolic) pH changes during nitrate uptake. As long as only transport is proceeding, cytosolic acidification by the strong ion, NO_3^-, will correspond to the alkalinization of the medium, whether protons are pumped back out or not. But, as soon as nitrate reduction, or more specifically the second step, nitrite reduction, starts within the cells, nitrate (a strong anion) and nitrite (a weaker anion, but also completely dissociated at pH 8) are converted to ammonium and amino groups (medium strong cations). Thus in the cytosol acidification may be diminished or prevented, and an overall intracellular alkalinization may result in addition to the extracellular alkalinization. Cell sap pH measurements as carried out by Blevins *et al.* (1974), in our experiments performed in the absence of extracellular K^+, showed an increase in pH usually 20 min after the addition of NO_3^- [as $Ca(NO_3)_2$] (unpublished results). In root hairs of *Limnobium*, measurements with pH micro-electrodes also revealed a rapid pH rise by about 0.2 units in the cytosol, whereas there was a significant acidification when Cl^- or $H_2PO_4^-$ had been added to the plants (Ullrich and Novacky 1990). This indicates that, in contrast to Cl^- and $H_2PO_4^-$, NO_3^- is rapidly metabolized and converted to the amino cations. Interference with pH by extracellular K^+, even if occasionally released from the cells, was excluded by using a flow-through system with $CaSO_4/Ca(NO_3)_2$. Differences in the immediate metabolization of nitrate may also contribute to the differences in the electrical response to addition of nitrate, because changes in pH may affect the activity of H^+-ATPase and thus cause a delay in re-polarization.

Ammonium inhibition of nitrate uptake

A problem currently under discussion is the inhibition of nitrate uptake, to a much lesser extent than that of nitrite uptake, by ammonium. As mentioned earlier, in algae this inhibition is often complete, in *Lemna* a certain component remains uninhibited, and in roots, even when they are severely nitrogen-starved, there is usually only a moderate inhibition in short-term experiments (see MacKown *et al.* 1982). For *Lemna* it was shown that extent and duration of nitrate uptake inhibition depended on nitrogen starvation (Ullrich *et al.* 1984). Uptake inhibition and E_m depolarization by ammonium correlated well, both being especially strong in the dark. As phosphate uptake was inhibited by ammonium in the same way as nitrate uptake, it was concluded that membrane potentials and driving forces for protons become limiting in those cases. This may easily explain the effect in nitrogen-starved *Lemna* and various other plants, but in algae the inhibition by ammonia can be completely relieved by either absence of carbon skeletons for amino acid formation, or by artificially blocking glutamine synthetase with methionine sulfoximine, or glutamate synthase with azaserine (Syrett and Leftley 1976; Rigano *et al.* 1979; Flores *et al.* 1983; Ullrich 1983; Vega and Menacho 1990). In cyanobacteria inhibition of nitrate uptake could also be induced by some amino acids such as, for example, arginine (Flores *et al.* 1983), in *Monoraphidium* also with caseine hydrolysate (Jarczyk, unpublished). Hence, the most common explanation of this effect is an inhibition of the transport step, not by ammonium itself, but by assimilation products of ammonium via glutamine synthetase–glutamate synthase. Although this is plausible, the physical explanation, i.e. by reduction of driving forces, may apply to the algal system as well: inhibition of glutamine synthetase not only prevents formation of assimilation products, but also influx of ammonium. This is the essential depolarizing process, not the mere presence of ammonium or its formation within the cells. In addition to depolarization, influx of NH_4^+/NH_3 may induce pH changes (Bertl et al. 1984) which will affect the activity of the plasmalemma H^+-ATPase. Another argument for this explanation is that ammonium inhibits nitrate transport instantly, i.e. within a few seconds (Syrett and Leftley 1976; Pistorius et al. 1976; Ullrich 1983), a time interval that seems to be too short for the formation of more than trace amounts of assimilation products. The processes of inactivation and repression of nitrate reductase by ammonium, on the other hand, are too slow to account for this rapid inhibition.

Mechanisms of ammonium transport

As pointed out earlier the uptake of ammonium as a cation, in a wide range of external concentrations, does not need the driving force of another cation such as H^+. Thus, ammonium uptake may be compared to K^+ uptake (Kochian and Lucas 1982) but the ammonium concentrations in the cytosol are usually much lower than those of K^+. Accumulation has been found, for example, in barley roots (Lee and Rudge 1986), but rarely to the concentrations of K^+, and seems to be mainly located in the vacuoles of the tissue (Glass 1988). Hence, $\Delta\tilde{\mu}H^+$ in H^+ co-transport may often be necessary for accumulation of K^+ (Rodríguez-Navarro et al. 1986), rarely for NH_4^+. In any case its demonstration is difficult, since pH changes cannot easily be attributed to H^+ transport, and since NH_4^+ causes depolarization itself. Concomitant pH changes in the cytosol depend on form and concentration of NH_4^+/NH_3 transported (Bertl et al. 1984), and on the ions used for charge balance.

In various organisms ammonium can be replaced by its analogue, methylammonium, for transport studies (reviewed by Glass 1988), but in others (for example, *Monoraphidium*) the uptake rates are very low and do not cause inhibition of nitrate uptake (W. R. Ullrich, unpublished results). It is assumed, though not proven for many species, that both ions are transported by the same carrier.

Nitrate transport proteins and their possible identity with nitrate reductase

Doddema and Telkamp (1979) reported on mutants of *Arabidopsis thaliana* in which nitrate reductase had very little activity, while nitrate uptake showed the normal pattern of kinetics. Together with other physiological data (Ullrich 1987; Ullrich *et al*. 1990) this questioned the hypothesis that membrane-bound nitrate reductase may also be in charge of nitrate transport across the plasmalemma, a hypothesis proposed by Butz and Jackson (1977), but based on data from a wide range of organisms. More recently, a small but constant amount of membrane-bound nitrate reductase has been localized in the plasmalemma of roots and algal cells by immuno-staining, a nitrate reductase apparently not (or much less) subjected to induction and decomposition by proteases. Ward *et al*. (1988) report on inhibition of nitrate transport with anti-nitrate reductase IgG fragments. This special form (and apparently small fraction) of nitrate reductase is presently being characterized: it is a low molecular weight protein different from the cytosolic form but with all the partial activities of the cytosolic enzyme (Tischner 1990). Its role is yet unclear, since it does not represent the inducible components of either nitrate reductase or nitrate transport. Warner and Huffaker (1989) denied a connection between nitrate transport and nitrate reduction, although there was still a small amount of nitrate reductase activity left in NR^- mutants of barley in their experiments. Identification of the new, membrane-bound, nitrate reductase with the constitutive, low-affinity, nitrate uptake system may also be difficult, since this system, at least in the diatom *Skeletonema costatum*, shows no sensitivity to tungstate (Serra *et al*. 1978).

In cyanobacteria another type of mutant, lacking the high-affinity nitrate transport protein, has been studied (Sivak *et al*. 1989; Omata *et al*. 1989). In these studies a 45 kDa protein (or 47 kDa, in the work of Sivak *et al*.) could be identified as the essential carrier protein for high-affinity nitrate transport in *Synechococcus* (= *Anacystis*). Nitrate reductase activity was completely independent of the 45 kDa protein in these cyanobacteria, which could even be cultivated on nitrate, but only at much higher concentrations.

Recently another possible function of membrane-bound nitrate reductase has been proposed: electron transport across the plasmalemma for the reduction of extra-cellular substrates such as iron (Jones and Morel 1988). This has again been supported by immuno-staining and by inhibition of ferricyanide reduction with anti-nitrate reductase IgG. Also, in this case only a small part of nitrate reductase, if any, could be responsible, since ammonium-cultivated *Monoraphidium braunii* showed the usual high rates of ferricyanide reduction together with only 6 per cent of nitrate reductase activity (Corzo *et al*. 1991). An electron transport protein from maize root plasmalemma, in addition, was found to readily reduce ferricyanide and other oxidants with NADH but not nitrate, as electron donor (Luster and Buckhout 1989).

Conclusion

The data on uptake of inorganic nitrogen compounds now available suggest that various mechanisms may be responsible for their transport across the plasmalemma. They are schematically presented in Fig. 9.2.

Nitrate is taken up by two systems:

1. A constitutive, low-affinity, system that appears even linear in the range of physiological concentrations. It could be a carrier system but could also reflect the action of anion channels and thus mediate a passive nitrate influx and, under other conditions, especially in a low-energy state, nitrate efflux.

2. An inducible, high-affinity, transport system regulated by $\Delta\tilde{\mu}H^+$, by cellular energy supply, and by intracellular nitrate consumption. It is pre-

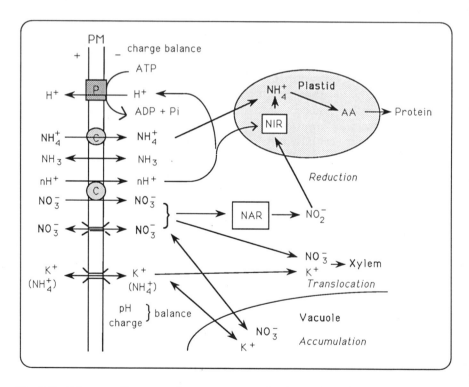

Fig. 9.2. Scheme of proposed transport mechanisms for nitrate (nitrite) and ammonium at the plasmalemma of a plant cell. PM = plasma membrane (= plasmalemma); P = proton pump (H^+-ATPase), acting mainly in charge balance; C = carrier protein; channels shown by symbols, serving for pH and charge balance. For further explanation see text.

vented from functioning at low E_m in the order of the K^+ diffusion potential. It is regarded as an H^+/anion co-transport carrier mechanism that produces transient plasmalemma depolarization upon addition of nitrate. The depolarization is counteracted by the plasmalemma H^+-ATPase. Immediate reduction of the nitrate taken up will cause alkalinization of the cytosol. This may inhibit or delay re-polarization and lead to a more pronounced primary depolarization as observed regularly in *Lemna* and *Limnobium*, whereas acidification by accumulation of nitrate may stimulate the H^+-ATPase. This may reduce or even conceal the primary depolarization, as is often observed in roots. Induction of the inducible nitrate transport system and of nitrate reductase may thus contribute to the expression of depolarization responses. The inducible nitrate transport system is closely related to a 45 (or 47) kDa protein isolated from cyanobacteria. A plasmalemma-bound low-molecular-weight nitrate reductase seems to be involved in nitrate transport but cannot be clearly associated with the two systems as yet.

Nitrite is also transported by two systems, of which the low-affinity system may play a greater role than in nitrate uptake. In some, but not all cases, the high-affinity system has been shown to be identical with that of nitrate transport, by uptake competition as well as by studies of voltage changes.

Ammonium as a cation has its own $\Delta\mu$ directed into the cytosol, at least for a wide range of external concentrations. There is again a saturable, but apparently constitutive, carrier system with high substrate affinity, which may carry out ammonium uniport as long as H^+-ATPases restore E_m and thus $\Delta\bar{\mu}NH_4^+$. The low-affinity component of transport is stimulated by high external pH and probably reflects diffusion of uncharged NH_3 across the lipid phase of the plasmalemma. If ammonium enters the cells at high rates, it causes strong membrane depolarization and will block anion/H^+ co-transport.

References

Azuara, M. P. and Aparicio, P. J. (1984). Spectral dependence of photoregulation of inorganic nitrogen metabolism in *Chlamydomonas reinhardii*. *Plant Physiology*, **77**, 95–8.

Behl, R., Tischner, R., and Raschke, K. (1988). Induction of a high-capacity nitrate-uptake mechanism in barley roots prompted by nitrate uptake through a constitutive low-capacity mechanism. *Planta*, **176**, 235–40.

Bertl, A., Felle, H., and Bentrup, F. W. (1984). Amine transport in *Riccia fluitans*. Cytoplasmic and vacuolar pH recorded by a pH-sensitive microelectrode. *Plant Physiology*, **76**, 76–8.

Blevins, D. G., Hiatt, A. J., and Lowe, R. H. (1974). The influence of nitrate and chloride uptake on expressed sap pH, organic acid synthesis, and potassium accumulation in higher plants. *Plant Physiology*, **54**, 82–7.

Butz, R. G. and Jackson, W. A. (1977). A mechanism for nitrate transport and reduction. *Phytochemistry*, **16**, 409–17.

Calero, F., Ullrich, W. R., and Aparicio, P. J. (1980). Regulation by monochromatic light of nitrate uptake in *Chlorella fusca*. In *The blue light syndrome* (ed. H. Senger), pp. 411–21. Springer Verlag, Berlin.

Corzo, A., Plasa, R., and Ullrich, W. R. (1991). Extracellular ferricyanide reduction and nitrate reductase activity in the green alga *Monoraphidium braunii*. *Plant Science*, **75**, 221–8.

Deane-Drummond, C. E. (1982). Mechanisms for nitrate uptake into barley (*Hordeum vulgare* L. cv. Fergus) seedlings grown at controlled nitrate concentrations in the nutrient medium. *Plant Science Letters*, **24**, 79–89.

Deane-Drummond, C. E. (1984). Mechanism of nitrate uptake into *Chara corallina* cells: lack of evidence for obligatory coupling to proton pump and a new NO_3^-/NO_3^- exchange model. *Plant Cell and Environment*, **7**, 317–23.

Dijkshoorn, W. (1962). Metabolic regulation of the alkaline effect of nitrate utilization in plants. *Nature*, **194**, 165–7.

Doddema, H. and Telkamp, G. P. (1979). Uptake of nitrate by mutants of *Arabidopsis thaliana*, disturbed in uptake or reduction of nitrate. II. Kinetics. *Physiologia Plantarum*, **45**, 332–8.

Eisele, R. and Ullrich, W. R. (1975). Stoichiometry between photosynthetic nitrate reduction and alkalinization by *Ankistrodesmus braunii in vivo*. *Planta*, **123**, 117–23.

Eisele, R. and Ullrich, W. R. (1977). Effect of glucose and CO_2 on nitrate uptake and coupled OH^- flux in *Ankistrodesmus braunii*. *Plant Physiology*, **59**, 18–21.

Flores, E., Romero, J. M., Guerrero, M. G., and Losada, M. (1983). Regulatory interaction of photosynthetic nitrate utilization and carbon dioxide fixation in the cyanobacterium *Anacystis nidulans*. *Biochimica et Biophysica Acta*, **725**, 529–32.

Fuggi, A. (1990). Uptake of inorganic nitrogen compounds in an acidophilic alga. In *Inorganic nitrogen in plants and microorganisms: uptake and metabolism* (ed. W. R. Ullrich, C. Rigano, A. Fuggi, and P. J. Aparicio), pp. 66–72. Springer Verlag, Berlin.

Glass, A. D. M. (1988). Nitrogen uptake by plant roots. *ISI Atlas of Science: Plants and Animals*, **1**, 151–61.

Good, N. E. (1988). Active transport, ion movements, and pH changes. I. The chemistry of pH changes. *Photosynthesis Research*, **19**, 225–36.

Jones, G. J. and Morel, F. M. M. (1988). Plasmalemma redox activity in the diatom *Thalassiosira*. A possible role for nitrate reductase. *Plant Physiology*, **87**, 143–7.

Keller, B. U., Hedrich, R., and Raschke, K. (1989). Voltage-dependent anion channels in the plasma membrane of guard cells. *Nature*, **341**, 450–3.

Kirkby, E. A. and Mengel, K. (1967). Ionic balance in different tissues of the tomato plant in relation to nitrate, urea, and ammonium nutrition. *Plant Physiology*, **42**, 6–14.

Kochian, L. V. and Lucas, W. J. (1982) Potassium transport in corn roots. 1. Resolution of kinetics into a saturable and linear component. *Plant Physiology*, **70**, 1723–31.

Larsson, M., Larsson, C.-M., and Ullrich, W. R. (1982) Regulation by amino acids of photorespiratory ammonia and glycolate release from *Ankistrodesmus* in the presence of methionine sulfoximine. *Plant Physiology*, **70**, 1637–40.

Lee, R. B. and Rudge, K. A. (1986). Effects of nitrogen deficiency on the absorption of nitrate and ammonium by barley plants. *Annals of Botany*, **57**, 471–86.

Luster, D. G. and Buckhout, T. J. (1989). Isolation and characterization of an electron transport protein from maize plasma membranes. *Plant Physiology*, **89**, 44 (Suppl. 262).

MacKown, C. T., Jackson, W. A., and Volk, R. J. (1982). Restricted nitrate influx and reduction in corn seedlings exposed to ammonium. *Plant Physiology*, **69**, 353–9.

Maldonado, J. M., Agüera, E., and de la Haba, P. (1990). Nitrate and nitrite utilization by sunflower plants. In *Inorganic nitrogen in plants and microorganisms: uptake and metabolism* (ed. W. R. Ullrich, C. Rigano, A. Fuggi, and P. J. Aparicio), pp. 159–64. Springer Verlag, Berlin.

Marigo, G., Bouyssou, H., Pennarun, A. M., and Rona, J. P. (1990). Accumulation and vacuolar remobilization of nitrate in *Catharanthus roseus* cells. In *Inorganic nitrogen in plants and microorganisms: uptake and metabolism* (ed. W. R. Ullrich, C. Rigano, A. Fuggi, and P. J. Aparicio), pp. 54–9. Springer Verlag, Berlin.

McClure, P. R., Omholt, T. E., Pace, G. M., and Bouthyette, P.-Y. (1987). Nitrate-induced changes in protein synthesis and translation of RNA in maize roots. *Plant Physiology*, **84**, 52–7.

McClure, P. R., Shaff, J. E., Spanswick, R. M., and Kochian, L. V. (1989). Response of the membrane potential of maize roots to nitrate. *Plant Physiology*, **89**, 45 (Suppl. 267).

Morgan, M. A., Volk, R. J., and Jackson, W. A. (1973). Simultaneous influx and efflux of nitrate during uptake by perennial ryegrass. *Plant Physiology*, **51**, 267–72.

Novacky, A., Fischer, E., Ullrich-Eberius, C. I., Lüttge, U., and Ullrich, W. R. (1978). Membrane potential changes during transport of glycine as a neutral amino acid and nitrate in *Lemna gibba* G 1. *FEBS Letters*, **88**, 264–7.

Omata, T., Ohmori, M., Arai, N., and Ogawa, T. (1989). Genetically engineered mutant of the cyanobacterium *Synechococcus* PCC7942 defective in nitrate transport. *Proceedings of the National Academy of Science, USA*, **86**, 6612–6.

Oscarson, P., Ingemarsson, B., af Ugglas, M., and Larsson, C.-M. (1987). Short-term studies of NO_3^- uptake in *Pisum* using $^{13}NO_3^-$. *Planta*, **170**, 550–7.

Pistorius, E. K., Gewitz, H. S., Voss, H., and Vennesland, B. (1976). Reversible activation of nitrate reductase in *Chlorella vulgaris in vivo*. *Planta*, **128**, 73–80.

Quiñones, M. A. and Aparicio, P. J. (1990). Blue light activation of nitrate reductase and blue light promotion of the biosynthesis of nitrite reductase in *Monoraphidium braunii*. In *Inorganic nitrogen in plants and microorganisms: uptake and metabolism* (ed. W. R. Ullrich, C. Rigano, A. Fuggi, and P. J. Aparicio), pp. 171–7. Springer Verlag, Berlin.

Rigano, C., Di Martino Rigano, V., Vona, V., and Fuggi, A. (1979). Glutamine synthetase activity, ammonia assimilation and control of nitrate reduction in the unicellular red alga *Cyanidium caldarium*. *Archives of Microbiology*, **121**, 117–20.

Rodríguez-Navarro, A., Blatt, M.R., and Slayman, C.L. (1986). A potassium-proton symport in *Neurospora crassa*. *Journal of General Physiology*, **87**, 649–74.

Rona, J.P., Monestiez, M., Pennarun, A.M., Convert, M., Cornel, D., Bousquet, U., Kiolle, R., and Marigo, G. (1990). Nitrate uptake in *Catharanthus roseus* cells: electrophysiological effects. In *Inorganic nitrogen in plants and microorganisms: uptake and metabolism* (ed. W.R. Ullrich, C. Rigano, A. Fuggi, and P.J. Aparicio), pp. 60–5. Springer Verlag, Berlin.

Roos, W. (1989) Cytosolic control of vacuolar amino acid transport in *Penicillium cyclopium*. In *Plant membrane transport: the current position* (ed. J. Dainty, M.I. De Michelis, E. Marrè, and F. Rasi-Caldogno), pp. 393–6. Elsevier, Amsterdam.

Sanders, D., Slayman, C.L., and Pall, M.L. (1983). Stoichiometry of H^+/amino acid co-transport in *Neurospora crassa* revealed by current-voltage analysis. *Biochimica et Biophysica Acta*, **735**, 67–76.

Serra, J.L., Llama, M.J., and Cadenas, E. (1978). Nitrate utilization by the diatom *Skeletonema costatum*. II. Regulation of nitrate uptake. *Plant Physiology*, **62**, 991–4.

Sivak, M.N., Lara, C., Romero, J.M., Rodríguez, R., and Guerrero, M.G. (1989). Relationship between a 47 kDa cytoplasmic membrane polypeptide and nitrate transport in *Anacystis nidulans*. *Biochemical and Biophysical Research Communications*, **158**, 257–62.

Spanswick, R.M. (1989). Some problems concerning proton co-transport at the plasma membrane. In *Plant membrane transport: the current position* (ed. J. Dainty, M.I. De Michelis, E. Marrè, and F. Rasi-Caldogno), pp. 13–8. Elsevier, Amsterdam.

Stewart, P.A. (1983). Modern quantitative acid-base chemistry. *Canadian Journal of Physiology and Pharmacology*, **61**, 1444–61.

Syrett, P.J. (1981). Nitrogen metabolism of microalgae. *Canadian Bulletin of Fisheries and Aquatic Sciences*, **210**, 182–210.

Syrett, P.J. and Leftley, J.W. (1976). Nitrate and urea assimilation by algae. In *Perspectives in experimental biology* (ed. N. Sunderland), Vol. II, pp. 221–34. Pergamon Press, Oxford.

Thibaud, J.-B. and Grignon, C. (1981). Mechanism of nitrate uptake in corn root. *Plant Science Letters*, **22**, 279–89.

Tischner, R. (1990). New regulatory steps in nitrate assimilation of lower and higher plants. In *Inorganic nitrogen in plants and microorganisms: uptake and metabolism* (ed. W.R. Ullrich, C. Rigano, A. Fuggi, and P.J. Aparicio), pp. 51–3. Springer Verlag, Berlin.

Touraine, B. and Grignon, C. (1982). Potassium effect on nitrate secretion into the xylem of corn roots. *Physiologie Végétale*, **20**, 23–31.

Ullrich, C.I. and Novacky, A.J. (1990). Extra- and intracellular pH and membrane potential changes induced by K^+, Cl^-, $H_2PO_4^-$ and NO_3^- uptake and fusicoccin in root hairs of *Limnobium stoloniferum*. *Plant Physiology*, **94**, 1561–7.

Ullrich, W.R. (1974). Die nitrat- und nitritabhängige photosynthetische O_2-Entwicklung in N_2 bei *Ankistrodesmus braunii*. *Planta*, **116**, 143–52.

Ullrich, W.R. (1983). Uptake and reduction of nitrate: Algae and fungi. In *Inorganic plant nutrition* (ed. A. Läuchli and R.L. Bieleski), *Encyclopedia of Plant Physiology*, Vol. 15 A, pp. 376–97. Springer Verlag, Berlin.

Ullrich, W.R. (1987). Nitrate and ammonium uptake in green algae and higher plants: Mechanism and relationship with nitrate metabolism. In *Inorganic nitrogen metabolism* (ed. W.R. Ullrich, P.J. Aparicio, P.J. Syrett, and F. Castillo), pp. 32–8. Springer Verlag, Berlin.

Ullrich, W.R. and Novacky, A. (1981). Nitrate-dependent membrane potential changes and their induction in *Lemna gibba* G1. *Plant Science Letters*, **22**, 211–17.

Ullrich, W.R., Larsson, M., Larsson, C.-M., Lesch, S., and Novacky, A. (1984) Ammonium uptake in *Lemna gibba* G1, related membrane potential changes, and inhibition of anion uptake. *Physiologia Plantarum*, **61**, 369–76.

Ullrich, W.R., Jaenicke, H., and Brandl, G. (1989). Nitrate uptake in roots: induction and charge balance. In *Plant membrane transport: the current position* (ed. J. Dainty, M.I. De Michelis, E. Marrè, and F. Rasi-Caldogno), pp. 335–8. Elsevier, Amsterdam.

Ullrich, W.R., Lesch, S., Jarczyk, L., Herterich, M., and Trogisch, G.D. (1990). Transport of inorganic nitrogen compounds: physiological studies on uptake and assimilation. In *Inorganic nitrogen in plants and microorganisms: uptake and metabolism* (ed. W.R. Ullrich, C. Rigano, A. Fuggi, and P.J. Aparicio), pp. 44–50. Springer Verlag, Berlin.

Vega, J.M. and Menacho, A. (1990). Regulation of inorganic nitrogen metabolism in *Chlamydomonas reinhardtii*. In *Inorganic nitrogen in plants and microorganisms: uptake and metabolism* (ed. W.R. Ullrich, C. Rigano, A. Fuggi, and P.J. Aparicio), pp. 73–8. Springer Verlag, Berlin.

Ward, M.R., Tischner, R., and Huffaker, R.C. (1988). Inhibition of nitrate transport by anti-nitrate reductase IgG fragments and the identification of plasma-membrane associated nitrate reductase in roots of barley seedlings. *Plant Physiology*, **88**, 1141–5.

Warner, R.L. and Huffaker, R.C. (1989). Nitrate transport is independent of NADH and NAD(P)H nitrate reductases in barley seedlings. *Plant Physiology*, **91**, 947–53.

10. Amino acid transport across the higher plant cell membrane

FRIEDRICH-WILHELM BENTRUP and BERND HOFFMANN

Botanisches Institut 1 der Justus-Liebig-Universität Giessen, Senckenbergstrasse 17–21, D-6300 Giessen, Germany

Introduction

This chapter first briefly outlines current aspects and problems of amino acid transport in the higher plant. Secondly, we present novel data from a photo-autotrophic suspension cell culture which demonstrates the operation of an amino acid exchange carrier in the plasmalemma by efflux analysis.

In higher plants, a flow of amino acids passes from the key site of amino acid synthesis in the photosynthetically active tissues to the heterotrophically growing meristematic and storage tissues of stem, fruits, and roots. Photosynthates in the phloem, including amino acids, cross the plasmalemma several times if passing through the apoplast while *en route* from the source mesophyll to sink tissues. Thus the plasmalemma is an established site for the recognition and specific transport of pertinent substrates, and is a plausible site of physiological control of the overall translocation process.

In passing we mention that, surprisingly, studies of amino acid transport into and out of the cellular storage organelle (the vacuole) are in their very infancy; only uptake of phenylalanine (Homeyer and Schultz 1988), and of alanine, leucine, and glutamine (Dietz *et al.* 1990) by vacuoles isolated from barley mesophyll protoplasts has been studied. Interestingly, uptake as well as release from the vacuole (Dietz *et al.* 1989) depends upon ATP, although ATP is not hydrolysed by an ATPase. A similar, intriguing, role of ATP in the release of betacyanin dyes by isolated *Chenopodium* vacuoles has also been demonstrated (Lommel *et al.* 1989).

The experimental design of transport studies of higher plant tissues and organs at the molecular level, in terms of the specificity and kinetics of individual transport proteins, is cumbersome, and progress in this field has been slow compared with what is known from comparable animal tissues or the isolated higher plant cell. Moreover, even segments or discs cut from roots and leaves respectively differ from *in situ* plant material, so that experiments are hampered by well-known ageing effects. For obvious reasons, aquatic higher plants which can absorb nutrients through the whole leaf surface from their environment have a long experimental record; for example, protonmotive force-driven amino acid carriers have recently been identified in the plasmalemma of *Egeria* (Petzold *et al.* 1985) and *Lemna* (Datko and Mudd 1985); amino acid transport in the aquatic liverwort *Riccia fluitans* has been analysed in considerable detail by Felle's group (Felle 1981, 1984; Johannes and Felle 1985, 1987).

Recent progress from *in situ* studies may be illustrated by three cases. First, concerning the symplast–apoplast–symplast pathway concept of phloem-loading, a careful study of amino acid and sugar uptake by mesophyll cells (without veins) and leaf discs containing veins, respectively, from *Commelina* leaves led van Bel *et al.* (1986) to conclude that high affinity carrier systems are restricted to the mesophyll cells. This keeps apoplastic concentration rather low there, whereas near the veins photosynthate exits to the apoplast and is then loaded by low affinity, high capacity, carrier systems to the sieve-tube/companion cell complexes.

Secondly, we refer to work on the castor bean, where the cotyledons actively take up amino acids, preferably glutamine, from the endosperm (Mengel and Haeder 1977; Robinson and Beevers 1981). Recently Schobert and Komor (1989) showed that the sieve-tube sap of isolated cotyledons contained 150 mM amino acids with 50 mM glutamine, and 10–15 mM each of valine, isoleucine, lysine, and arginine. Notably, the transport specificity of phloem-loading differs from the amino acid uptake specificity of the cotyledonar tissue. Hence the occurrence of tissue-specific amino acid carriers has been proposed.

Thirdly, the reader is encouraged to follow recent progress on the transport of peptides, purine, and pyrimidine bases, and nucleosides in germinating seeds. In the plasmalemma of scutellum cells of germinating barley grains there are different carriers that actively import oligopeptides (di- to pentapeptides) and amino acids respectively (Higgins and Payne, 1978*a, b*). Interestingly, peptide uptake shows higher rates than amino acid uptake. First attempts to isolate the peptide transport carrier from a scutellar plasmalemma preparation have been reported (Payne and Walker-Smith 1987).

In closing this introductory outline, we note that in the plasmalemma of higher plant cells constitutive amino acid carrier systems operate. (Inducible amino acid carriers have been described for *Chlorella* by Sauer *et al.* 1983.) They are generally believed to be energized by the protonmotive force (proton/amino acid co-transport). However, there is need for reliable, independent measurement of both proton and amino acid fluxes to verify kinetic details of this proton/amino acid co-transport. In fact, Johannes and Felle (1987) recorded the cytoplasmic pH during amino acid uptake and found that it was kinetically much more closely coupled to the proton pump than had been anticipated: amino acid-induced depolarization activates the proton pump which, in turn, over-rules the acidification of the cytoplasm expected from the proton/amino acid co-transport process.

Furthermore, in addition to isolation of carrier proteins by the direct biochemical approach or by molecular cloning, reliable functional identification of carrier systems (proteins?) in an individual membrane is needed. The information available is to date contradictory, possibly due to overlapping carrier substrate spectra. Tests of carrier specificity commonly rely upon competitive (*cis*) inhibition of suspected amino acids. Thus Berry *et al.* (1981) propose that a general L-amino acid carrier operates in the plasmalemma of cultured tobacco cells, accepting also arginine, whereas separate uptake systems for neutral and basic amino acids, respectively, have been reported repeatedly, including in sugar cane suspension cells (Wyse and Komor 1984) and *Lemna* leaves (Datko and Mudd, 1985). In *Lemna*, uptake systems for further nitrogenous compounds have been proposed, i.e. for choline, ethanolamine, tyramine, urea, and purine bases. The frequent observation of various concentration-dependent amino acid uptake patterns have been incorporated into various kinetic models; for a critical discussion see Borstlap and Schuurmans (1988), who studied four amino acid-transport mutants of tobacco, and concluded that in the mesophyll cells of these lines at least three physically distinct transport systems operate.

Cell suspension cultures from higher plants have been a favourite system to study transport processes of plasmalemma and tonoplast (*see* Reinhold and Kaplan 1984). These heterotrophically grown suspension cells may be regarded as a storage cell model system. However, the commonly used rich nutritional media hardly simulate the probably less opulent milieu in the apoplast of leaf, stem, or root tissues. In fact, a suspension culture from parsley under investigation in our laboratory from 1970 to 1978 revealed unusual membrane transport properties (Pfrüner and Bentrup 1978).

Therefore, we have changed to a strictly photo-autotrophically growing suspension cell culture of *Chenopodium rubrum* established by Hüsemann and Barz (1977); these cells indeed resemble mesophyll cells in respect to pertinent transport properties, including electrogenic transport of hexoses (Ohkawa *et al.* 1981; Gogarten and Bentrup 1989*a*, *b*) and amino acids (Steinmüller and Bentrup 1981, and this study).

Below we report trans-membrane stimulation of efflux of a labelled amino acid (α-amino-isobutyric, AIB, in this study) by external amino acids. Contrary to the popular competitive *cis*-inhibition test, *trans*-stimulation of efflux of A by substrate B directly proves whether A and B are indeed transported (exchanged) by the same carrier molecule (see Stein 1986). It relies on the common observation that the rate of transport by a given carrier molecule is limited by the trans-membrane movement of its empty binding site, and is accelerated if occupied by a transportable substrate.

We have worked out this approach for analysis of the proton/hexose co-transporter in the plasmalemma of *Riccia* (Gogarten and Bentrup 1983) and *Chenopodium* (Gogarten and Bentrup 1989*a*, *b*). Previous observations have indicated that in these plants, large vacuoles quickly refill (and re-label) the cytoplasm back to steady-state tracer content after the flux equilibrium at the plasmalemma has been perturbed by intermittent stimulation. We report data suggesting that the neutral amino acids AIB and tryptophan, and the basic amino acids lysine, arginine, and histidine are transported by the same carrier molecule, which is probably identical to the carrier found previously to mediate the electrogenic uptake of α-AIB, L-alanine, and L-serine (Steinmüller and Bentrup 1981).

Material and methods

Cultivation of the photo-autotrophic *Chenopodium rubrum* cell suspension has been described previously (Hüsemann and Barz 1977; Steinmüller and Bentrup, 1981). Our test medium contained 2.0 mM NaCl, 0.1 mM KCl, 0.2 mM CaCl$_2$, 0.1 mM MgSO$_4$, and 20 mM Mes/NaOH to adjust the pH to 5.1. The analogue α-amino-isobutyric acid (α-AIB) was not measurably metabolized by the cells even after 72 h of incubation, as indicated by thin layer chromatography of ^{14}C-labelled α-AIB.

The experimental design of radiotracer uptake, counting, and efflux compartmental analysis with ^{14}C-labelled α-AIB closely followed the previous study of *Chenopodium* hexose transport using ^{14}C-labelled 3-O-methyl glucose (Gogarten and Bentrup 1989*b*). Accordingly, for efflux experiments the cells were preloaded with ^{14}C-α-AIB for 40 h (0.2–2.5 μCi ^{14}C-α-AIB in 20 ml samples). Efflux was started by removal of the radiotracer from the test medium by washing the cells in the perfusion chamber (α-AIB concentration remained unchanged).

Tryptophan exit was continuously assayed from the absorbance at 280 nm of the test medium after it had passed the cell suspension in a perfusion chamber. The experimental set-up is described by Hoffmann (1988). Similarly, L-arginine left in the perfusion medium was assayed by fluorimetry of a pyrrolinone arising from the coupling of the amino acid to fluorescamine (Stein *et al.* 1973).

Results and discussion

Uptake of ^{14}C-α-AIB yielded the commonly observed saturating and linear components; initial uptake rates plotted versus ^{14}C-α-AIB concentration over the tested range from 10 to 1000 μM yielded $k_{1/2}$ values from 8 μM (pH 4.6) to 39 μM (pH 6.9), and are consistent with comparable values for L-alanine for this material (4 μM, Steinmüller and Bentrup 1981) and the generally observed range for so-called high affinity amino acid uptake by higher plant cells (see Reinhold and Kaplan 1984).

Efflux kinetics showed two components quite similar to those from previous hexose flux compartmental analysis (Gogarten and Bentrup 1989*b*); that is, a slowly exchanging large component (half-time of exchange 40 h) attributable to the vacuole, and a fast component (half-time about 8 min), reflecting the amino acid transport activity of the plasmalemma.

At 2.5 μM α-AIB in the medium, average cytoplasmic and vacuolar concentrations of 0.3 mM and 5 mM α-AIB, respectively, have been derived from the flux analysis. Evidently, the α-AIB transporting carrier in the plasmalemma accumulates 120-fold, and further 17-fold vacuolar accumulation indicates active transport at the tonoplast. Accumulation (112-fold) of L-alanine across the plasmalemma of these cells has been reported by Steinmüller and Bentrup (1981).

Figure 10.1 illustrates that tracer efflux drastically rises, if the external α-AIB concentration is temporarily increased from the steady-state concentration of 5 μM to the values given at the efflux peaks. (Correspondingly, efflux decreases if external α-AIB is withdrawn.) Following the concept and

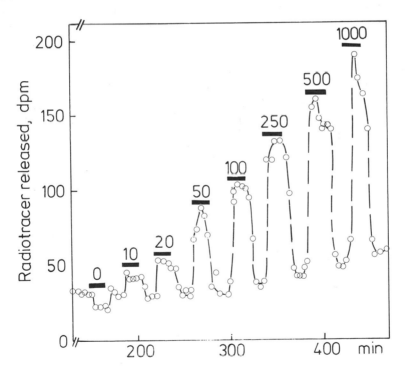

Fig. 10.1. Kinetics of efflux of ^{14}C-labelled-α-AIB from a *Chenopodium* cell suspension at pH 5.2. Tracer efflux was started at $t = 0$ by removing the tracer, but keeping 5 μM α-AIB in the perfusion medium. During the indicated time intervals (16 min) the α-AIB concentration was raised to the given values (μM).

experimental design of efflux stimulation (Gogarten and Bentrup 1983, 1989*b*), the rising external α-AIB concentration increasingly refills the carrier binding site emptied after release of ^{14}C-labelled α-AIB, and thus accelerates the rate-limiting recycling of the binding site to the cytoplasmic face of the plasmalemma. Efflux stimulation (EST) may be expressed through the Michaelis–Menten rate equation:

$$EST = EST_{max} \times [A] / (k_{1/2} + [A])$$ (10.1)

where [A] is the external amino acid concentration and $k_{1/2}$ gives [A] for half-maximal efflux stimulation. Figure 10.2 shows that the data of Fig. 10.1 fairly fit this formalism; the Hofstee plot yields a $k_{1/2}$ value of 30 μM α-AIB.

Efflux stimulation obviously saturates, if carrier binding site recycling is no longer rate-limiting. These features cogently demonstrate that the stimulating substrate indeed enters the cell by those carrier molecules mediating tracer efflux. In our efflux experiments, prior to perturbance by a stimulating substrate, a steady-state carrier occupancy holds with stationary concentrations, on both sides of the plasmalemma, of the putative ternary complex CSH of carrier C$^-$, substrate S, and co-substrate H$^+$ (Gogarten and Bentrup 1989*a*, *b*); addition of a stimulating substrate will accelerate

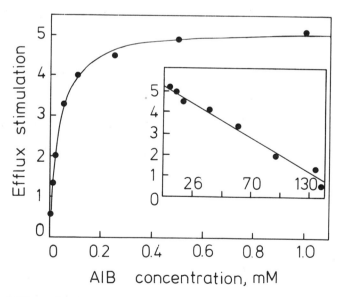

Fig. 10.2. Efflux stimulation of Fig. 10.1 plotted in relative units versus α-AIB concentration (unit efflux at [α-AIB] = 0). The data have been fitted to eqn (10.1). The Hofstee plot (inset) yields a $k_{1/2}$ of 30 μM α-AIB.

exchange diffusion of CSH, and thus not necessarily cause a net influx of charge and protons.

We note that the observed maximum stimulation exceeding a factor of five excludes that it is simulated by the well-known label re-uptake syndrome. Efflux stimulation could be feigned by competitive (*cis*)-inhibition, that is, sudden addition of substrate (unlabelled α-AIB or any other amino acid) could decrease the permanently extant re-uptake of ^{14}C-labelled-α-AIB by competition for the binding site. We have previously checked this well-known re-uptake problem on the *Chenopodium* cells by independent changes of specific radioactivity and chemical concentration of the transported molecules: during efflux of ^{14}C-labelled 3-O-methyl-D-glucose re-uptake by the hexose carrier operating in parallel and at comparable rates in the plasmalemma of *Chenopodium rubrum*, amounted to 21 per cent or less (Gogarten and Bentrup 1989*b*). Therefore, radiotracer re-uptake may be ignored if efflux stimulation exceeds a factor of, say, two.

It is useful then to test other amino acids. Figure 10.3 demonstrates that

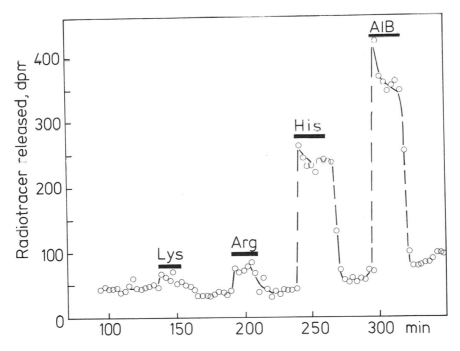

Fig. 10.3. Stimulation of tryptophan exit from a *Chenopodium* cell suspension by intermittent addition (12 min) of α-AIB to the test medium. Over the test periods used (1 h) a steady-state tryptophan concentration of approx. 2 μM resulted from the rates of net tryptophan efflux (= exit) and perfusion of the test chamber, respectively.

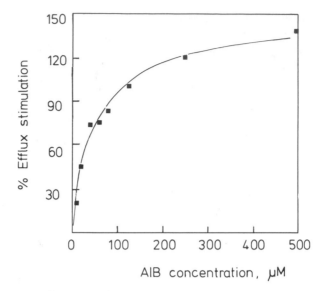

Fig. 10.4. Tracer efflux kinetics as in Fig. 10.1. During the indicated time intervals 1 mM each of L-lysine, L-arginine and L-histidine, respectively, have been added to the perfusion medium.

α-AIB stimulates the exit of L-tryptophan. Stimulation half-saturates at 53 μM α-AIB; maximum stimulation approximates 150 per cent. Thus both of the neutral amino acids α-AIB and tryptophan share the same carrier.

Figure 10.4 shows an experiment where efflux of ^{14}C-labelled-α-AIB is stimulated by addition of the basic amino acids lysine, arginine, and histidine. The kinetics show that lysine, arginine, histidine, and α-AIB at 1 mM each stimulate α-AIB-efflux differently, i.e. by a factor of 1.4, 2.1, 6.2, and 6.7, respectively; only histidine clearly stimulates α-AIB efflux, suggesting a common carrier usage at this acid pH.

This conclusion does not imply that the other basic amino acids are not taken up actively: for instance, evidence for saturable uptake of L-arginine by the *Chenopodium* cells is given by Fig. 10.5. Over the tested range up to 1 mM, initial uptake rates are satisfactorily described by a single hyperbola which gives a reasonably small $k_{1/2}$ of 36.1 μM to indicate high-affinity uptake. The maximum uptake is comparable to the rate accomplished by the carrier handling α-AIB and tryptophan. Assuming the presence, therefore, of a second carrier species for amino acid transport in *Chenopodium*, does not rule out overlapping substrate spectra, depending, for instance, on the pH-dependent charge. Johannes and Felle (1985) have demonstrated in *Riccia* distinct carrier species for neutral and basic amino acids, although the former also accepts basic amino acids.

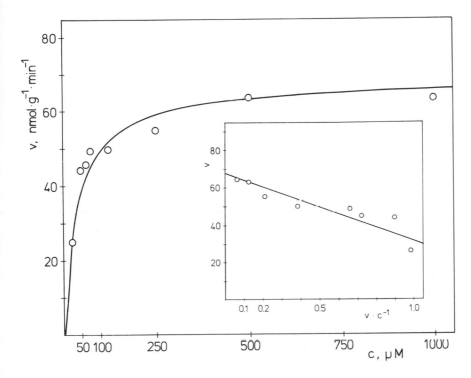

Fig. 10.5. Uptake of L-arginine by a *Chenopodium* cell suspension at pH 5 as a function of substrate concentration. Ordinate values give initial rates (10 min). The data are fitted by a hyperbola which gives $k_{1/2} = 36.1\ \mu M$ according to the Michaelis–Menten rate equation.

Changes of the protonmotive force, i.e. membrane depolarization and changes of the transmembrane pH-gradient, due to addition of proton-driven amino acid (or sugar) can be incorrectly interpreted as showing sharing by two substrates of a common carrier. By contrast to the conventional competitive *cis*-inhibition experiment, this pitfall is bypassed by the experimental design of the efflux experiment, where the addition of the stimulating substrate accelerates carrier cycling but does not necessarily cause net influx of charge or protons.

There is ample evidence for this notion; it is a frequent observation that depolarization by a proton-driven co-transported substrate is transient, i.e. by Felle and Bentrup (1980) for hexoses, or Johannes and Felle (1985) for amino acids. In fact, the latter authors have identified the quoted two transport systems also by the observation that after pre-incubation of *Riccia* thalli with a neutral amino acid (1 mM α-AIB, or alanine), 0.1 mM serine did not, but arginine, lysine, ammonia, and glucose markedly depolarized the plasmalemma, because they obviously engage carrier systems other than the one pertinent to neutral amino acids.

We were not surprised, therefore, that addition of 1 mM D-glucose, which depolarizes the *Chenopodium* plasmalemma by approximately 60 mV (Ohkawa *et al.* 1981), i.e. reduces the protonmotive force by about 20 per cent, indeed measurably stimulates ^{14}C-labelled-α-AIB efflux by a factor of 1.6 (compare Fig. 10.4). Thus heterologous efflux stimulation due to tapping the protonmotive force by a physically different proton-driven transport is possible but may be ruled out quantitatively. Histidine, but not arginine, shares the carrier with α-AIB, as Fig. 10.4 indicates, because at pH 5.2 a substantial fraction of the zwitterionic neutral form of histidine is available to the carrier accepting neutral amino acids like α-AIB (the imidazole group of histidine protonates with pK = 6.1). In fact, at pH 7.2, 1 mM histidine stimulates α-AIB efflux to a comparable extent as 1 mM α-AIB.

Finally, we include the α-AIB-dependent alkalinization of the test medium (Fig. 10.6). According to the generally favoured proton/amino acid co-transport mechanism, a neutral (zwitterionic) amino acid S and H^+ bind to the carrier C^- removing a proton from the medium. Evidence for the negatively charged carrier and the thus uncharged ternary complex CHS has been seen for an amino acid carrier in *Riccia* by Felle (1984) and for the hexose carrier in *Chenopodium* by Gogarten and Bentrup (1989*a*). Saturation of the alkalization with AIB and half-saturation at 25 μM α-AIB may be

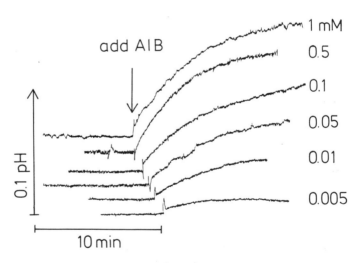

Fig. 10.6. pH Recordings from *Chenopodium* cell suspensions in unbuffered medium. α-AIB of noted concentration was added as indicated (subsequent traces have been shifted for better visibility). Ordinate scale denotes rising pH, which was 4.35 $\pm$ 0.02 at the time of α-AIB addition.

abstracted. By contrast, histidine clearly acidifies the medium (data not shown); obviously its de-protonation at the cytosolic neutral pH will lead to enhanced proton pump activity, as has been shown by Johannes and Felle (1987).

It seems instructive to compile the $k_{1/2}$ values from the different measured transport quantities involving α-AIB:

27 ± 5 μM α-AIB, uptake of α-AIB, initial rate at pH 6.0 (not shown)
30 ± 8 μM α-AIB, stimulation of α-AIB efflux at pH 5.2 (Fig. 10.2)
25 μM α-AIB, alkalinization of the medium, initial rate (Fig. 10.6)
53 ± 7 μM α-AIB, stimulation of tryptophan exit at pH 5.2 (Fig. 10.3)

The small range covered by these $k_{1/2}$ figures suggests that they indeed reflect the activity of a distinct amino acid carrier in the plasmalemma of these cells.

Concluding remarks

We conclude that trans-membrane stimulation experiments are a reliable tool for solving the current problem of identifying in the intact cell amino acid carriers operating in parallel in a given plasmalemma. It should be kept in mind, that in bacterial and animal membranes, numerous carriers for absorption of amino acids and other amino compounds have been identified (see Stein 1986).

Secondly, although proton-driven transport is the strongly indicated active transport mechanism, reliable experimental control of the cytosolic pH, in fact of both substrates at either side of the membrane, is required, as has been emphasized in the introduction to this chapter and demonstrated by Johannes and Felle (1987). On the one hand, 'simple' test systems, such as the quasi-cellular liverwort *Riccia* or a higher plant suspension cell, are still powerful experimental systems. On the other hand, transport within particular plant organs addressed in the introduction, clearly shows that a given suspension cell, reflecting a particular differentiated state, may not be expected to have expressed the complete array of amino acid transporters operative in the whole plant.

Acknowledgments

We are grateful to Dieter Klingelhöfer for carrying out the arginine uptake experiment and to Dr Wilfried Blum for a critical reading of the manuscript. Supported by the Deutsche Forschungsgemeinschaft (Grant Be 466/21).

References

Berry, S. L., Harrington, H. M., Bernstein, R. L., and Henke, R. R. (1981). Amino acid transport into cultured tobacco cells. III. Arginine transport. *Planta*, **153**, 511–18.

Borstlap, A. C. and Schuurmans, J. (1988). Kinetics of valine uptake in tobacco leaf discs. Comparision of wild-type, the digenic mutant Val^r-2, and its monogenic derivatives. *Planta*, **176**, 42–50.

Datko, A. H. and Mudd, S. H. (1985). Uptake of amino acids and other organic compounds by *Lemna paucicostata* Hegelm. 6746. *Plant Physiology*, **77**, 770–8.

Dietz, K.-J., Martinoia, E., and Heber, U. (1989). Mobilisation of vacuolar amino acids in leaf cells as affected by A T P and the level of cytosolic amino acids: A T P regulates but appears not to energize vacuolar amino-acid release. *Biochimica et Biophysica Acta*, **984**, 57–62.

Dietz, K.-J., Jäger, R., Kaiser. G., and E. Martinoia (1990). Amino acid transport across the tonoplast of vacuoles isolated from barley mesophyll protoplasts. *Plant Physiology*, **92**, 123–9.

Felle, H. (1981). Stereospecificity and electrogenicity of amino acid transport in *Riccia fluitans*. *Planta*, **152**, 505–12.

Felle, H. (1984). Steady-state current-voltage characteristics of amino acid transport in rhizoid cells of *Riccia fluitans*. Is the carrier negatively charged? *Biochimica et Biophysica Acta*, **772**, 307–50.

Felle, H. and Bentrup, F.-W. (1980). Hexose transport and membrane depolarization in *Riccia fluitans*. *Planta*, **147**, 471–6.

Gogarten, J. P. and Bentrup, F.-W. (1983). Fluxes and compartmentation of 3-O-methyl-D-Glucose in *Riccia fluitans*. *Planta*, **159**, 423–31.

Gogarten, J. P. and Bentrup, F.-W. (1989*a*). Substrate specifity of the hexose carrier in the plasmalemma of *Chenopodium* suspension cells probed by exchange diffusion. *Planta*, **178**, 52–60.

Gogarten, J. P. and Bentrup, F.-W. (1989*b*). The electrogenic proton/hexose carrier in the plasmalemma of *Chenopodium rubrum* suspension cells: effects of delta C, delta pH, and delta Φ on hexose exchange diffusion. *Biochimica et Biophysica Acta*, **978**, 43–50.

Higgins, C. F. and Payne, J. W. (1978*a*). Peptide transport by germinating barley embryos: uptake of physiological di-and oligopeptides. *Planta*, **138**, 211–15.

Higgins, C. F. and Payne, J. W. (1978*b*). Peptide transport by germinating barley embryos: Evidence for a single common carrier for di- and oligopeptides. *Planta*, **138**, 217–21.

Hoffmann, B. (1988) Untersuchungen zum Aminosäure- und Protonentransport an Zellen und isolierten Vakuolen einer photoautotrophen Zellkultur von *Chenopodium rubrum* L. Unpublished Ph.D. dissertation, Giessen.

Homeyer, U. and Schultz, G. (1988). Transport of phenylalanine into vacuoles isolated from barley mesophyll protoplasts. *Planta*, **176**, 378–82.

Hüsemann, W. and Barz, W. (1977). Photo-autotrophic growth and photosynthesis in cell suspension cultures of *Chenopodium rubrum*. *Physiology of Plants*, **40**, 77–81.

Johannes, E. and Felle, H. (1985). Transport of basic amino acids in *Riccia fluitans*. Evidence for a second binding site. *Planta*, **166**, 244–51.

Johannes, E. and Felle, H. (1987). Implications for cytoplasmic pH, protonmotive force, and amino-acid transport across the plasmalemma of *Riccia fluitans*. *Planta*, 172, 53–9.

Lommel, C., Hoffmann, B., Weintraut, H., and Bentrup, F.-W. (1989). ATP-dependent betacyanin release from isolated vacuoles measured by means of computer aided microphotometry. In *Plant membrane transport: the current position*, (ed. J. Dainty, M. I. de Michelis, E. Marrè, and F. Rasi-Caldogno), pp. 201–4. Elsevier, Amsterdam.

Mengel, K. and Haeder, H. E. (1977). Effect of potassium supply on the rate of phloem sap exudation and the composition of phloem sap of *Ricinus communis*. *Plant Physiology*, 59, 282–4.

Ohkawa, T.-A., Koehler, K., and Bentrup, F.-W. (1981). Electrical membrane potential and resistance in photo-autotrophic suspension cells of *Chenopodium rubrum* L. *Planta*, 151, 88–94.

Payne, J. W. and Walker-Smith, D. J. (1987). Isolation and identification of proteins from the peptide-transport carrier in the scutellum of germinating barley (*Hordeum vulgare*, L.) embryos. *Planta*, 170, 263–71.

Petzold, U., Dahse, I., and Müller, E. (1985). Light-stimulated electrogenic uptake of neutral amino acids by leaf cells of *Egeria densa* and *Vicia faba*. *Biochemie. Physiologie Pflanzen*, 180, 655–66.

Pfrüner, H. and Bentrup. F.-W. (1978). Fluxes and compartmentation of K^+, Na^+ and Cl^-, and action of auxins in suspension-cultured *Petroselinum* cells. *Planta*, 143, 213–23.

Reinhold, R. and Kaplan, A. (1984). Membrane transport of sugars and amino acids. *Annual Review of Plant Physiology*, 35, 45–83.

Robinson, S. P. and Beevers, H. (1981). Amino acid transport in germinating castor bean seedlings. *Plant Physiology*, 68, 560 6.

Sauer, N., Komor, E., and Tanner, W. (1983). Regulation and characterization of two inducible amino-acid transport systems in *Chlorella vulgaris*. *Planta*, 159, 404–10.

Schobert, C. and Komor, E. (1989). The differential transport of amino acids into the phloem of *Ricinus communis* L. seedlings as shown by the analysis of sieve-tube sap. *Planta*, 177, 324–9.

Stein, S., Böhlen, P. Stone, J., Dairman, W., and Udenfried, S. (1973). Amino acid analysis with fluorescamine at the picomole level. *Archives of Biochemistry and Biophysica*, 155, 202–12.

Stein, W. D. (1986). *Transport and diffusion across cell membranes*. Academic Press, London.

Steinmüller, F. and Bentrup, F.-W. (1981). Amino acid transport on photo-autotrophic suspension cells of *Chenopodium rubrum* L.: Stereospecificity and interaction with potassium ions. *Zeitschrift für Pflanzenphysiologie*, 102, 353–61.

van Bel, A. J. E., Ammerlaan, A., and Blaauw-Jansen, G. (1986). Preferential accumulation by mesophyll cells at low, and by veins at high, exogeneous amino acid and sugar concentrations in *Commelina communis* L. leaves. *Journal of Experimental Botany*, 37, 1899–910.

Wyse, R. E. and Komor, E. (1984). Mechanism of amino acid uptake by sugarcane suspension cells. *Plant Physiology*, 76, 865–70.

11. Ammonia assimilation in higher plants

PETER J LEA, RAYMOND D BLACKWELL, and
KENNETH W JOY*

*Division of Biological Sciences, The University of Lancaster,
Lancaster LA1 4YQ, UK*
**Department of Biology, Carleton University, Ottawa K1S 5B6,
Canada*

Introduction

Since 1974, the pathway of the assimilation of ammonia in higher plants has been considered to be via the combined action of glutamine synthetase and glutamate synthase, as shown in Fig 11.1.

The pathway has been termed the glutamate synthase cycle by Rhodes *et al.* (1980), and there have been numerous review articles discussing the operation of the cycle (Miflin and Lea 1980; Miflin and Lea 1982; Joy 1988; Lea *et al.* 1990*b*). The reactions of the pathway are outlined below.

Glutamine synthetase (GS; EC 6.3.1.2)

$$\text{L-glutamate} + \text{ATP} + \text{NH}_3 \longrightarrow \text{L-glutamine} + \text{ADP} + \text{Pi} + \text{H}_2\text{O}$$

Glutamate synthase; ferredoxin dependent (EC 1.4.7.1)

$$\text{2-oxoglutarate} + \text{L-glutamine} + \text{ferredoxin (red)} \longrightarrow 2\text{ L-glutamate} + \text{ferredoxin (ox)}$$

NAD(P) H-dependent (EC 1.4.1.13)

$$\text{2-oxoglutarate} + \text{L-glutamine} + \text{NAD(P) H}_2 \longrightarrow 2\text{ L-glutamate} + \text{NAD(P)}$$

A third enzyme, glutamate dehydrogenase (GDH; EC 1.4.1.3),

$$\text{L-glutamate} + \text{H}_2\text{O} + \text{NAD(P)} \rightleftharpoons \text{2-oxoglutarate} + \text{NH}_3 + \text{NAD(P)H}_2$$

has also been reported to be involved in ammonia assimilation (Loyola-Vargas and Jimenez 1984; Srivastava and Singh 1987; Yamaya and Oaks 1987; Muñoz-Blanco and Cardenas 1989). There has been considerable controversy over the last 16 years concerning the importance of GDH in ammonia assimilation.

In this chapter we will review the evidence supporting the hypothesis that over 95 per cent of ammonia is assimilated via the glutamate synthase cycle,

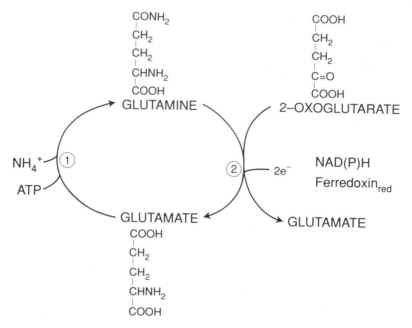

Fig. 11.1. The glutamate synthase cycle. (1) Glutamine synthetase; (2) Glutamate synthase.

and propose possible roles for the high levels of GDH activity frequently found in plant tissues.

Properties of enzymes

Glutamine synthetase

Higher plant GS is an octameric protein with a native molecular weight of 350–400 kDa (Stewart *et al.* 1980). In size and quaternary structure it strongly resembles the enzyme isolated from mammals but is distinct from the bacterial enzyme which consists of twelve subunits. The active site of the *Salmonella typhimurium* enzyme has been shown to be located between two adjacent subunits (Almassy *et al.* 1986). Eisenberg *et al.* (1987) have proposed a model of how a similar mechanism could also operate in eukaryotic octameric GS. Full details of the assay and purification of GS have been described by Lea *et al.* (1990*a*).

   In leaves, two major isoenzymes of GS have been isolated (Mann *et al.* 1979; McNally *et al.* 1983) and shown to be localized in the cytoplasm (GS$_1$) and chloroplast (GS$_2$) respectively. The proportion of the two isoenzymic forms may vary with the species studied (McNally *et al.* 1983), and is dependent upon the developmental age of the tissue (Tobin *et al*. 1985, 1988), and the presence of light (McNally and Hirel 1983; Edwards and Coruzzi 1989). Cytosolic and plastid forms of GS have also been found in roots (Vezina *et al*. 1987).

Extremely high levels of GS activity have been isolated from actively nitrogen-fixing legume root nodules (Cullimore and Bennett 1988), the majority of which is located in the plant cytoplasm. Lara *et al*. (1983) showed that *Phaseolus vulgaris* nodule GS could be separated into two isoenzymes designated GS_{n1} and GS_{n2}, the latter being very similar to the non-nodulated root enzyme. Analysis of GS subunits in *P. vulgaris* has shown that there are three distinct isoelectric forms (designated α, β, and γ) with a molecular weight of 43 kDa. GS_{n1} was shown to be composed of both β- and γ-subunits, whilst GS_{n2} contained mainly β subunits (Lara *et al*. 1984). The possibility of the occurrence of various combinations of β and γ-subunits has been discussed (Robert and Wong 1986; Bennett and Cullimore 1989). Multiple cytosolic GS polypeptides have also been observed in nodules of alfalfa (Groat and Schrader 1982), soybean (Hirel *et al*. 1987), and pea (Tingey *et al*. 1987).

The kinetic and regulatory properties of the root and leaf cytosolic enzymes are very similar (Hirel and Gadal 1980; Mann *et al*. 1980). In *P. vulgaris*, both α- and β-subunits are found in the leaf and root cytosolic enzymes. It was originally thought that the γ-subunit was only synthesized in the nodule of *P. vulgaris*, however, expression of the gene has been detected in petioles, stems, and green cotyledons (Bennett *et al*. 1989; Swarup *et al*. 1990).

The chloroplast isoenzyme of GS comprises one larger subunit of 43–45 kDa (Hirel *et al*. 1984; Ericson 1985; Tingey *et al*. 1988) termed δ in *P. vulgaris* (Lara *et al*. 1984). The polypeptide is synthesized on cytoplasmic ribosomes and transported into the chloroplast following the removal of a 4–5 kDa peptide. The primary amino acid sequences of the chloroplastic and cytoplasmic GS subunits are very similar, with an additional 60 and 16 residues being located at the N-terminus and C-terminus respectively (Forde and Cullimore 1989). The chloroplast isoenzyme does however contain two additional cysteine residues, which may account for the sensitivity of the isoenzymes to *N*-ethylmaleimide and other sulphydryl reagents (Sumar *et al*. 1984; Stewart *et al*. 1986).

It is now well established in *P. vulgaris* and *Pisum sativum* that GS is coded for by a small multigene family and much of the previously described subunit structure has been explained at the gene level. A simplified model of the mechanism of the genetic regulation of GS synthesis in *P. vulgaris* is shown in Fig. 11.2. A full description of the molecular biology of GS is beyond the remit of this chapter. The reader is referred to two excellent recent review articles, Coruzzi *et al*. (1988), Forde and Cullimore (1989).

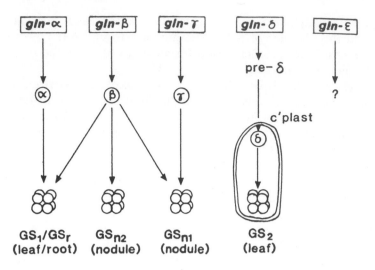

Fig. 11.2. Genetic control of glutamine synthetase isoenzymes in *Phaseolus vulgaris* (from Forde and Cullimore 1989).

Glutamate synthase

Ferredoxin dependent glutamate synthase can be separated from the NAD(P)H-dependent activity by ion exchange or molecular exclusion chromatography (Cullimore and Sims 1981*a*; Marquez *et al*. 1984; Avilla *et al*. 1984, 1987; Hecht *et al*. 1988). Suzuki *et al*. (1982) showed that antisera raised against the rice leaf ferredoxin-dependent enzyme did not cross-react with NAD(P)H-dependent glutamate synthase located in etiolated leaf and root tissue. Similar results were obtained with glutamate synthase activity in maize (Suzuki *et al*. 1987) and tomato (Botella *et al*. 1988). Full details of the assay and purification of the enzymes have been described by Lea *et al*. (1990*a*).

Ferredoxin-dependent   The enzyme is an iron–sulphur flavoprotein (Galvan *et al*. 1984; Marquez *et al*. 1988) and may represent up to 1 per cent of the protein content of leaves (Marquez *et al*. 1988). Molecular weight determinations of the native enzyme, using a variety of techniques have given values ranging from 141 to 230 kDa in different plant species. The barley leaf enzyme is a single polypeptide of molecular weight 154 kDa, reasons for the wide range of values reported have been discussed by Marquez *et al*. (1988).

Tissue fractionation studies have shown that ferredoxin-dependent glutamate synthase is localized in the chloroplasts of leaves (Wallsgrove *et al*.

1979, 1983), and the green algae *Chlamydomonas reinhardtii* (Cullimore and Sims 1981*b*; Fischer and Klein 1988). Using immuno-gold antibody labelling techniques on EM sections of tomato, Botella *et al.* (1988) were able to show that the enzyme was localized solely in the chloroplast stroma of mesophyll, xylem parenchyma, and epidermal cells. Enzyme activity increases considerably during leaf development in the light (Matoh and Takahashi 1982; Suzuki *et al.* 1982; Wallsgrove *et al.* 1982; Tobin *et al.* 1985). Suzuki *et al.* (1987) suggested that in maize, during the greening period, there was an initial activation of the enzyme followed by new synthesis of the glutamate synthase protein. In mustard cotyledons the induction of the ferredoxin-dependent enzyme has been shown to be due to the high irradiance reaction of phytochrome (Hecht *et al.* 1988).

Very recently Sakakibara *et al.* (1991) have succeeded in cloning the enzyme from maize leaves: 42 per cent of the amino acid sequence was shown to be identical to the NADPH-dependent enzyme in *E. coli*. Levels of mRNA were shown to increase markedly following illumination of etiolated maize seedlings.

NAD(P)H-dependent   Early reports indicated that the enzyme was able to use either NADH or NADPH as a coenzyme (Beevers and Storey 1976; Chiu and Shargool 1979; Kang and Titus 1980). However, it is now established that the NADH-dependent form predominates in higher plant tissue. In green leaves the activity is low in comparison to ferredoxin-dependent activity (Matoh and Takahashi 1981; Wallsgrove *et al.* 1982; Avilla *et al.* 1984; Hecht *et al.* 1988). Enzyme activity has been detected in a range of non-green tissues, for example roots (Suzuki *et al.* 1981; Matoh and Takahashi 1982; Oaks and Hirel 1985), developing cotyledons (Matoh *et al.* 1979; Beevers and Storey 1976) young seedlings (Singh and Srivastava 1986), and tissue culture cells (Jain and Shargool 1987; Levee and Chupeau 1989).

The NADH-dependent enzyme has been shown to have a molecular weight of 240 kDa in *Chlamydomonas* (Cullimore and Sims 1981*a*), 158 kDa in tomato leaves (Avilla *et al.* 1987), and 200 kDa in *P. vulgaris* (Chen and Cullimore 1989) and alfalfa (Anderson *et al.* 1989) root nodules. The enzyme from both species of root nodules has been shown to be a monomer and is probably one of the highest molecular weight enzyme subunits known.

The NADH-dependent enzyme appears to play a major role in ammonia assimilation in nitrogen-fixing legume root nodules where the activity increases dramatically following the onset of nitrogen fixation (Robertson *et al.* 1975; Boland and Benny 1977; Awonaike *et al.* 1981; Suzuki *et al.* 1984; Chen and Cullimore 1988; Anderson *et al.* 1989; Chen *et al.* 1990). Chen and Cullimore (1989) were able to isolate two isoenzymic forms of NADH glutamate synthase in *P. vulgaris* both of which were located in the plastid. Antisera raised against alfalfa root nodule glutamate synthase did not detect the presence of a similar protein in the roots or leaves (Anderson *et al.* 1989).

Glutamate dehydrogenase

The structure of the enzyme has been covered by Stewart *et al.* (1980) and more recently by Srivastava and Singh (1987). Scheid *et al.* (1980) compared the NADH-dependent enzyme from *Lemna minor* and *P. sativum*, and showed that both were tetramers of molecular weight of 230 kDa. However, in a similar study Kindt *et al.* (1980) demonstrated that the enzyme had a molecular weight of 260 kDa, and comprised six subunits. In *Sinapis alba* NADH-dependent GDH has a molecular weight of 270 kDa but comprises four polypeptides of 19, 21, 23 and 25 kDa (Lettgen *et al.* 1989). In contrast, Itagaki *et al.* (1988) isolated the enzyme from turnip mitochondria, and determined a molecular weight of 300–310 kDa with subunits of 43 kDa, but were unable to state whether the enzyme was a hexamer or octamer. Considering the remarkable consistency of the molecular structure data obtained with GS, there is clearly a need to clarify the position with GDH.

Electrophoresis of the native enzyme from a wide variety of sources has frequently shown the presence of seven isoenzymic forms (Thurman *et al.* 1965; Nagel and Hartmann 1980; Nauen and Hartmann 1980; Pahlich *et al.* 1980; Cammaerts and Jacobs 1985). The isoenzymic forms have similar kinetic properties (Nagel and Hartmann 1980) and could not be distinguished immunologically by Ouchterlony (Pahlich *et al.* 1980) or western blot analysis (Lettgen *et al.* 1989).

The addition of ammonium ions to a plant tissue has frequently been shown to stimulate an increase in GDH activity (Joy 1971, 1973; Kanamori *et al.* 1972; Shepard and Thurman 1973; Barash *et al.* 1975; Postius and Jacobi 1976; Lauriere and Daussant 1983; Cammaerts and Jacobs 1985; Lettgen *et al.* 1989; Muñoz-Blanco and Cardenas 1989; Zink 1989). New isoenzymic forms of GDH have been detected following the application of ammonia (Kanamori *et al.* 1972; Barash *et al.* 1975; Lettgen *et al.* 1989). It is possible that the new isoenzymes are formed as a result of senescence-induced changes (Nauen and Hartmann 1980; Lauriere and Daussant 1983). GDH isolated from higher plants has a high K_m for ammonia, with values between 5 and 70 mM quoted (Stewart *et al.* 1980). These values have been used by Miflin and Lea (1980) as one of a number of pieces of evidence to suggest that GDH is unlikely to have a role in ammonia assimilation, a hypothesis that has been challenged by a number of workers (Murray and Kennedy 1980; Loyola-Vargas and Sanchez de Jimenez 1984; Yamaya and Oaks 1987; Muñoz-Blanco and Cardenas 1989). It should be noted that the apparent K_m for ammonia can be influenced by pH and the concentration of substrates. Dramatic decreases in the K_m value for ammonia have been detected at low ammonia concentrations (25-fold) by Pahlich and Gerlitz (1980) and at low NADH concentrations (tenfold) by Nagel and Hartmann (1980).

Most of the NADH-dependent GDH activity in a plant cell is located in the mitochondria (Miflin 1970; Davies and Teixeira 1975; Nauen and Hartmann 1980; Yamaya *et al.* 1984; Itagaki *et al.* 1988). There is limited evidence to suggest that separate isoenzymic forms are present in the cytoplasm (Kanamori *et al.* 1972; Postius and Jacobi 1976; Gasparikova *et al.* 1978). An NADP-dependent form of GDH has been detected in leaf chloroplasts (Lea and Thurman 1972) and root plastids (Washitami and Sato 1977). The specific properties of NADPH-dependent GDH in the green algae will be discussed in the next section.

NADPH-dependent GDH in algae

In *Chlorella sorokiniana* there is a chloroplast-located ammonia inducible NADPH-GDH composed of six identical subunits of 53 kDa, which has been studied in detail by Schmidt and his colleagues (Gronostajski *et al.* 1978; Prunkard *et al.* 1986) and Tischner (1984). The subunit is synthesized on cytosolic ribosomes as a precursor protein of 58.5 kDa (Prunkard *et al.* 1986; Bascomb *et al.* 1986). The NADPH-dependent enzyme was shown to have different physical, chemical, and antigenic properties when compared to the mitochondrial NADH-GDH which was composed of four subunits of 45 kDa (Gronostajski *et al.* 1978; Meredith *et al.* 1978).

More recently two forms of NADPH-GDH have been detected in *C. sorokiana*. The α-isoenzyme comprising 55 kDa subunits was induced at low ammonia (2 mM) concentrations. The β-isoenzyme comprising the previously described 53 kDa subunits was only fully induced at higher (29 mM) ammonia concentrations. The K_m for ammonia for the β-isoenzyme remained constant at 75 mM, but the K_m for the α-isoenzyme ranged from 0.02 to 3.5 mM depending on the NADPH concentration (Bascomb and Schmidt 1987; Bascomb *et al.* 1987).

Origin of ammonia within plant

Ammonia can be generated by a range of different processes within the plant, the more important of which are described below and shown in Fig. 11.3. There is now considerable evidence to suggest that ammonia is recycled on a number of occasions from initial entry into the plant to final deposition in protein (Joy 1988; Lea *et al.* 1990*b*; Lea 1991).

Nitrate reduction All higher plants are able to utilize nitrate as a source of nitrogen. Nitrate may be either reduced in the root or transported in the xylem for reduction in the leaf (Andrews 1986; Pate and Layzell 1990). Nitrate is reduced by the enzyme NADH-dependent nitrate reductase (EC 1.6.6.1) to nitrite in the leaf cytoplasm, which is then further reduced in the chloroplast by ferredoxin-dependent nitrite reductase (EC 1.7.7.1) to yield ammonia directly (Wray and Kinghorn 1989; Kleinhofs and Warner 1990).

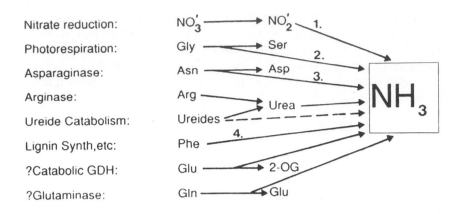

Fig. 11.3. Metabolic reactions that produce ammonia in plants. 1. Nitrite reductase; 2. Photorespiratory metabolism; 3. Asparaginase; 4. Phenylalanine ammonia lyase (from Joy 1988).

Nitrogen fixation   The symbiotic association of *Rhizobium* with legume root nodules has been the subject of recent review articles (Pate 1989; Vance 1990). Nitrogenase located in the bacteroid converts nitrogen gas to ammonia, which is transported into the plant nodule cells and rapidly assimilated. In the cyanobacteria nitrogenase is located in specialized cells termed heterocysts (Gallon and Chaplin 1988; Haselkorn 1986).

Metabolism of transport compounds   Asparagine is a major transport and storage compound in higher plants (Lea and Miflin 1980; Sieciechowicz *et al.* 1988). Asparaginase (E C 3.5.1.1), which converts asparagine to aspartate and ammonia has been shown to function in young expanding leaves and in maturing legume seeds during protein synthesis. The leaf enzyme is subject to complex light-dependent regulation (Sieciechowicz *et al.* 1988, 1989) and the seed enzyme may be either potassium-dependent or independent (Sodek *et al.* 1980; Joy and Ireland 1990). Asparagine may also be metabolized via trans-amination to 2-oxosuccinamic acid. The enzyme responsible co-purifies with serine:glyoxylate aminotransferase and is probably linked with photorespiration (see below). 2-Oxosuccinamic acid is metabolized to hydroxysuccinamic acid, both compounds may be deaminated to yield ammonia (Sieciechowicz *et al.* 1988).

   In a number of tropical legumes, ureides contain a large proportion of the nitrogen entering the shoot and pod (Pate 1989; Schubert and Boland 1990). It was originally thought that allantoin was metabolized through urea by the action of urease in a similar manner to arginine (Kerr *et al.* 1983; Polacco *et al.* 1985). However, it has recently been proposed that ureide catabolism involves the action of allantoinate amidohydrolase and ureidoglycollate amidohydrolase, without urea as an intermediate (Meyer-Bothling *et al*. 1987; Winkler *et al*. 1988; Blevins 1989). Whichever pathway operates, a large amount of free ammonia is still liberated.

Amino acid catabolism Proteins are broken down in senescing tissue (for example, germinating seedlings, and deciduous leaves) with the subsequent transport of the nitrogen in the form of amides (Thomas 1978; Lea and Joy 1982). Release of the amino group in the form of ammonia can be catalysed by amino acid oxidases or dehydrogenases (Mazelis 1980). The most well-characterized of these enzymes, glutamate dehydrogenase, will be discussed in detail later in this chapter. Phenylalanine is the precursor of a range of secondary products, including lignins and flavonoids. The first step is catalysed by the enzyme phenylalanine ammonia lyase with a subsequent release of ammonia (Grisebach 1985; Rhodes 1985).

Photorespiration In the process of photorespiration, ammonia is released from the mitochondria of C_3 plants, during the conversion of glycine to serine:-

$$2 \; Glycine + H_2O \rightarrow Serine + CO_2 + NH_3 + 2H^+ + 2e^-$$

The CO_2 liberated in the reaction is that lost in the photorespiration process. It has been estimated that this loss may account for 25–40 per cent of the net rate of photosynthesis (Somerville and Somerville 1983; Keys 1986; Gerbaud and André 1987). A key point about the reaction shown above is that ammonia is also released in stoichiometric amounts equal to that of CO_2 which is equivalent to a rate of at least ten times the rate of nitrogen assimilation.

Glycine is synthesized as a result of the oxygenase reaction of RuBP carboxylase/oxygenase (Keys 1990; Gutteridge and Lorimer 1990). The initial product, phosphoglycollate, is de-phosphorylated and the glycollate transported to the peroxisome where it is oxidized and transaminated to yield glycine. The full series of reactions involving the chloroplast, peroxisome, and mitochondria is termed the photorespiratory carbon and nitrogen cycle and is shown in Fig. 11.4. It was originally thought (Keys *et al*. 1978) that the nitrogen cycle was closed and that only serine and glutamate were able to act as nitrogen donors for the synthesis of glycine. However, work by Joy and his colleagues has clearly shown that alanine, and to a lesser extent asparagine, are able to enter the cycle (Ta and Joy 1986; Ta *et al*. 1987). The three carbon atoms of serine are returned to the chloroplast as glycerate, thus ensuring that 75 per cent of the carbon in glycollate is recovered. Evidence for the re-assimilation in the chloroplast of the ammonia released in the glycine to serine conversion will be presented in later sections.

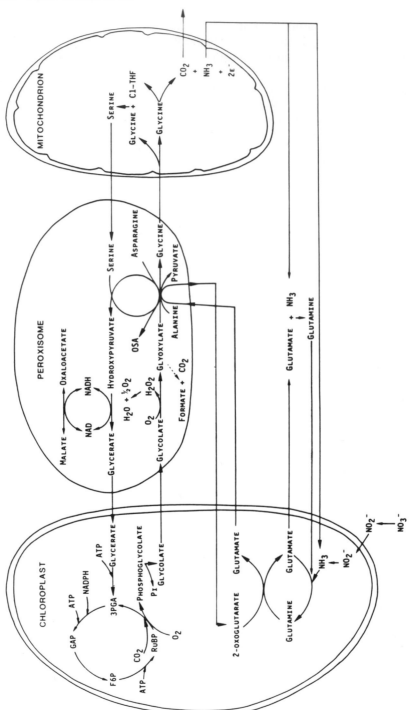

Fig. 11.4. The photorespiratory nitrogen and carbon cycle.

Inhibition of the glutamate synthase pathway

The use of inhibitors of GS has gone a long way to demonstrate the importance of the enzyme in plant metabolism. (Lea and Ridley 1989; Lea 1991). L-methionine-*S*-sulphoximine (MSO) was originally shown to be a potent inhibitor of sheep brain GS (Meister 1980). Detailed analyses have been carried out on the mode of inhibition of the *E. coli* (Rhee *et al.* 1981), *Saccharomyces* (Kim and Rhee 1987), sheep brain (Logusch *et al.* 1989), pea seed (Wedler and Horn 1976), pea leaf (Leason *et al.* 1982), and spinach chloroplast (Ericson 1985) enzyme. In conjunction with detailed X-ray crystallographic studies Almassey *et al.* (1986) were able to show that MSO binds to an active site situated at the interface of two adjacent subunits in *Salmonella typhimurium* GS.

A second GS inhibitor (phosphinothricin PPT), initially isolated as a tripeptide antibiotic (Bayer *et al.* 1972), is now marketed as a non-selective herbicide as the ammonium salt of gluphosinate under the trade name of 'Basta' (Schwerdtle *et al.* 1981; Koecher 1989). PPT is an even more potent inhibitor of bacterial, (Bayer *et al.* 1972), animal (Logusch *et al.* 1989), and in particular, plant GS (Ridley and McNally 1985; Mandersheid and Wild 1986; Lea and Ridley 1989; Lea 1991).

When MSO and PPT are applied to plant tissues, they cause a dramatic increase in the level of free ammonia almost without fail. Lea and Ridley (1989) and Lea (1991) have produced lists of the large range of plant material tested. A particularly valuable use of MSO has been to try to establish the rate of ammonia production in the photorespiratory nitrogen cycle. However, in most cases, rates of ammonia evolution were not as high as had been measured by other techniques following CO_2 evolution, due to the rapid inhibition of the photosynthetic rate (Lea 1991). Similar results have been obtained with PPT (Sauer *et al.* 1987; Wild *et al.* 1987; Ziegler and Wild 1989).

The correlation between the evolution of ammonia and photorespiration has been elegantly demonstrated by Tobin *et al.* (1988) using developing wheat leaves. The rate of light and MSO-dependent ammonia accumulation increased 15-fold from the leaf base and reached a maximum at 4–5 cm, remaining constant in mature tissue (Fig. 11.5a). The ammonia level corresponded to the photosynthetic capacity of the leaf segments. Ammonia production could be prevented by the addition of potassium bicarbonate, which would increase the CO_2 level and hence decrease the rate of photorespiration (Fig. 11.5b). The addition of the glycollate oxidase inhibitor HPMS, also reduced MSO-dependent ammonia evolution by reducing the synthesis of glycine (Fig. 11.5c).

It was originally thought that ammonia had a direct effect on photosynthesis probably by uncoupling photosynthetic phosphorylation. However,

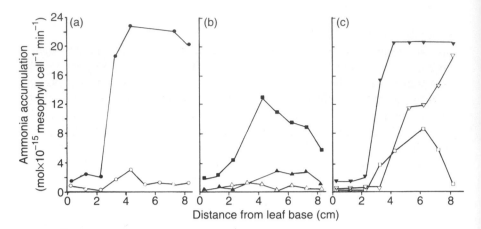

Fig. 11.5. Light-dependent ammonia accumulation in serial transverse 5 mm sections of 8-day-old wheat primary leaves. Note that the x-axis refers to distance of the section from the basal meristem, i.e. cells increase in age towards the leaf tip. Each point represents the mean of duplicate assays. Leaf sections were treated with: (a) water ($\bigcirc-\bigcirc$), 2 mol m^{-3} MSO ($\bullet-\bullet$); (b) 10 mol m^{-3} KHCO$_3$ ($\triangle-\triangle$), 10 mol m^{-3} KHCO$_3$ plus 2 mol m^{-3} MSO ($\blacksquare-\blacksquare$), 10 mol m^{-3} glycine ($\blacktriangle-\blacktriangle$); (c) 2 mol m^{-3} MSO plus 10 mol m^{-3} HPMS ($\square-\square$), 2 mol m^{-3} MSO plus 10 mol m^{-3} glycine ($\blacktriangledown-\blacktriangledown$, 2 mol m^{-3} MSO plus 10 mol m$^-$ HPMS plus 10 mol m^{-3} glycine ($\triangledown-\triangledown$) (from Tobin *et al.* 1988).

Walker *et al.* (1984) showed that wheat leaves supplied with 30 mM ammonium chloride accumulated more ammonia than leaves treated with MSO, but showed less inhibition of photosynthesis. Ikeda *et al.* (1984), also using wheat leaves, could reverse the MSO-induced fall in the photosynthetic rate by the addition of glutamine, despite the presence of high internal concentrations of ammonia. Similar results have been obtained using both PPT and the tripeptide bialaphos on *Sinapis alba* (Sauer *et al.* 1987; Ziegler and Wild 1989). A fuller discussion of the mechanisms of action of MSO and PPT on the photosynthetic rate will be presented on p. 167, when mutants lacking chloroplastic GS are described.

Tabtoxinine-β-lactam (T-β-L) is produced in a tripeptide by a number of strains of *Pseudomonas syringae* (Lea 1991). T-β-L is also a potent inhibitor of plant GS and acts in a similar manner to MSO (Langston-Unkefer *et al.* 1987). The toxin is responsible for the occurrence of 'wild fire' disease in tobacco (Turner and Debbage 1982). Knight and Langston-Unkefer (1988) have shown that infection of nodulated alfalfa plants with a T-β-L-producing strain of *P. syringae* caused an increase in plant growth, nitrogenase activity, nodule number and weight. The nodule-specific GS iso-enzyme was found to be insensitive to T-β-L, whilst the root GS isoenzyme was totally inhibited. Ammonia levels were found to be considerably higher in the infected, but actively fixing nodules. Clearly the results obtained by Knight and Langston-Unkefer (1988) have extremely important implications in identifying possible mechanisms for improving the nitrogen fixing capacities of legumes.

Three inhibitors of glutamate synthase have been used in a limited number of studies on plant metabolism; azaserine (*O*-diazoacetylserine), DON (6-diazo-5-oxonorleucine), and albizziine. These inhibitors block the assimilation of ammonia and cause a build-up of glutamine within the plant tissue (van der Meulen and Bassham 1959; Stewart and Rhodes 1976; Stewart 1979; Fentem *et al.* 1983; Berger *et al.* 1986; Martin *et al.* 1986).

Labelling studies using isotopes of nitrogen

The short lived isotope ^{13}N has been used to study the pathway of ammonia assimilation in cyanobacteria (Thomas *et al.* 1977), soybean nodules (Meeks *et al.* 1978), tobacco cells (Skokut *et al.* 1978), and *Alnus glutinosa* nodules (Schubert *et al.* 1981). The experiments have all shown a rapid accumulation of radioactivity into glutamine followed by glutamate being formed secondarily. In most cases the use of MSO and azaserine indicated that assimilation by any other mechanism than the glutamate synthase pathway (Fig. 11.1), was at a low level. However, Schubert *et al.* (1981) were able to demonstrate MSO and azaserine-insensitive assimilation of ^{13}NH$_3$ into glutamate in nodules of *A. glutinosa*.

Due to the high cost of producing ^{13}N and the difficulties involved in carrying out experiments for very short time periods, ^{15}N is now the predominant isotope for studying nitrogen assimilation. The reader should refer to the excellent work carried out by Rhodes and his colleagues utilizing ^{15}N, some of which is referred to below (Rhodes *et al.* 1980, 1986, 1989*a,b*).

It has been mentioned previously that the major ammonia generating system in the leaves of a C$_3$ plant is the conversion of glycine to serine during photorespiration. Mitochondria which contain GDH but not GS (Wallsgrove *et al.* 1980) are only able to re-assimilate a very small percentage of the ^{15}N-ammonia released when supplied with [^{15}N]-glycine (Keys *et al.* 1978; Hartmann and Ehmke 1980; Yamaya *et al.* 1986; Yamaya and Oaks 1987). Detailed studies with intact leaves have demonstrated that [^{15}N]-glycine is converted to [^{15}N]-serine and that ^{15}N also enters the amide group of glutamine followed by the amino group of glutamate, thus confirming the operation of the glutamate synthase cycle (Woo *et al.* 1982).

In the original photorespiratory nitrogen cycle proposed by Keys *et al.* (1978) ammonia was first shown to be assimilated into glutamine in the cytoplasm followed by transport into the chloroplast for the glutamate synthase reaction. However, in plants with low or zero GS in the cytoplasm, ammonia and 2-oxoglutarate must enter the chloroplast at the same rate and glutamate must be exported (Fig. 11.4). Woo and his colleagues have studied the system in detail and proposed a two-translocator model for 2-oxoglutarate and glutamate transport during ammonia assimilation in chloroplasts (Woo and Osmond 1982; Woo *et al.* 1987*a, b*). More recently it has been suggested that in plants containing a leaf cytosolic GS, glutamine is transported into the chloroplast on a specific translocator (Barber and Thurman 1978; Yu and Woo 1988).

In a series of detailed experiments using ^{15}N, Rhodes *et al.* (1989*a, b*) attempted to highlight some of the technical difficulties associated with assigning upper and lower boundaries to the flux via GDH *in vivo*, when superimposed upon an, admittedly, much larger flux via the GS/glutamate synthase pathway.

Utilizing tomato (*Lycopersicon esculentum*) roots, they were able to show that in the presence of 1 mM MSO, glutamate, alanine and γ-aminobutyrate continued to become labelled with ^{15}N, even when the synthesis of gluta-mine was completely inhibited. However, the rate of ammonia assimilation was only 1 per cent of that in the absence of MSO. Rhodes *et al.* (1989*a*) argued that whilst such data was convincing evidence for the operation of the glutamate synthase pathway (Fig. 11.1), the possibility that MSO was having a secondary effect on carbon metabolism and inhibiting the synthesis of 2-oxoglutarate could not be ruled out.

In a second series of experiments, Rhodes *et al.* (1989*a*) made use of a mutant of maize (82–137S) which contained only 5–10 per cent of the root mitochondrial GDH. The rate of root ammonia assimilation by the mutant was 3.12 μmol h^{-1} g^{-1} fresh weight as compared to a rate of 5.7 μmol h^{-1} g^{-1} fresh weight for the wild-type, suggesting that GDH could carry out up to 30 per cent of the total ammonia assimilation of a maize plant. However, MSO completely inhibited incorporation of ^{15}NH$_4^+$ into all the amino groups of both genotypes, suggesting that GDH was playing no role in ammonia assimilation. These conflicting results could be explained by the observation that the GDH-deficient mutant had a 20–30 per cent lower shoot/root ratio than the wild-type regardless of the nitrogen source utilized.

A further mutant of maize lacking 90 per cent of the total glutamate dehydrogenase activity was shown to grow normally, except when exposed to low night temperatures (Pryor 1990).

In an attempt to get away from the problems of using potentially 'unspe-cific' inhibitors, Rhodes *et al.* (1989*a*) carried out a complex analysis of the ^{15}NH$_4^+$ labelling kinetics of amino acids in *Lemna minor* and tobacco tissue culture cells. Computer simulation models were constructed that revealed the presence of multiple pools of glutamine and glutamate. The analysis of the data was further complicated by protein turnover and metabolite exchange between compartments. The authors concluded that the upper limit that could be placed on the flux of ammonia through GDH was only 4–5 per cent of the total nitrogen assimilated.

NMR spectroscopy is a useful method of following nitrogen metabolism in plant cells in a non-destructive manner (Martin 1985; Roberts 1988). It is possible to use both ^{14}N and ^{15}N in high resolution NMR experiments (Neeman *et al.* 1985; Monselise *et al.* 1987). Using shoot-forming cultures of white spruce buds, Thorpe *et al.* (1989) were able to follow the metabolism of ^{15}N-labelled NH_4^+ and NO_3 in an NMR spectrometer. ^{15}N was first incorporated into the amide group of glutamine and then into α-amino acids. Arginine and alanine were shown to contain considerable amounts of ^{15}N-label.

Studies on mutant lines lacking enzymes of ammonia accumulation

Following an initial suggestion by Somerville and Ogren (1979), a wide range of mutant plant lines have been isolated that are deficient in key enzymes of the photorespiratory nitrogen and carbon cycle (Somerville 1986; Blackwell *et al.* 1988a; Lea *et al.* 1989; Lea and Blackwell 1990). Such mutants have been selected on the basis that they grow normally under non-photorespiratory conditions (0.7 per cent CO_2), but show severe stress symptoms and chlorotic lesions on exposure to air.

Glutamine synthetase A number of lines of barley have been isolated that lack the chloroplastic isoenzyme GS_2, (Blackwell *et al.* 1987a; Wallsgrove *et al.* 1987), whilst still containing normal levels of GS_1, the cytoplasmic isoenzyme. Classical genetic studies have demonstrated that the lack of GS_2 activity behaves as a single nuclear recessive gene. All of the lines tested so far have been shown to be allelic (Wallsgrove *et al.* 1987) confirming that only one gene (*gln* δ. Fig. 11.2) is responsible for the synthesis of the chloroplastic isoenzyme (Lightfoot *et al.* 1988; Tingey *et al.* 1988). The amount of the δ subunit of GS_2 present in the different mutant lines of barley has been shown to be variable (Wallsgrove *et al.* 1987; Fig. 11.6). In a later study, Freeman *et al.* (1990) were able to identify three classes of barley mutants lacking GS_2 activity: Class I in which the absence of GS_2 protein was correlated with low levels of mRNA; Class II which had normal or increased levels of GS_2 mRNA but very little protein; and Class III, which had significant amounts of mRNA and protein. Mutant lines of plant species lacking GS_2, other than barley have not been isolated. This may represent the fact that plants such as *A. thaliana*, lack the cytoplasmic iso-enzyme GS_1, which is essential for growth under non-photorespiratory conditions.

Plants deficient in GS_2 accumulate high levels of ammonia in the leaves when placed in the air. The maximum rate of ammonia evolution has been calculated as 120 μmol h^{-1} g^{-1} fresh weight (Lea *et al.* 1989). This value should theoretically be equivalent to the rate of photorespiratory CO_2 loss from the conversion of glycine to serine. The rate of ammonia release is in good agreement with the current estimates of photorespiration (40 per cent

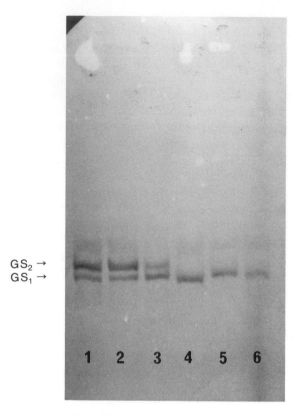

$GS_2 \rightarrow$
$GS_1 \rightarrow$

1 2 3 4 5 6

Fig. 11.6. Western blot with antibodies raised against *Phaseolus* glutamine synthetase for: **1.** wild-type barley; **2.** L_AP_r 85/14 (12 per cent); **3.** 85/38 (38 per cent); **4.** 85/80 (8 per cent); **5.** 86/38 (8 per cent) and **6.** 85/67 (16 per cent) mutants of barley. Figures in brackets indicate GS activity as a percentage of wild-type. (GS was generously supplied by Dr J. V. Cullimore.)

of the net rate of CO_2 fixation) in air, as obtained by other methods (Somerville and Somerville 1983; Gerbaud and André 1987).

Following the increase in ammonia levels in the air-exposed leaves, there is a concomitant decrease in the key amino acids of the photorespiratory nitrogen cycle (Blackwell *et al.* 1987*a*). The rate of photosynthetic CO_2 assimilation decreases slowly following exposure to air. The characteristics of the GS-deficient plants are in fact very similar to those treated with the GS inhibitors, MSO and PPT (Lea and Ridley 1989; Lea 1991).

Recently, evidence has been presented to show that the accumulation of ammonia does not have a direct inhibitory effect on photosynthesis (Ikeda *et al.* 1984; Walker *et al.* 1984; Blackwell *et al.* 1988*b*). The fall in the rate of photosynthesis can be prevented by feeding the amino acids alanine and

Table 11.1 Recovery of ^{15}N (nmol ^{15}N g^{-1}) fresh weight in the leaves from $^{15}NO_3$ feeding to wild-type (WT), glutamine synthetase deficient (85/80), or glutamate synthase deficient (85/73) barley plants for 4 or 8 h in high CO_2 (0.7 per cent)

	WT(4)	WT(8)	85/80(4)	85/80(8)	85/73(4)	85/73(8)
Asp	99.5	271.8	39.1	87.1	7.0	9.9
Glu	255.1	628.6	120.1	224.8	9.8	47.7
Asn (an)	0.5	0.8	3.1	10.3	2.7	0.8
Asn (ad)	4.3	5.9	7.7	51.7	6.8	10.9
Ser	28.3	86.2	8.6	14.2	14.2	2.6
Gln (an)	53.3	102.0	51.1	82.8	109.0	789.8
Gln (ad)	124.1	141.1	71.6	151.9	346.9	1463.8
Gly	0.4	0.1	0.3	Na	Na	Na
Ala	92.9	221.0	35.9	71.2	0.3	1.6

asparagine to the leaves (Lea *et al.* 1989). It is clear that a functional photorespiratory nitrogen cycle is necessary, in order to achieve maximum photosynthetic rates. In the absence of suitable amino donors to carry out the conversion of glyoxylate to glycine, it would be predicted that glyoxylate would leak out of the peroxisome into other organelles in the leaf cell. Glyoxylate has been shown to inhibit the activation of Rubisco at micromolar concentrations in intact, lysed and reconstituted chloroplasts (Chastain and Ogren 1989; Campbell and Ogren 1990).

When $^{15}NO_3$ was fed to whole barley plants in liquid culture there was very little difference in the amount of ^{15}N recovered in either the GS deficient or the wild-type leaves (Table 11.1). This shows quite clearly that the 10 per cent cytosolic GS remaining in the mutant is capable of maintaining normal growth under high CO_2 (non-photorespiratory) conditions. This table also shows that most of the label (80 per cent) remains locked up in glutamine for the glutamate synthase deficient plant, indicating that despite the non-photorespiratory conditions this plant is struggling to carry out basic metabolism. When plants were fed $^{15}NH_4$ + MSO there was no recovery of label in amino acids indicating that there was no residual GS activity in the GS mutant and no GDH bypass.

Ferredoxin-dependent glutamate synthase In 1980, Somerville and Ogren reported the first isolation of a mutant of *A. thaliana* deficient in chloroplastic ferredoxin-dependent glutamate synthase. The mutant has since been studied by a number of workers (Grumbles 1987; Morris *et al.* 1988, 1989). Glutamate synthase deficient mutants have also been isolated from barley (Kendall *et al.* 1986; Blackwell *et al.* 1987*b*, 1988*b*) and pea (Blackwell *et al.* 1987*b*). The chloroplastic ferredoxin-dependent enzyme is virtually absent in the leaves of the barley mutants. However, an NADH-dependent

Table 11.2 Amino acid content of leaves from wild-type and glutamate synthase mutant (86/33) peas after transfer to air from 0.7 per cent CO_2 (per cent total)

	WT		86/33	
	0 h	3 h	0 h	3 h
Asp	2.1	5.5	1.7	0.5
Glu	7.5	14.1	10.6	1.1
Asn	55.0	43.9	10.7	7.0
Ser	1.5	1.8	0.6	0.3
Gln	4.6	4.4	65.3	83.7
HSer	16.5	24.4	9.1	4.0
Ala	12.8	5.9	2.0	3.4
Gly	0	0	0	0
Total	105390	55195	62402	62446
(nmol g^{-1} fresh weight)				

enzyme (5 per cent of the leaf ferredoxin-dependent activity) is present in all mutant lines so far tested at normal wild-type levels. Characterization of the mutant lines has shown that the ferredoxin-dependent and NADH-dependent enzymes are under separate gene control and that the chloroplast ferredoxin-dependent enzyme is coded for by a single nuclear gene (Kendall *et al.* 1986; Blackwell *et al.* 1988*b*). This hypothesis was confirmed by the cloning of the gene for glutamate synthase in maize (Sakakibara *et al.* 1991).

The rate of photosynthesis in the glutamate synthase deficient mutants falls rapidly following transfer from 1 per cent O_2 to air. This inhibition is far greater than that seen in the glutamine synthetase mutants under the same conditions (Kendall *et al.* 1986; Wallsgrove *et al.* 1987; Blackwell *et al.* 1988*b*). Levels of glutamine increase rapidly in the leaves following exposure to air, with a corresponding decrease in other amino acids, whilst there is only a small increase in the ammonia concentration. Alanine and to a lesser extent glutamate were able to prevent the sharp decline in photorespiratory rate (Blackwell *et al.* 1988*b*). Once again the inhibition of photosynthesis could be explained by the accumulation of glyoxylate, due to the lack of availability of amino donors in the peroxisome (Chastain and Ogren 1989; Campbell and Ogren 1990). Previous work on the characterization of the mutants lacking ammonia-assimilating enzymes has tended to concentrate on the analysis of soluble amino acid levels. There would clearly be some merit in carrying out an analysis of the concentrations of organic (in particular 2-oxo) acids during the early stages of exposure to air.

In a screen of air-sensitive peas one plant (86/33) has been isolated that lacks Fd-glutamate synthase activity. Transfer to air of WT pea leaves

Table 11.3 Recovery of ^{15}N from ^{15}N-glutamine feeding to wild-type and glutamate synthase mutant (86/33) peas in photorespiratory and non-photorespiratory conditions (nmol $N^{15}g^{-1}$ fresh weight)

	WT		86/33	
	2% O_2	Air	2% O_2	Air
Asp	40.1	147.5	0.0	0.8
Glu	259.8	328.4	12.1	6.7
Asn (an)	*	12.6	9.1	3.0
Asn (ad)	12.4	12.6	9.1	3.0
Ser	18.0	8.2	0.2	0.3
Gln (an)	51.9	70.3	291.6	324.1
Gln (ad)	203.6	228.1	1077.6	1262.7
HSer	281.0	314.8	0.0	0.2
Ala	117.8	119.7	1.6	1.8

results in a decline in the levels of asparagine with a maintenance of levels of the other amino acids (Table 11.2). As with the barley glutamate synthase mutant (Blackwell *et al.* 1988*b*), the pea mutant has a high concentration of glutamine after transfer to air. It is interesting to note that as the glutamine level increases, all other amino donors decline to almost zero, including asparagine, which has an important role as a transport form of nitrogen in a large number of plants especially nitrogen fixers.

An interesting point about the amino acid composition of WT peas is the large amount of homoserine, the role of which is not clear at present. Homoserine can be derived from aspartate, via aspartyl phophate and aspartyl semialdehyde, in chloroplasts and may also be produced by transamination (Joy and Prabha 1986). In the WT homoserine comprises about 24 per cent of the total amino acid pool after 3 h in air (an increase from the basal level in 0.7 per cent CO_2 of 16 per cent. In the glutamate synthase pea mutant this basal level is 9 per cent which falls on transfer to air to 4 per cent of the total pool. Presumably in this situation there are insufficient amino donors or aspartate present for homoserine to accumulate, or if it is being imported it is being used as an amino donor.

The WT and mutant pea were fed ^{15}N from amide labelled glutamine for 4 h in either photorespiratory or non-photorespiratory conditions. The results found are similar to those seen with the barley glutamate synthase mutant (Table 11.3), although for 86/33 there was a much larger incorporation of label into the amino group of glutamine. Also the incorporation of label into homoserine in the mutant was minimal suggesting perhaps that peas do not have an absolute requirement for homoserine.

By normal genetic crossing techniques it has been possible to construct mutant lines of barley lacking both chloroplastic glutamine synthetase and ferredoxin-dependent glutamate synthase (Kendall *et al.* 1986; Blackwell *et al.* 1988*b*). The double mutant plants are just capable of growth in elevated levels of CO_2, but show severe stress symptoms immediately following transfer to air. The plants have the properties of both parent lines, exhibiting very low rates of CO_2 fixation combined with high levels of ammonia when exposed to air. The amino acid levels of the double mutant are low and slowly decrease under photorespiratory conditions (Blackwell *et al.* 1988*b*).

The massive increase in ammonia concentration in the leaves of the GS deficient plants exposed to air clearly demonstrates the high flux of nitrogen through the photorespiratory cycle. It has frequently been argued that GDH could assimilate ammonia at concentrations above the K_m value (Srivastava and Singh 1987; Yamaya and Oaks 1987). As concentrations of ammonia have frequently been recorded at over 50 mM in the GS deficient mutants of barley (Lea *et al.* 1989), it must be assumed that there is some other reason to explain the apparent inoperation of GDH in the aminating direction.

A role for glutamate dehydrogenase?

It is generally assumed that in most micro-organisms grown in an ammonia rich medium, the GDH pathway operates in the direction of glutamate synthesis (Hander and Dijkhuizen 1983). The enzyme has also been shown to function in nitrogen fixing *Bacillus* spp. (Kanamori *et al.* 1987*a*, *b*) and in the fungi (Sims and Folkes 1964; Folkes and Sims 1974; Genetet *et al.* 1984). However, it has been clearly demonstrated in the previous sections that in higher plants, it is unlikely that GDH could be responsible for more than 5 per cent of the flux of ammonia into amino acids.

Observations that GDH levels are higher in root tissues and in the senescing organs has come from many groups of workers (Storey and Beevers 1978; Murray and Kennedy 1980; Groat and Vance 1981, 1982; Loyola-Vargas and Jimenez 1984; Smirnoff and Stewart 1987). During the senescence of germinating cotyledons (Lea and Fowden 1975; Lea and Joy 1982) or leaves (Thomas 1978), protein is hydrolysed and redistributed throughout the plant. Asparagine is frequently a major constituent of the transported nitrogen and it has been proposed that the ammonia required for asparagine synthesis is derived from the breakdown of glutamate by glutamate dehydrogenase (Lea and Fowden 1975; Thomas 1978). Oxidation of glutamate could provide the cell with reduced nucleotides and carbon in circumstances where chloroplasts are non-functional (Cammaerts and Jacobs 1985). Kar and Feierabend (1984) obtained similar results using wheat leaves and concluded that GDH represents the main path for the liberation of ammonia from amino acids, since no L-amino acid oxidase activity was found in the leaves. In ^{15}N labelling studies on senescing wheat leaves, Berger and Fock (1985) were able to provide evidence for the deamination of [^{15}N]-glutamate. This deamination converts amino nitrogen to ammonia, which would then be available for synthesis of amides, a more efficient form for export of nitrogen from the senescent leaf.

There is an increasing volume of data to suggest that the availability of soluble carbohydrates rather than ammonia may effect the level of GDH activity. Increases in activity have been shown to be a response to carbon limitation and could be reversed by the addition of various sugars (Oaks *et al.* 1980; Nauen and Hartmann 1980; Sahulka and Lisa 1980; Tassi *et al.* 1984). Such information is circumstantial evidence in favour of GDH supplying carbon skeletons from proteins during periods of carbohydrate shortage. In fact, the stimulation of GDH by the addition of ammonia could be explained by the rapid removal of 2-oxo acids required for the synthesis of amino acids.

Very recently Robinson *et al.* (1991) have carried out a series of studies on the nitrogen metabolism of cell suspension cultures of carrot. The experiments using *in vivo* nuclear magnetic resonance spectroscopy, *in vitro* gas chromatography, mass spectrometry and automated $^{15}N/^{13}C$ mass spectrometry demonstrate that glutamate dehydrogenase is active in the oxidation of glutamate, but provide no evidence to show that reductive amination occurs *in vivo*. The results confirm suggestions that the function of glutamate dehydrogenase (at least in tissue culture cells) is the oxidation of glutamate, thus ensuring sufficient carbon skeletons for effective functioning of the TCA cycle. This catabolic role for glutamate dehydrogenase implies an important regulatory function in carbon and nitrogen metabolism.

Dedication

This chapter is dedicated to the memory of Dr A. P. Sims, whose research on ^{15}N-amonium assimilation in yeast served as a platform for all the later work on higher plants.

References

Almassy, R. J., Janson, C. A., Hamlin, R., Xuong, N. H., and Eisenberg, D. (1986). Novel subunit–subunit interactions in the structure of glutamine synthetase. *Nature*, **323**, 304–9.

Anderson, M. P., Vance, C. P., Heichel, G. H., and Miller, S. S. (1989). Purification and characterisation of NADH-glutamate synthase from alfalfa root nodules. *Plant Physiology*, **90**, 351–8.

Andrews, M. (1986). The partitioning of nitrate assimilation between root and shoot of higher plants. *Plant, Cell and Environment*, **9**, 511–19.

Avilla, C., Canovas, F., Nunez de Castro, I., and Valpuesta, V. (1984). Separation of two forms of glutamate synthase in leaves of tomato (*Lycopersicon esculentum*). *Biochemical and Biophysical Research Communications*, **122**, 1125–30.

Avilla, C., Botella, J. R., Canovas, F. M., de Castro, I. N., and Valpuesta, V. (1987). Different characteristics of the two glutamate synthases in the green leaves of *Lycopersicon esculentum. Plant Physiology*, **85**, 1036–39.

Awonaike, K. O., Lea, P. J., and Miflin, B. J. (1981). The localisation of the enzymes of ammonia assimilation in root nodules of *Phaseolus vulgaris. Plant Science Letters*, **23**, 189–95.

Barash, I., Mor, H., and Sadon, T. (1975). Evidence for ammonia dependent *de novo* synthesis of glutamate dehydrogenase in detached oat leaves. *Plant Physiology*, **56**, 856–8.

Barber, D. J. and Thurman, D. A. (1978). Transport of glutamine into isolated pea chloroplasts. *Plant, Cell and Environment*, **1**, 1–15.

Bascomb, N. F. and Schmidt, R. R. (1987). Purification and partial kinetic and physical characterisation of two chloroplast-localised NADP-specific glutamate dehydrogenase isoenzymes and their preferential accumulation in *Chlorella Sorokiana* cells cultured at low or high ammonium levels. *Plant Physiology*, **83**, 75–84.

Bascomb, N. F., Turner, K. J., and Schmidt, R. R. (1986). Specific polysome immunoadsorption to purify an ammonium-inducible glutamate dehydrogenase mRNA from *Chlorella sorokiana* and sythesis of full length double stranded cDNA from the purified mRNA. *Plant Physiology*, **81**, 527–32.

Bascomb, N. F., Prunkard, D. E., and Schmidt, R. R. (1987). Different rates of synthesis and degradation of two chloroplastic ammonium-inducible NADP-specific glutamate dehydrogenase isoenzymes during induction and deinduction in *Chlorella sorokiana* cells. *Plant Physiology*, **83**, 85–91.

Bayer, E., Gugel, K. K., Haegele, K., Hagenmaier, H., Jessipow, S., Konig, W. A., and Zahner, H. (1972). Stoffwechselprodukte von mikrootganismen. Phosphinothricin and phoshinothrycyl-alanyl-alanin. *Helvetica Chimica Acta*, **55**, 224–39.

Beevers, L. and Storey, R. (1976). Glutamate synthase in developing cotyledons of *Pisum sativum. Plant Physiology*, **57**, 862–6.

Bennett, M. J. and Cullimore, J. V. (1989). Glutamine synthetase iso-enzymes of *Phaseolus vulgaris* L.: subunit composition in developing root nodules and plumules. *Planta*, **179**, 433–40.

Bennett, M. J., Lightfoot, D. A., and Cullimore, J. V. (1989). cDNA sequence and differential expression of the gene encoding the glutamine synthetase polypeptide of *Phaseolus vulgaris* L. *Plant Molecular Biology*, **12**, 553–65.

Berger, M. G. and Fock, H. P. (1985). Comparative studies on the photorespiratory nitrogen metabolism in wheat and maize leaves. *Journal of Plant Physiology*, **119**, 257–67.

Berger, M. G., Sprengart, M. L., Kusnan, M., and Fock, H. P. (1986). Ammonia fixation via glutamine synthetase and glutamate synthase in the CAM plant *Cissus guadrangularis* L. *Plant Physiology*, **81**, 356–60.

Blackwell, R. D., Murray, A. J. S., and Lea, P. J. (1987*a*). Inhibition of photosynthesis in barley with decreased levels of glutamine synthetase activity. *Journal of Experimental Botany*, **38**, 1799–809.

Blackwell, R. D., Murray, A. J. S., and Lea, P. J. (1987*b*). The isolation and characterisation of photorespiratory mutants of barley and pea. In *Progress in photosynthesis research* (ed. J. Biggins), pp. 625–8. Martinus Nijhoff, Dordrecht.

Blackwell, R. D., Murray, A. J. S., Lea, P. J., Kendall, A. C., Hall, N. P., Turner,

J.C., and Wallsgrove, R.M. (1988*a*). The value of mutants unable to carry out photorespiration. *Photosynthesis Research*, **16**, 155–76.

Blackwell, R.D., Murray, A.J.S., Lea, P.J., and Joy, K.W. (1988*b*). Sucrose synthesis and N metabolism in a photorespiratory mutant of barley deficient in both chloroplastic glutamine synthetase and ferredoxin-dependent glutamate synthase. *Journal of Experimental Botany*, **39**, 845–58.

Blevins, D.G. (1989). An overview of nitrogen metabolism in higher plants. In *Plant nitrogen metabolism*, (ed. J.E. Poulton, J.T. Romeo, and E.E. Conn), pp. 1–41. Plenum Press, New York.

Boland, M.J. and Benny, A.G. (1977). Enzymes of nitrogen metabolism in legume nodules. Purification and properties of NADH-dependent glutamate synthase from lupin nodules. *European Journal of Biochemistry*, **79**, 355–62.

Botella, J.R., Verbelen, J.P., and Valpuesta, V. (1988). Immunological localisation of glutamine synthetase in green leaves and cotyledons of *Lycopersicon esculentum*. *Plant Physiology*, **87**, 255–7.

Cammaerts, D. and Jacobs, M. (1985). A study of the role of glutamate dehydrogenase in the nitrogen metabolism of *Arabidopsis thaliana*. *Planta*, **163**, 517–26.

Campbell, W.J. and Ogren, W.L. (1990). Glyoxylate inhibition of ribulosebisphosphate carboxylase/oxygenase activation in intact, lysed and reconstituted chloroplasts. *Photosynthesis Research*, **23**, 257–68.

Chastain, C.J. and Ogren, W.L. (1989). Glyoxylate inhibition of ribulosebisphosphate carboxylase/oxygenase activation state *in vivo*. *Plant Cell Physiology*, **30**, 937–44.

Chen, F-L. and Cullimore, J.V. (1988). Two isoenzymes of NADH-dependent glutamate synthase in root nodules of *Phaseolus vulgaris* L. *Plant Physiology*, **88**, 1411–17.

Chen, F-L. and Cullimore, J.V. (1989). Location of two isoenzymes of NADH-dependent glutamate synthase in root nodules of *Phaseolus vulgaris* L. *Planta*, **179**, 441–7.

Chen, F-L., Bennett, M.J., and Cullimore, J.V. (1990). Effect of nitrogen supply on the activities of NADH-dependent glutamate synthase and glutamine synthetase iso-enzymes in root nodules *Phaseolus vulgaris* L. *Journal of Experimental Botany*, **41**, 1215–21.

Chiu, J.Y. and Shargool, P.D. (1979). Importance of glutamate synthase in glutamate synthesis by soybean cell suspension cultures. *Plant Physiology*, **63**, 409–15.

Coruzzi, G.M., Edwards, J.W., Tingey, S.V. Tsai, F.Y., and Walker, E.L. (1988). Glutamine synthetase: molecular evolution of an eclectic multi-gene family. In *The molecular basis of plant development*, U.C.L.A. Symposia on Molecular and Cell Biology (ed. R. Goldberg), Vol. 92, pp. 223–32. Alan R. Liss, New York.

Cullimore, J.V. and Bennett, M.J. (1988). The molecular biology and biochemistry of plant glutamine synthetase from root nodules of *Phaseolus vulgaris* L. and other legumes. *Journal of Plant Physiology*, **132**, 387–93.

Cullimore, J.V. and Sims, A.P. (1981*a*). Occurrence of two forms of glutamate synthase in *Chlamydomonas reinhardii*. *Phytochemistry*, **20**, 597–600.

Cullimore, J.V. and Sims, A.P. (1981*b*). Pathway of ammonia assimilation in illuminated and darkened *Chlamydomonas reinhardii*. *Phytochemistry*, **20**, 933–40.

Davies, D. D. and Teixeira, A. W. (1975). The synthesis of glutamate and control of glutamate dehydrogenase in pea mitochondria. *Phytochemistry*, **14**, 647–56.

Edwards, J. W. and Coruzzi, G. M. (1989). Photorespiration and light act in concert to regulate the expression of the nuclear gene for chloroplast glutamine synthetase. *The Plant Cell*, **1**, 241–48.

Eisenberg, D., Almassy, R. J., Janson, C. A., Chapman, M. S., Suh, S. W., Cascio, D., and Smith, W. W. (1987). Some evolutionary relationships of the primary biological catalysts glutamine synthetase and Rubisco. *Cold Spring Harbor Symposium on Quantitative Biology*, **52**, 483–90.

Ericson, M. C. (1985). Purification and properties of glutamine synthetase from spinach leaves. *Plant Physiology*, **79**, 923–7.

Fentem, P. A., Lea, P. J., and Stewart, G. R. (1983). Action of inhibitors of ammonia assimilation on amino acid metabolism in *Hordeum vulgare* L. *Plant Physiology*, **71**, 496–501.

Fischer, P. and Klein, U. (1988). Localisation of nitrogen-assimilating enzymes in the chloroplast of *Chlamydomonas reinhardii*. *Plant Physiology*, **88**, 947–52.

Folkes, B. F. and Sims, A. P. (1974). The significance of amino acid inhibition of NADP-linked glutamate dehydrogenase in the physiological control of glutamate synthesis in *Candida utilis*. *Journal of General Microbiology*, **82**, 77–95.

Forde, B. G. and Cullimore, J. V. (1989). The molecular biology of glutamine synthetase in higher plants. In *Oxford surveys of plant molecular and cell biology* (ed. B. J. Miflin), Vol. 6, pp. 247–96. Oxford University Press, Oxford.

Freeman, J., Marquez, A. J., Wallsgrove, R. M., Saarelainen, R., and Forde, B. G. (1990). Molecular analysis of barley mutants deficient in chloroplast glutamine synthetase. *Plant Molecular Biology*, **14**, 297–311.

Gallon, J. R. and Chaplin, A. E. (1988). Nitrogen fixation. In *Biochemistry of the algae and cyanobacteria* (ed. L. J. Rogers and J. R. Gallon), pp. 147–73. Oxford University Press, Oxford.

Galvan, E., Marquez, A. J., and Vega, J. M. (1984). Purification and molecular properties of ferredoxin-glutamate synthase from *Chlamydomonas reinhardii*. *Plant Physiology*, **162**, 180–7.

Gasparikova, O., Psenakova, T., and Niznanska, A. (1978). Location of nitrate reductase, nitrite reductase and glutamate dehydrogenase in *Zea mays* roots. *Biologia*, **33**, 35–42.

Genetet, I., Martin, F., and Stewart, G. R. (1984). Ammonium assimilation in the N-starved ectomyccorrhizal fungus *Cerococcum graniforme*. *Plant Physiology*, **76**, 395–9.

Gerbaud, A. and André, M. (1987). An evaluation of recycling in measurements of photorespiration. *Plant Physiology*, **83**, 933–7.

Grisebach, H. (1985). Topics in flavonoid biosynthesis. In *The biochemistry of plant phenolics* (ed. C. F. Van Sumere and P. J. Lea), pp. 183–98. Oxford University Press, Oxford.

Groat, R. G. and Schrader, L. E. (1982). Isolation and immunochemical characterisation of plant glutamine synthetase in alfalfa (*Medicago sativa* L.) nodules. *Plant Physiology*, **70**, 1759–61.

Groat, R. G. and Vance, C. P. (1981). Root nodule enzymes of ammonia assimilation in alfalfa, (*Medicago sativa* L.). *Plant Physiology*, **67**, 1198–203.

Groat, R. G. and Vance, C. P. (1982). Root and nodule enzymes of ammonia assimilation, in two plant-conditioned symbiotically ineffective genotypes of alfalfa. *Plant Physiology*, **69**, 614–18.

Gronostajski, R.M., Yeung, A.T., and Schmidt, R.R. (1978). Purification and properties of the inducible NADP-specific glutamate dehydrogenase from *Chlorella sorokiana*. *Journal of Bacteriology*, **134**, 621–8.

Grumbles, R.M. (1987). The effects of glutamate deficiency and ammonia on *Arabidopsis* metabolism. *Journal of Plant Physiology*, **130**, 368–71.

Gutteridge, S. and Lorimer, G.H. (1990). The *in vitro* mutagenesis of ribulose bisphosphate carboxylase. In *Perspectives in biochemical and genetic regulation of photosynthesis* (ed. I. Zelitch), pp. 225–38. Alan R. Liss, New York.

Hander, W. and Dijkhuizen, L. (1983). Physiological responses to nutrient limitation. *Annual Review of Microbiology*, **37**, 1–23.

Hartmann, T. and Ehmke, A. (1980). Role of mitochondrial glutamate dehydrogenase in the reassimilation of ammonia produced by glycine-serine transformation. *Planta*, **149**, 207–8.

Haselkorn, R. (1986). Organisation of the genes for nitrogen fixation in photosynthetic bacteria and cyanobacteria. *Annual Review of Microbiology*, **40**, 525–47.

Hecht, U., Oelmüller, R., Schmidt, S., and Mohr, H. (1988). Action of light, nitrate and ammonium on the levels of NADH- and ferredoxin-dependent glutamate synthases in the cotyledons of mustard seedlings. *Planta*, **175**, 130–8.

Hirel, B. and Gadal, P. (1980). Glutamine synthetase in rice. A comparative study of the enzymes from roots and leaves. *Plant Physiology*, **66**, 619–23.

Hirel, B., Bouet, C., King, B., Layzell, D., Jacobs, F., and Verma, D.P.S. (1987). Glutamine synthetase genes are regulated by ammonia provided externally or by symbiotic nitrogen fixation. *EMBO Journal*, **6**, 1167–71.

Hirel, B., Weatherley, C., Cretin, C., Bergounioux, C., and Gadal, P. (1984). Multiple subunit composition of chloroplastic glutamine synthetase of *Nicotiana tabacum* L. *Plant Physiology*, **74**, 448–50.

Ikeda, M. Ogren, W.L., and Hageman, R.H. (1984). Effect of methionine sulphoximine on photosynthetic carbon metabolism in wheat (*Triticum aestivum*) leaves. *Plant Cell Physiology*, **25**, 447–52.

Itagaki, T., Dry, I.B., and Wiskich, J.T. (1988). Purification and properties of NAD-glutamate dehydrogenase from turnip mitochondria. *Phytochemistry*, **27**, 3373–8.

Jain, J.C. and Shargool, P.D. (1987). Use of aneuploid soybean cell culture to examine the relative importance of the GS:GOGAT system and GDH in ammonia assimilation. *Journal Plant Physiology*, **130**, 137–46.

Joy, K.W. (1971). Glutamate dehydrogenase changes in *Lemna* not due to enzyme induction. *Plant Physiology*, **47**, 445–6.

Joy, K.W. (1973). Control of glutamate dehydrogenase from *Pisum sativum* roots. *Phytochemistry*, **12**, 1031–40.

Joy, K.W. (1988). Ammonia, glutamine and asparagine: a carbon–nitrogen interface. *Canadian Journal of Botany*, **66**, 2103–9.

Joy, K.W. and Ireland, R.J. (1990). Enzymes of asparagine metabolism. In *Methods in plant biochemistry* (ed. P.J. Lea), Vol. 3, pp. 287–96. Academic Press, New York.

Joy, K.W. and Prabha, C. (1986). The role of transamination in the synthesis of homoserine in peas. *Plant Physiology*, **82**, 99–102.

Kanamori, T., Konishi, S., and Takahashi, E. (1972). Inducible formation of glutamate dehydrogenase in rice plant roots by the addition of ammonia to the media. *Physiologia Plantarum*, **26**, 1–6.

Kanamori, K., Weiss, R.L., and Roberts, J.D. (1987*a*). Role of glutamate dehydrogenase in ammonia assimilation in nitrogen-fixing *Bacillus macerans*. *Journal of Bacteriology*, **169**, 4692–5.

Kanamori, K., Weiss, R.L., and Roberts, J.D. (1987*b*). Ammonia assimilation in *Bacillus polymyxa*. *Journal of Biological Chemistry*, **262**, 11038–45.

Kang, S-M. and Titus, J.S. (1980). Activity profiles of enzymes involved in glutamine and glutamate metabolism in the apple during autumnal senescence. *Physiologia Plantarum*, **50**, 291–7.

Kar, M. and Feierabend, J. (1984). Changes in the activities of enzymes involved in amino acid metabolism during the senescence of detached wheat leaves. *Plant Physiology*, **62**, 39–44.

Kendall, A.C., Wallsgrove, R.M., Hall, N.P., Turner, J.C., and Lea, P.J. (1986). Carbon and nitrogen metabolism in barley (*Hordeum vulgare*) mutants lacking ferredoxin-dependent glutamate synthase. *Planta*, **168**, 316–23.

Kerr, P.S., Blevins, D.G., Rapp, B., and Randall, D.D. (1983). Soybean leaf urease: comparison with seed urease. *Plant Physiology*, **57**, 339–45.

Keys, A.J. (1986). Rubisco, its role in photorespiration. *Philosophical Transactions of the Royal Society London*, Series B **313**, 325–36.

Keys, A.J. (1990). Biochemistry of ribulose bisphosphate carboxylase. In *Perspectives in biochemical and genetic regulation of photosynthesis* (ed. I. Zelitch), pp. 207–24. Alan R. Liss, New York.

Keys, A.J., Bird, I.F., Cornelius, M.J., Lea, P.J., Wallsgrove, R.M., and Miflin, B.J. (1978). The photorespiratory nitrogen cycle. *Nature*, **275**, 741–3.

Kim, K.H. and Rhee, S.G. (1987). Subunit interaction elicited by partial inactivation with L-methionine sulphoximine and ATP differently affects the biosynthetic and α-glutamyl transferase reactions catalysed by yeast glutamine syntetase. *Journal Biological Chemistry*, **262**, 13050–4.

Kindt, R., Pahlich, E., and Rasched, I. (1980). Glutamate dehydrogenase from peas: isolation, quaternary structure and influence of cations on activity. *European Journal of Biochemistry*, **112**, 533–40.

Kleinhofs, A. and Warner, R.L. (1990). Advances in nitrate assimilation. In *The biochemistry of plants* (ed. B.J. Miflin and P.J. Lea), Vol. 16, pp. 89–120. Academic Press, New York.

Knight, T.J. and Langston-Unkefer, P.J. (1988). Enhancement of symbiotic dinitrogen fixation by a toxin-releasing plant pathogen. *Science*, **241**, 951–4.

Koecher, H. (1989). Inhibitors of glutamine synthetase and their effects in plants. In *BCPC Monograph* (ed. L.G. Kopping, J. Dalziel, and A.D. Dodge), Vol. 42, pp. 173–82. BCPC, Farnham.

Langston-Unkefer, P.J., Robinson, A.C., Knight, T.J. and Durbin, R.D. (1987). Inactivation of pea seed glutamine synthetase by the toxin, tabtoxinine-β-lactam. *Journal of Biological Chemistry*, **262**, 1608–13.

Lara, M., Cullimore, J.V., Lea, P.J., Miflin, B.J., Johnston, A.W.B., and Lamb, J.W. (1983). Appearance of a novel form of plant glutamine synthetase during nodule development in *Phaseolus vulgaris* L. *Planta*, **157**, 254–8.

Ammonia assimilation in higher plants 179

Lara, M., Porta, H., Padilla, J., Falch, J., and Sanchez, F. (1984). Heterogeneity of glutamine synthetase polypeptides in *Phaseolus vulgaris*. *Plant Physiology*, **76**, 1019-23.

Lauriere, C. and Daussant, J. (1983). Identification of the ammonium-dependent isoenzyme of glutamate dehydrogenase as the form induced by senescence or darkness stress in the first leaf of wheat. *Plant Physiology*, **58**, 89-92.

Lea, P. J. (1991). The inhibition of ammonia assimilation — a mechanism of herbicide action. In *Topics in photosynthesis* (ed. N. R. Baker), Vol. 10, pp. 267-98. Elsevier, Amsterdam.

Lea, P. J. and Blackwell, R. D. (1990). In *Perspectives in biochemical and genetic regulation of photosynthesis* (ed. I. Zelitch), pp. 301-18. Alan R. Liss, New York.

Lea, P. J. and Fowden, L. (1975). Glutamine dependent asparagine synthetase in *Lupinus*. *Proceedings of the Royal Society, London*, B, **192**, 13-26.

Lea, P. J. and Joy, K. W. (1982). Amino acid interconversions in germinating seeds. In *Mobilisation of reserves in germination* (ed. C. Nozzolillo, P. J. Lea, and F. A. Loewus), pp. 77-109. Plenum Press, New York.

Lea, P. J. and Miflin, B. J. (1980). Transport and metabolism of asparagine and other nitrogen compounds within the plant. In *The biochemistry of plants* (ed. B. J. Miflin), Vol. 5, pp. 569-607. Academic Press, New York.

Lea, P. J. and Ridley, S. M. (1989). Glutamine synthetase and its inhibition. In *Herbicides and plant metabolism* (ed. A. D. Dodge), pp. 137-70. Cambridge University Press, Cambridge.

Lea, P. J. and Thurman, D. A. (1972). Intracellular location and properties of plant L-glutamate dehydrogenase. *Journal of Experimental Botany*, **23**, 440-9.

Lea, P. J., Blackwell, R. D., Murray, A. J. S., and Joy, K. W. (1989). The use of mutants lacking glutamine synthetase and glutamate synthase to study their role in plant nitrogen metabolism. In *Plant nitrogen metabolism* (ed. J. E. Poulton, J. T. Romeo, and E. E. Conn), pp. 157-189. Plenum Press, New York.

Lea, P. J., Blackwell, R. D., Chen, F-L., and Hecht, U. (1990a). Enzymes of ammonia assimilation. In *Methods in plant biochemistry* (ed. P. J. Lea), Vol. 3, pp. 257-76. Academic Press, New York.

Lea, P. J., Robinson, S. A., and Stewart, G. R. (1990b). The enzymology and metabolism of glutamine, glutamate and asparagine. In *The biochemistry of plants* (ed. B. J. Miflin and P. J. Lea), Vol. 16, pp. 121-59. Academic Press, New York.

Leason, M., Cunliffe, D., Parkin, D., Lea, P. J., and Miflin, B. J. (1982). Inhibition of pea leaf glutamine synthetase by methionine sulphoximine, phosphinothricin and other glutamate analogues. *Phytochemistry*, **21**, 855-7.

Lettgen, W., Britsch, L., and Kasemir, H. (1989). The effect of light and exogenously supplied ammonium ions on glutamate dehydrogenase activity and isoforms in young mustard (*Sinapis alba* L.) seedlings. *Botanica Acta*, **102**, 189-95.

Levee, P. and Chupeau, Y. (1989). Development of nitrogen assimilating enzymes during growth of cells derived from protoplasts of sunflower and tobacco. *Plant Science*, **59**, 109-17.

Lightfoot, D. A., Green, N. K., and Cullimore, J. V. (1988). The chloroplast located glutamine synthetase of *Phaseolus vulgaris* L: Nucleotide sequence, expression in different organs and uptake into isolated chloroplast. *Plant Molecular Biology*, **11**, 191-202.

Logusch, E. W., Walker, D. M., McDonald, J. F., and Franz, J. E. (1989). Substrate variability as a factor in enzyme inhibitor design. *Biochemistry*, **28**, 3043–51.

Loyola-Vargas, V. M. and Jimenez, E. S. (1984). Differential roles of glutamate dehydrogenase in nitrogen metabolism of maize tissues. *Plant Physiology*, **76**, 536–40.

Manderscheid, R. and Wild, A. (1986). Studies on the mechanism of inhibition by phosphinothricin of glutamine synthetase isolated from *Triticum aestivum* L. *Journal of Plant Physiology*, **123**, 135–42.

Mann, A. F., Fentem, P. A., and Stewart, G. R. (1979). Identification of two forms of glutamine synthetase in barley. *Biochemical and Biophysical Research Communications*, **88**, 515–21.

Mann, A. F., Fentem, P. A., and Stewart, G. R. (1980). Tissue localisation of barley (*Hordeum vulgare*) glutamine synthetase isoenzymes. *FEBS Letters*, **110**, 265–7.

Marquez, A. J., Avila, C., Forde, B. G., and Wallsgrove, R. M. (1988). Ferredoxin-glutamate synthase from barley leaves: Rapid purification and partial characterisation. *Plant Physiology Biochemistry*, **26**, 645–51.

Marquez, A. J., Galvan, F., and Vega, J. M. (1984). Purification and characterisation of the NADH-glutamate synthase from *Chlamydomonas reinhardii*. *Plant Science Letters*, **34**, 305–14.

Martin, F. (1985). Monitoring plant metabolism by ^{13}C, ^{14}N and ^{15}N NMR spectroscopy. A review of the application to algae, fungi and higher plants. *Physiologie Vegetale*, **23**, 463–90.

Martin, F., Stewart, G. R., Genetet, I., and Le Tacon, F. (1986). Assimilation of ^{15}NH$_4^+$ by beech (*Fagus sylvatica* L.) ectomycorrhizas. *New Phytologist*, **102**, 85–94.

Matoh, T. and Takahashi, E. (1981). Glutamate synthase in greening pea shoots. *Plant Cell Physiology*, **22**, 727–31.

Matoh, T. and Takahashi, E. (1982). Changes in the activity of ferredoxin and NADH-glutamate synthase during seedling development of peas. *Planta*, **154**, 289–94.

Matoh, T., Takahashi, E., and Ida, S. (1979). Glutamate synthase in developing pea cotyledons. Occurrence of NADH-dependent and ferredoxin-dependent enzymes. *Plant Cell Physiology*, **20**, 1455–9.

Mazelis, M. (1980). Amino acid catabolism. In *The biochemistry of plants* (ed. B. J. Miflin), Vol. 5, pp. 541–67. Academic Press, New York.

McNally, S. F. and Hirel, B. (1983). Glutamine synthetase isoforms in higher plants. *Physiologie Vegetale*, **21**, 761–74.

McNally, S. F., Hirel, B., Gadal, P., Mann, A. F., and Stewart, G. R. (1983). Glutamine synthetases of higher plants. Evidence for a specific isoform content related to their possible physiological role and their compartmentation within the leaf. *Plant Physiology*, **72**, 22–5.

Meeks, J. C., Wolk, C. P., Schilling, N., Schaffer, P. W., Avissar, Y., and Chien, W. S. (1978). Initial organic products of fixation of [^{13}N] dinitrogen by root nodules of soybean (*Glycine max*). *Plant Physiology*, **61**, 980–3.

Meister, A. (1980). Glutamine synthetase. In *Glutamine: metabolism, enzymology and regulation* (ed. J. Mora and R. Palacios), pp. 1–40. Academic Press, New York.

Meredith, M. J., Gronostajski R. M., and Schmidt, R. R. (1978). Physical and kinetic properties of the N A D-specific glutamate dehydrogenase purified from *Chlorella sorokiana*. *Plant Physiology*, **61**, 969–74.

Meyer-Bothling, L. E., Polacco, J. C., and Cianzio, S. R. (1987). Pleiotropic soybean mutants defective in both urease isoenzymes. *Molecular and General Genetics*, **209**, 432–8.

Miflin, B. J. (1970). Studies on the subcellular location of particulate nitrate and nitrite reductase, glutamic dehydrogenase and other enzymes in barley roots. *Planta*, **93**, 160–70.

Miflin, B. J. and Lea, P. J. (1980). Ammonia assimilation. In *The biochemistry of plants* (ed. B. J. Miflin), Vol. 5, pp. 169–202. Academic Press, New York.

Miflin, B. J. and Lea, P. J. (1982). Ammonia assimilation and amino acid metabolism. In *Encyclopaedia of plant physiology* (ed. D. Boulter and B. Parthier), Vol. 14A, pp. 5–64. Springer Verlag, Berlin.

Monselise, E. B-I., Kost, D., Porath, D., and Tal, M. (1987). A ^{15}N-N M R study of ammonium ion assimilation by *Lemna gibba*. *New Phytologist*, **102**, 341–5.

Morris, P. F., Layzell, D. B., and Canvin, D. T. (1988). Ammonia production and assimilation in glutamate synthase mutants of *Arabidopsis thaliana*. *Plant Physiology*, **87**, 148–54.

Morris, P. F., Layzell, D. B., and Canvin, D. T. (1989). Photorespiratory ammonia does not inhibit photosynthesis in glutamate synthase mutants of *Arabidopsis*. *Plant Physiology*, **89**, 498–500.

Muñoz-Blanco, J. and Cardenas, J. (1989). Changes in glutamate dehydrogenase activity of *Chlamydomonas reinhardtii* under different trophic and stress conditions. *Plant, Cell and Environment*, **12**, 173–82.

Murray, D. R. and Kennedy, I. R. (1980). Changes in activities of enzymes of nitrogen metabolism in seedcoats and cotyledons during embryo development in pea seeds. *Plant Physiology*, **66**, 782–5.

Nagel, M. and Hartmann, T. (1980). Glutamate dehydrogenase from *Medicogo sativa*. *Zeitschrift Naturforschaften*, **35c**, 406–15.

Nauen, W. and Hartmann, T. (1980). Glutamate dehydrogenase from *Pisum sativum*. *Planta*, **148**, 7–16.

Neeman, M., Aviv, D., Degani, H., and Galvin, E. (1985). Glucose and glyine metabolism in regenerating tobacco protoplasts. *Plant Physiology*, **77**, 374–8.

Oaks, A. and Hirel, B. (1985). Nitrogen metabolism in roots. *Annual Review of Plant Physiology*, **36**, 345–65.

Oaks, A., Stulen, I., Jones, K., Winspear, M. J., Misra, S., and Boesel, I. L. (1980) Enzymes of nitrogen assimilation in maize roots. *Planta*, **148**, 477–84.

Pahlich, E. and Gerlitz, C. H. R. (1980). Deviations from Michaelis–Menten behaviour of plant glutamate dehydrogenase. *Phytochemistry*, **19**, 11–13.

Pahlich, E., Ott, W., and Schad, B. (1980). Immunochemical investigations with highly purified glutamate dehydrogenase from pea seeds by means of the Ouchterlony test. *Journal of Experimental Botany*, **31**, 419–23.

Pate, J. S. (1989). Synthesis, transport, and utilisation of products of symbiotic nitrogen fixation. In *Plant nitrogen metabolism* (ed. J. E. Poulton, J. T. Romeo, and E. E. Conn), pp. 65–115. Plenum Press, New York.

Pate, J. S. and Layzell, D. B. (1990). Energetics and biological costs of nitrogen assimilation. In *The biochemistry of plants* (ed. B.J. Miflin and P.J. Lea), Vol. 16, pp. 1–42. Academic Press, New York.

Polacco, J.C., Kruger, R.W. and Winkler, R.G. (1985). Structure and possible ureide degrading function of the ubiquitous urease of soybean. *Plant Physiology*, **79**, 794–800.

Postius, C. and Jacobi, G. (1976). Dark starvation and plant metabolism. Biosynthesis of glutamic acid dehydrogenase in detached leaves from *Cucurbita maxima. Zeitschrift Pflanzenphysiology*, **78**, 133–40.

Prunkard, D.E., Bascomb, N.F., Robinson, R.W., and Schmidt, R.R. (1986). Evidence for chloroplast localisation of ammonia-inducible glutamate dehydrogenase and synthesis of its subunits from a cytosolic precursor-protein in *Chlorella sorokiana. Plant Physiology*, **81**, 349–55.

Pryor, A. (1990). A maize glutamic dehydrogenase null mutant is cold temperature sensitive. *Maydica*, **35**, 367–72.

Rhee, S.G., Chock, P.B., Wedler F.C., and Sugiyama, Y. (1981). Subunit interaction in unadenylated glutamine synthetase from *Eschericia coli. Journal Biological Chemistry*, **256**, 644–8.

Rhodes, M.J.C. (1985). The physiological significance of plant phenolic compounds. In *The biochemistry of plant phenolics* (ed. C.F. Van Sumere and P.J. Lea), pp. 99–118. Oxford University Press, Oxford.

Rhodes, D., Sims, A.P., and Folkes, B.F. (1980). Pathway of ammonia assimilation in illuminated *Lemna minor. Phytochemistry*, **19**, 357–65.

Rhodes, D., Deal, L., Haworth, P., Jameson, G.C., Reuter, C.C., and Ericson, M.C. (1986). Amino acid metabolism of *Lemna minor*, L. *Plant Physiology*, **82**, 1057–62.

Rhodes, D., Brunk, D.G., and Magalhaes, J.R. (1989a). Assimilation of ammonia by glutamate dehydrogenase. In *Plant nitrogen metabolism* (ed. J.E. Poulton, J.T. Romeo, and E.E. Conn), pp. 191–226. Plenum Press, New York.

Rhodes, D., Rich, P.J., and Brunk, D.G. (1989b). Amino acid metabolism of *Lemna minor* L. *Plant Physiology*, **89**, 1161–71.

Ridley, S.M. and McNally, S.F. (1985). Effects of phosphinothricin on the isoenzymes of glutamine synthetase isolated from plant species which exhibit varying degrees of susceptibility to herbicides. *Plant Science*, **39**, 31–36.

Roberts, J.K.M. (1984). A study of plant metabolism *in vivo*, using NMR spectroscopy. *Annual Review Plant Physiology*, **35**, 375–86.

Robert, F.M. and Wong, P.P. (1986). Isoenzymes of glutamine synthetase in *Phaseolus vulgaris* L. and *Phaseolus lunatus* L. root nodules. *Plant Physiology*, **81**, 142–8.

Robertson, J.G., Warburton, M.P., and Farnden, K.J.F. (1975). Induction of glutamate synthase during nodule development in lupin. *FEBS Letters*, **55**, 33–7.

Robinson, S.A., Slade, A.P., Fox, G.G. Phillips, R., Ratcliffe, R.G., and Stewart, G.R. (1991). The role of glutamate dehydrogenase in plant nitrogen metabolism. *Plant Physiology*, **95**, 509–16.

Sahulka, J. and Lisa, L. (1980). Effect of some disaccharides, hexoses and pentoses on nitrate reductase, glutamine synthetase and glutamate dehydrogenase in excised pea roots. *Plant Physiology*, **50**, 32–6.

Sakakibara, H., Watanabe, M., Hase, T., and Sugiyama, T. (1991). Molecular cloning and characterisation of complementary DNA encoding for ferredoxin-dependent glutamate synthase in maize leaf. *Journal of Biological Chemistry*, **266**, 2028-34.

Sauer, H., Wild, A., and Ruehle, W. (1987). The effect of phosphinothricin (glufosinate) on photosynthesis. *Zeitschrift Naturforschaften*, **42**, 270-8.

Scheid, H.-W., Ehmke, A., and Hartmann, T. (1980). Plant NAD-dependent glutamate dehydrogenase purification, molecular properties and metal ion activation of the enzymes from *Lemna minor* and *Pisum sativum*. *Zeitschrift Naturforschaften*, **35c**, 213-21.

Schubert, F. R. and Boland M. J. (1990). The Ureides. In *The biochemistry of plants* (ed. B. J. Miflin and P. J. Lea), Vol. 16, pp. 197-282. Academic Press, New York.

Schubert, K. R., Coker, G. T., and Firestone, R. B. (1981). Ammonia assimilation in *Alnus glutinosa* and *Glycine max*. *Plant Physiology*, **67**, 662-5.

Schwerdtle, F., Bieringer, H., and Finke, M. (1981). Hoe 39866—ein neues nicht selektives blattherbizid. *Zeitschrift Pflanzeniology*, **IX**, 431-40.

Shepard, D. V. and Thurman, D. A. (1973). Effect of nitrogen source upon the activity of glutamate dehydrogenase of *Lemna gibba*. *Phytochemistry*, **12**, 1937-46.

Sieciechowicz, K., Ireland, R. J., and Joy, K. W. (1988). The metabolism of asparagine in plants. *Phytochemistry*, **27**, 663-71.

Sieciechowicz, K. A., Joy, K. W., and Ireland, R. J. (1989). Effect of methionine sulfoximine on asparaginase activity and ammonium levels in pea leaves. *Plant Physiology*, **89**, 192-6.

Sims, A. P. and Folkes, B. F. (1964). A kinetic study of the assimilation of ^{15}N-ammonia and the synthesis of amino acids in an exponentially growing culture of *Candida utilis*. *Proceedings of the Royal Society, London*, Series B, **159**, 479-502.

Singh, P. and Srivastava, H. S. (1986). Increase in glutamate synthase (NADH) activity in maize seedlings in response to nitrate and ammonium nitrogen. *Plant Physiology*, **66**, 413-6.

Skokut, T. A., Wolk, C. P., Thomas, J., Meeks, J. C., Schatten, P. W., and Chuen, W. A. (1978). Initial organic products of assimilation of ^{13}N ammonia and ^{13}N nitrate by tobacco cells cultured on different sources of nitrogen. *Plant Physiology*, **62**, 299-304.

Smirnoff, N. and Stewart, G. R. (1987). Glutamine synthetase and ammonium assimilation in roots of zinc tolerant and non-tolerant clones of *Anthoxanthum odoratum* L. *New Phytologist*, **107**, 659-70.

Sodek, L., Lea, P. J., and Miflin, B. J. (1980). Distribution and properties of a potassium dependent asparaginase isolated from developing seeds of *Pisum sativum* and other plants. *Plant Physiology*, **65**, 22-6.

Somerville, C. R. (1986). Analysis of photosynthesis and photorespiration using mutants of higher plants and algae. *Annual Review Plant Physiology*, **37**, 467-507.

Somerville, C. R. and Ogren, W. L. (1979). A phosphoglycollate phosphatase-deficient mutant of *Arabidopsis*. *Nature*, **280**, 833-6.

Somerville, C. R. and Ogren, W. L. (1980). The inhibition of photosynthesis in *Arabidopsis* mutants lacking in glutamate synthase activity. *Nature*, **286**, 257-9.

Somerville, S. C. and Somerville, C. R. (1983). Effect of oxygen and carbon dioxide on photorespiratory flux determined from glycine accumulation in a mutant of *Arabidopsis*. *Journal of Experimental Botany*, **34**, 415-21.

Srivastava, H.S. and Singh, R.P. (1987). Role and regulation of L-glutamate dehydrogenase activity in higher plants. *Phytochemistry*, **26**, 597–610.

Stewart, C.R. (1979). The effect of ammonium, glutamine, methionine sulphoximine and azaserine on asparagine synthesis in soybean leaves. *Plant Science Letters*, **14**, 169–73.

Stewart, G.R. and Rhodes, D. (1976). Evidence for the assimilation of ammonia via the glutamine pathway in nitrate grown *Lemna minor*. *FEBS Letters*, **64**, 296–9.

Stewart, G.R., Mann, A.F., and Fentem, P.A. (1980). Enzymes of glutamate formation: glutamate dehydrogenase, glutamine synthetase and glutamate synthase. In *The biochemistry of plants* (ed. B.J. Miflin), Vol. 5, pp. 271–327. Academic Press, New York.

Stewart, G.R., Popp, M., Holzapfel, I., Stewart, J.A., and Dickie-Eskew, A. (1986). Localisation of nitrate reduction in ferns and its relationship to environment and physiological characteristics. *New Phytologist*, **104**, 373–84.

Storey, R. and Beevers, L. (1978). Enzymology of glutamine metabolism related to senescence and seed development in pea (*Pisum sativum* L.). *Plant Physiology*, **61**, 494–500.

Sumar, N., Casselton, P.J., McNally, S.F., and Stewart, G.R. (1984). Occurrence of isoenzymes of glutamine synthetase in the alga *Chlorella kessleri*. *Plant Physiology*, **74**, 204–7.

Suzuki, A., Gadal, P., and Oaks, A. (1981). Intracellular distribution of enzymes associated with nitrogen assimilation in roots. *Planta*, **151**, 457–61.

Suzuki, A., Vidal, J., and Gadal, P. (1982). Glutamate synthase isoforms in rice. Immunological studies in green leaf, etiolated leaf and root tissues. *Plant Physiology*, **70**, 827–32.

Suzuki, A., Vidal, J., Nguyen, J., and Gadal, P. (1984). Occurrence of ferredoxin-dependent glutamate synthase in plant cell fractions of soybean root nodules (*Glycine max*). *FEBS Letters*, **173**, 204–8.

Suzuki, A., Audet, C. and Oaks, A. (1987). Influence of light on the ferredoxin-dependent glutamate synthase in maize leaves. *Plant Physiology*, **84**, 578–81.

Swarup, R., Bennett, M.J., and Cullimore, J.V. (1990). Expression of glutamine synthetase genes in cotyledons of germinating *Phaseolus vulgaris* L. *Planta*, **183**, 51–6.

Ta, T.C. and Joy, K.W. (1986). Metabolism of some amino acids in relation to the photorespiratory nitrogen cycle of pea leaves. *Planta*, **169**, 117–22.

Ta, T.C., Joy, K.W., and Ireland, R.J. (1987). Role of asparagine in the photorespiratory nitrogen metabolism of pea leaves. *Plant Physiology*, **78**, 334–7.

Tassi, F., Restivo, F.M., Puglisi, P.P. and Cacco, G. (1984). Effects of glucose on glutamate dehydrogenase and acid phosphatase and its reversal by cyclic adenosine 3′:5′-monophosphate in single cell cultures of *Asparagus officianalis*. *Plant Physiology*, **60**, 61–4.

Thomas, H. (1978). Enzymes of nitrogen metabolism in detached leaves of *Lolium temulentum* during senescence. *Planta*, **142**, 161–9.

Thomas, J., Meeks, J.C., Wolk, C.P., Shaffer, P.W., Austin, S.M., and Chien, W.S. (1977). Formation of glutamine from [^{13}N] ammonia, [^{13}N] dinitrogen and [^{14}C] glutamate by heterocysts of *Anabaena cylindrica*. *Journal of Bacteriology*, **129**, 1545–55.

Thorpe, T. A., Bagh, K., Cutler, A. J., Dunstan, D. I., Mcintyre, D. D., and Voge, H. J. (1989). A ^{14}N and ^{15}N nuclear magnetic resonance study of nitrogen metabolism in shoot-forming cultures of white spruce (*Picea glavca*) buds. *Plant Physiology*, **91**, 193-202.

Thurman, D. A., Palin, C., and Laycock, M. V. (1965). Isoenzymatic nature of L-glutamic dehydrogenase of higher plants. *Nature*, **207**, 193-4.

Tingey, S. V., Walker, E. L., and Coruzzi, G. M. (1987). Glutamine synthetase genes of pea encode distinct polypeptides which are differentially expressed in leaves, roots and nodules. *EMBO Journal*, **6**, 1-9.

Tingey, S. V., Tsai, F-Y., Edwards, J. W., Walker, E. L., and Coruzzi, G. M. (1988). Chloroplast and cytosolic glutamine synthetases are encoded by homologous nuclear genes which are differentially expressed *in vivo*. *Journal of Biological Chemistry*, **263**, 9651-7.

Tischner, R. (1984). Evidence for the participation of NADP-glutamate dehydrogenase in the ammonium assimilation of *Chlorella sorokiana*. *Plant Science Letters*, **34**, 60-73.

Tobin, A. K., Ridley, S. M., and Stewart, G. R. (1985). Changes in the activities of chloroplast and cytosolic iso-enzymes of glutamine synthetase during normal leaf growth and plastid development in wheat. *Planta*, **163**, 544-8.

Tobin, A. K., Sumar, N., Patel, M., Moore, A. L., and Stewart, G. R. (1988). Development of photorespiration during chloroplast biogenesis in wheat leaves. *Journal of Experimental Botany*, **39**, 833-43.

Turner, J. G. and Debbage, J. M. (1982). Tabtoxin-induced symptoms are associated with the accumulation of ammonia formed during photorespiration. *Physiological Plant Pathology*, **20**, 223-33.

Van der Meulen, P. Y. F. and Bassham, J. A. (1959). Study of inhibition of azaserine and diazooxo-norleucine (DON) on the algae *Scendesmus* and *Chlorella*. *Journal of the American Chemical Society*, **81**, 2233-9.

Vance, C. P. (1990). Symbiotic nitrogen fixation: recent genetic advances. In *The biochemistry of plants* (ed. B. J. Miflin and P. J. Lea), Vol. 16, pp. 43-88. Academic Press, New York.

Vezina, L. P., Hope, H. J., and Joy, K. W. (1987). Iso-enzymes of glutamine synthetase in roots of pea (*Pisum sativum* L. cv. Little Marvel) and alfalfa (*Medicago media* Pers. cv. Saranac). *Plant Physiology*, **83**, 58-62.

Walker, K. A., Keys, A. J., and Givan, C. V. (1984). Effect of L-methionine sulphoximine on the products of photosynthesis in wheat (*Triticum aestivum*) leaves. *Journal of Experimental Botany*, **35**, 1800-10.

Wallsgrove, R. M., Lea, P. J. and Miflin, B. J. (1979). Distribution of the enzymes of nitrogen assimilation with the pea leaf. *Plant Physiology*, **63**, 232-6.

Wallsgrove, R. M., Keys, A. J., Bird, I. F., Cornelius, M. J., Lea, P. J., and Miflin, B. J. (1980). The location of glutamine synthetase in leaf cells and its role in the assimilation of ammonia released in photorespiration. *Journal of Experimental Botany*, **31**, 1005-17.

Wallsgrove, R. M., Lea, P. J., and Miflin, B. J. (1982). The development of NAD(P)H-dependent and ferredoxin-dependent glutamate synthase in greening barley and pea leaves. *Planta*, **154**, 473-6.

Wallsgrove, R. M., Keys, A. J., Lea, P. J., and Miflin, B. J. (1983). Photosynthesis, photorespiration and nitrogen metabolism. *Plant, Cell and Environment*, **6**, 301-9.

Wallsgrove, R.M., Turner, J.C., Hall, N.P., Kendall, A.C., and Bright, S.W.J. (1987). Barley mutants lacking chloroplast glutamine synthetase-biochemical and genetical analysis. *Plant Physiology*, **83**, 155-8.

Washitami, I. and Sato, S. (1977). Studies on the function of proplastids in the metabolism of *in vitro* cultured tobacco cells. *Plant Cell Physiology*, **18**, 117-26.

Wedler, F.C. and Horn, B.R. (1976). Catalytic mechanisms of glutamine synthetase enzymes. *Journal of Biological Chemistry*, **251**, 7530-8.

Wild, A., Sauer, H., and Ruehle, W. (1987). The effect of phosphinothricin (glufosinate) on photosynthesis. *Zeitschrift Naturforschaften*, **42**, 263-9.

Winkler, R.G., Blevins, D.G., Polacco, J.C., and Randall, D.D. (1988). Ureide catabolism in nitrogen fixing legumes. *Trends in biochemical science*, **13**, 97-100.

Woo, K.C. and Osmond, B.C. (1982). Stimulation of ammonia and 2-oxoglutarate dependent O_2 evolution in isolated chloroplasts by dicarboxylates and the role of the chloroplast in photorespiratory nitrogen recycling. *Plant Physiology*, **69**, 591-6.

Woo, K.C., Boyle, F.-A., Flugge, I.U., and Heldt, H.W. (1987a). ^{15}N-ammonia assimilation, 2-oxoglutarate transport and glutamate export in spinach chloroplasts in the presence of dicarboxylates in the light. *Plant Physiology*, **85**, 621-5.

Woo, K.C., Flugge, V.I., and Heldt, H.W. (1987b). A two translocator model for the transport of 2-oxoglutarate and glutamate in chloroplasts during ammonia assimilation in the light. *Plant Physiology*, **84**, 624-32.

Woo, K.C., Morot-Gaudry, J.F., Summons, R.E., and Osmond, C.B. (1982). Evidence for the glutamine synthetase/glutamate synthase pathway during the photorespiratory nitrogen cycle in spinach leaves. *Plant Physiology*, **70**, 1514-7.

Wray, J.L. and Kinghorn, J.R. (1989). *Molecular and genetic aspects of nitrate assimilation*. Oxford University Press, Oxford.

Yamaya, T. and Oaks, A. (1987). Synthesis of glutamate by mitochondria—an anaplerotic function for glutamate dehydrogenase. *Physiologia Plantarum*, **70**, 749-56.

Yamaya, T., Oaks, A., and Matsumoto, H. (1984). Characteristics of glutamate dehydrogenase in mitochondria prepared from corn shoots. *Plant Physiology*, **76**, 1009-13.

Yamaya, T., Oaks, A., Rhodes, D., and Matsumoto, H. (1986). Synthesis of ^{15}N-glutamate from ^{15}NH$_4^+$ and ^{15}N-glycine by mitochondria isolated from pea and cornshoots. *Plant Physiology*, **81**, 754-7.

Yu, J. and Woo, K.C. (1988). Glutamine transport and the role of the glutamine translocator in chloroplasts. *Plant Physiology*, **88**, 1045-54.

Ziegler, C. and Wild, A. (1989). The effect of bialaphos on ammonium assimilation and photosynthesis. *Zeitschrift Naturforschaften*, **44c**, 103-8.

Zink, M.W. (1989). Regulation of ammonia assimilating enzymes by various sources in cultures of *Ipomoea* spp. *Canadian Journal of Botany*, **67**, 3127-33.

12. Environmentally induced adaptive mechanisms of plant amino acid metabolism

E. PAHLICH

Institut für Allgemeine Botanik und Pflanzenphysiologie, Heinrich Buff Ring 54-62, 6300 Giessen, Germany

Introduction

The basic pathways and interactions in plant metabolism have been uncovered, and this knowledge condensed in metabolic maps: these maps inform, at a qualitative level, which reactions occur, how pathways are composed, where ramifications exist, and which regulatory loops are operative. They answer the question 'what is'.

Unfortunately, these maps cannot inform us about actual fluxes, and their adaptation to a changing external and internal environment of cells. They provide no help in answering the question 'how much' of a transformation occurs under given circumstances and 'why' a respective flow rate is established under these conditions. These questions can be answered only with the help of the quantitative concepts of control of biochemical pathways which have been available since the early seventies (Kacser and Burns 1973; Heinrich and Rapoport 1974). From the quantitative concepts, it is apparent that the properties of metabolic pathway systems are characterized by (mainly genetically determined) sets of parameters (kinetic and thermodynamic constants) which specify the pools of the pathway variables (contents of metabolites) and the intensities of the respective fluxes. Only the changes of the parameters change the system's character.

Changes of parameters are usually thought to result from events of induction or repression. However, they can also be caused by an altered cell internal environment via conformative responses of catalysts (Citri 1973). In plants the latter case has to be considered important since the internal environment depends very much on the appropriate external conditions. These external variables change either in a periodic (day–night) or in a stochastic manner (temperature, water content, salt, toxic agents and so on). Therefore, the question arises of how, under natural conditions, a stable and steady motion of pathways is possible in plants. This problem is even more intricate because environmentally induced parameter changes are uncoordinated, individual, and reaction step-specific responses. How can metabolism work in an orderly manner in the presence of these uncoordinated parameter perturbations?

In my opinion this question comprises the major problem in examining adaptation in plant metabolism. With special regard to the environmentally switched dynamics of proline metabolism it is intended, in this chapter to discuss a concept for adaptation which hitherto has been scarcely considered. This might help to better understand processes like signalling, sensing, switching, tolerance, and even resistance.

Experimental

The experimental evidence for the mechanisms to be dicussed stems mainly from drought-induced switching of proline metabolism. Proline metabolism is a very valuable tool for these types of studies because:

(a) its metabolic map and biochemistry have been very well established;

(b) it is sensitive to drought and other environmental changes; and finally

(c) it may provide protecting effects in cells and therefore knowledge of its proper function and regulation could be of interest in plant breeding and engineering.

Most of the data used in the following discussion have been published recently (Pahlich *et al*. 1981, 1982, 1983; Krüger *et al*. 1986; Corcuera *et al*. 1989*a, b*; Pahlich 1990, Fricke and Pahlich 1990; Argandonia and Pahlich 1991). Some details concerning the quantitative treatment of data are given in the legends to the figures.

Drought-induced switching in proline synthesis

The problem of switching in metabolism is usually associated with on/off effects resulting from 'mono causal' signalling and causing all or none responses. For example, in mustard seedlings light switches 'on' the phenol propane pathway via induction of phenylalanine ammonia lyase. However, environmentally induced metabolic switching happens very often with pathways that are already operating, as is seen in drought-induced proline accumulation. This response is only one of many drought effects, but is typical (specific) for susceptible plants. The mechanistic concepts behind this way of switching are scarcely established and the 'primary effects' are hardly recognized.

It is well documented that the steady proline pool in cells is forced into a transient by withholding water. The transient phase is followed by an increased stationary rate of proline accumulation (Fig. 12.1A). Basically the

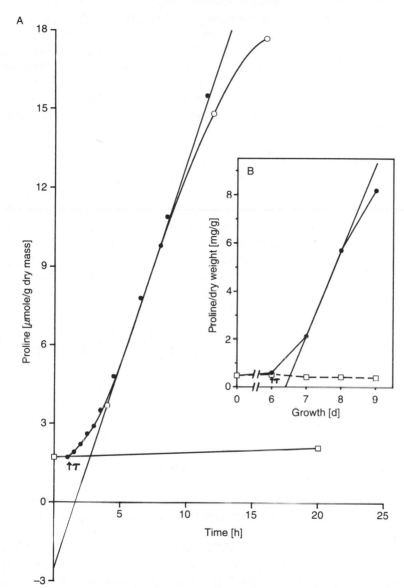

Fig. 12.1. The profiles of proline accumulation in potato cells after transfer of the culture into a PEG medium (Corcuera *et al.* 1989) and in barley seedlings (inset) after withholding water from day 0 onwards (Horn and Pahlich, unpublished). □—□, control; ○—○, water stressed; ●—●, calculated.

The simplified integrated rate law $P = v_o(t + \tau*\{\exp -t/\tau\} \exp -t/\tau - \tau)$ has been used for the calculation. $P =$ proline at time t; $v_o = 1.5$ μmol/g dry mass*t; $\tau = 1.3$ h is the sum of the relaxations involved in the accumulation process (see text for details).

same happens with proline in the barley seedlings (Fig. 12.1B). Analysis of the transient can give insight into the nature and time scale of the switching process. The quantitative concepts of Hess and of Easterby (Barwell and Hess 1970; Hess and Wurster 1970; Hess 1973; Easterby 1973, 1986) are followed in this chapter.

From the integrated rate laws of transients (Easterby 1986, also legend of Fig. 12.1) it becomes apparent that the velocity of the accumulation process (the steepness of the steady linear part) is determined only by the velocity of the initial, rate- limiting step. Drought-dependent changes of the steepness of this linear part, therefore, are indicative of the responsiveness of this process under varying stresses. Using this knowledge could help to elucidate questions concerning dose–response effects of stresses and problems of switching from pool to store mobilization by stresses (see later). Unfortunately no systematic surveys exist which use this simple principle to elucidate mechanisms of environmental effects.

Inspection of the lag phase helps to characterize the coupling enzymes following the initial process. The time interval between the very beginning of the transient (the moment of rate switching) and the intercept of the steady state line with the time axis comprises the sum of the relaxation times of these coupling enzymes. The respective lag is independent of the initial, rate-limiting step and controls the time, which makes apparent the pathway switching (time-limiting process) (Easterby 1986). A well developed lag phase shows that slowly working and 'ineffective' metabolic steps are involved in the transient, while very short lag times indicate high efficiency of these systems.

From the data in Fig. 12.1 it can be seen that the sum of the transients tau (τ) is at its minimum at 1.3 h in the potato cells (maximum 2.3 h), but lasts for a minimum of about 13 h (Fig. 12.1B, or even 2 days, and more under different conditions, not shown here) in barley seedlings. The minimum time intervals have been derived by curve fitting, since an exact determination of the start of the accumulation effect is nearly impossible. It is known that disturbed processes in the basic metabolism (glycolysis, respiration, citric acid cycle etc.) have relaxation times which are in the order of seconds or minutes (Heinrich and Rapoport 1977). Therefore it must be deduced from the measured and fitted data that the lag of environmentally induced proline accumulation most likely is dependent on additional and slower metabolic events. Most likely the draining of a depot, a step known to increase metabolic relaxation times (Reich and Sel'kov 1981), and mobilization of reserve stored materials (in the grains) and its transport contributes to the damping processes. The results also indicate that proline accumulation obviously depends on a much broader metabolic network than simply changes to its biosynthetic pathway.

The proline biosynthetic network

Proline in plants can be accumulated up to amounts of 15–20 per cent of the dry weight (Stewart and Lee 1974). These extreme amounts of a free amino acid can hardly be produced through re-grouping of existing pools of carbon and nitrogen precursors. In particular, the pools for the reductive potential and energy equivalents for the biosynthesis are too limited in the cell and need to be regenerated. Therefore proline biosynthesis has to be linked to regenerative processes. A 'minimum metabolic unit' which could be operating in the cytoplasm of heterotrophic cells is shown in Fig. 12.2. This is an open reaction network, apart from thermodynamic equilibrium, with carbohydrate as the fuel source and with CO_2 and proline as end products. According to the concept of Atkinson (1986) the interactions between the metabolic pathways are mediated and controlled by the nucleotide pairs NAD/NADH, NADP/NADPH, and ADP/ATP. At the moment of switching these pairs will have to be regarded as moiety conserved. What determines the dynamic patterns and the adaptive changes of this complex proline synthezising apparatus?

Key enzymes of metabolic network shown in Fig. 12.2 — phosphofructokinase, pyruvate kinase, glucose-6-phosphate (G-6-P) dehydrogenase, glutamate dehydrogenase (GDH), and pyrroline-5-carboxylate (P-5-C) reductase (Corcuera 1989a) — have been measured in control and in polyethylene glycol (PEG)-treated potato cell suspensions. Additionally, the activity of the anaplerotic enzyme phosphoenol pyruvate (PEP) carboxylase was determined (Bustamante and Pahlich, unpublished). While proline was accumulating in the treated cells, no changes in the activity ratios of the previously mentioned enzymes were observed. (Direct experimental evidence for flux rate changes in proline metabolism of PEG-treated suspension cells comes from Rhodes *et al.* 1986.)

However, a parameter change must have occurred *in vivo*. It was, therefore determined whether low water potentials (which must be expected in the cytoplasm of water stressed cells) affect these enzymes differentially *in vitro*. In fact it could be shown that glucose-6-phosphate dehydrogenase, glutamate dehydrogenase, and pyrroline-5-carboxylate reductase are inhibited differentially by this treatment (Corcuera *et al.* 1989a). The reaction environment specifically changes the enzymes' parameters, and comparable effects probably perturb the cytoplasm of the PEG-treated cells (Pahlich, 1990) resulting in those parametric changes which increase the fluxes of the proline synthezising network.

Switching in metabolic networks

Metabolic networks can be switched by mechanisms which have hitherto hardly been taken into account. The principle of this type of switching

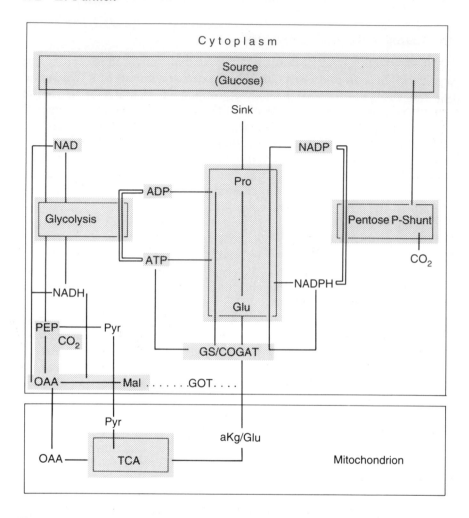

Fig. 12.2. The proline biosynthetic network of heterotrophic cells. The irreversible biosynthetic pathway of proline is linked through the nucleotide pairs ADP/ATP and NADP/NADPH with glycolysis and the pentose phosphate shunt. The anaplerotic phosphoenol pyruvate (PEP) carboxylase reaction supports the tricarboxylic acid cycle (TCA) with carbon which is exported to the cytoplasm as α-ketoglutarate or glutamic acid. The PEP carboxylase/malate dehydrogenase/transaminase reaction complex, which catalyses near-equilibrium reactions, is linked with the cytosolic NAD/NADH pair and interferes with the GS/GOGAT and the amino acid transaminases.

Table 12.1 Experimentally detemined parameters of glucose-6-phosphate dehydrogenase and pyrroline-5-carboxylate reductase. The data have been re-calculated from Krüger (1987).

Glucose-6-phosphate dehydrogenase		Pyrroline-5-carboxylate reductase	
K_s^{NADP}	44 μmol	K_s^{NADPH}	2.6 μmol
K_m^{NADP}	59 μmol	K_m^{NADPH}	8.3 μmol
K_i^{NADPH}	25 μmol	K_m^{P5C}	125.0 μmol
K_m^{G6P}	524 μmol	K_{eq}(pH 7.5)	10^5

mechanism will be discussed with the glucose-6-phosphate dehydrogenase/ pyrroline-5-carboxylate reductase circuit. Both enzymes have been studied intensively by our group. They exert Michaelis–Menten type kinetics, follow an ordered bi-bi mechanism (Clealand 1963*a*, *b*) and catalyse virtually irreversible reactions. The parameter characteristics of these two enzymes are summarized in Table 12.1

The network structure of the coupled two enzyme system considered is shown in the inset of Fig. 12.3. The system is fuelled by glucose-6-phosphate. Both enzyme reactions are linked through the NADP/NADPH couple. The nucleotides are assumed to be moiety conserved (the sum of NADP and NADPH is constant). Included in the reaction scheme is feedback inhibition of the glucose-6-phosphate dehydrogenase by its product NADPH and a mass action effect in the pyrroline-5-carboxylate reaction (see legend of Fig. 12.3).

The dynamic patterns of the complex reaction system are delineated in a parameter portrait (Fig. 12.3). The portrait of our system was calculated with the rate equations shown in the legend of Fig. 12.3 and shows the steady state redox charge of NADP as a function of the V_{max} of the pyrroline-5-carboxylate reductase. The redox charge is expressed as the NADP mole fraction (NADP/NADP + NADPH) at its measured capacity in potato cells (capacity = NADP + NADPH = 2.5×10^{-6} mol^{-1}). The enzyme parameters V_{max} are taken as arbitrary units, with the V_{max} of G-6-P dehydrogenase set at unity. In my laboratory, the V_{max} of the P-5-C reductase was chosen as the flexible parameter in the portrait because it changes with the respective environment. All our experimental results (activity ratio in cells, redox charge, and redox capacity gained with the two enzyme systems) obtained with the two enzymes produce a single point (o) in the parameter space (Fig. 12.3). This information is of limited value as far as the dynamic potential of the system is concerned.

The result included in the parameter portrait, however, is quite instructive. The steady states of the circuit at low V_{max} values of the P-5-C reductase

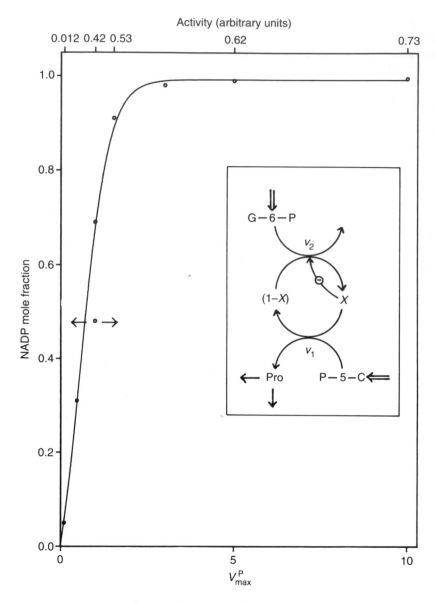

Activity (arbitrary units)

Fig. 12.3. The parameter portrait of the glucose-6-phosphate dehydrogenase/pyrroline-5-carboxylate reductase circuit. Both reactions are linked via the NADP/NADPH pair (inset). Moiety conservation of the coenzymes is assumed to exist. NADPH (X) is a strong product inhibitor of the G-6-P dehydrogenase; NADP = $(1-X)$. The graph shows the steady state redox charges (expressed as NADP mole fractions) as functions of the parameter V_{max} of the pyrroline-5-carboxylate reductase. The steady state fluxes of the coupled reaction system were calculated with the

would require a redox charge which is totally at its reductive side. At $V_{max} = 0.1$ for instance 90 per cent of the redox capacity must exist as NADPH. Small changes of the parameter under these conditions are accompanied by valuable changes of the redox charge (and hence the signalling character of the NADP/NADPH pair). Concomitantly the steady enzymatic activities operate far apart from their maxiumum rates (upper abscissa) and will change from 0.012 arbitrary units at $V_{max} = 0.1$ to 0.53 at $V_{max} = 1.5$. A fifteenfold increase of the parameter is accompanied by a 44-fold increase in activity. Obviously this region of the portrait uncovers a very sensitive part of the coupled process. A small parameter alteration will be followed by remarkable responses of the redox charges and the enzymatic activities.

The redox charge at $V_{max} = 1.5$ is 0.91 however, and far on the oxidative side. Of the total redox capacity 91 per cent exists as NADP and only 9 per cent as NADPH. A parameter shift to $V_{max} = 10$ changes the redox charge to 0.996 (99.6 per cent NADP and 0.4 per cent NADPH). The respective steady state activities then are 0.73 arbitrary units. A parameter shift by a factor of seven changes the steady state activity only 1.4-fold. The system does tolerate, and is insensitive to, the alterations in these parameters. It is keeping its redox charge (its signalling character), and the flux rate is hardly influenced by the parameter changes. Within this range the system exerts homoeostatic patterns.

With the insight received from the parameter portrait it is possible to interpret the dynamic potential of the experimentally determined G-6-P dehydrogenase/P-5-C reductase system from potato cell suspensions (○) (Fig. 12.3). It is apparent that the steady state of the reaction pair can belong to two different regimes of sensitivity. This depends on the V_{max} ratio of the

---

aid of the rate equations shown below, and the parameters summarized in Table 12.1. The extra ordinate (top) gives the actual fluxes under the respective conditions. The dynamics of the coupled reaction exerts two phases: the system is very sensitive towards parameter changes at V_{max} values up to 1.5 units. At higher V_{max} values it is insensitive and exerts homoeostatic properties. From the experimentally determined data from potato cells (○) it can be seen that the *in vivo* system is most probably located on the sensitive branch.

$$v_2 = \frac{V_{max}^G C}{K_s^{NADP} K_m^{G6P}\left(1 + \dfrac{[NADPH]}{K_i^{NADPH}}\right) + A + B + C}$$

$$A = K_m^{NADPH}[\text{G-6-P}]; B = K_m^{G6P}[NADP]; C = [\text{G-6-P}][NADP]$$

$$v_1 = \frac{v_{max}^P\left([NADPH][P5C]\right) - \left([Pro][NADP]/K_{eq}\right)}{\left(K_s^{NADPH} K_m^{P5C}\right) + \left(K_m^{NADPH}[P5C] + \left(K_m^{P5C}[NADPH]\right) + [NADPH][P5C]\right.}$$

two enzymes which is determined and can be altered genetically (gene dosage?), but which can also be shifted to a certain extent by environmental effects (Corcuera *et al.* 1989*a*). Preliminary results gained with this enzyme couple from barley epidermis in this laboratory might be taken as evidence that the ratio of P-5-C reductase/G-6-P dehydrogenase in stressed tissues is shifted to higher values, i.e. from 1.15 to 1.43 units. The problem is that epidermis isolation provides a big stress to tissues which cannot be differentiated from applied water stress effects.

The steady state of the system also depends on the redox charge, the stability and size of which is not well established in plants. The character of the reaction circuit in the system cannot be interpreted without detailed knowledge of the capacity, and the redox charge of the NADP/NADPH pair. A shift from the sensitive to the insensitive branch of the portrait is possible via small changes of the V_{max} if the system is in the vicinity of the transition zone. The absolute change of a V_{max}, therefore, does not elucidate which effect it really has in a coupled system. The effect of a parameter change on the system's flux depends on the specific shape of the parameter portrait and cannot be deduced intuitively.

A motion in the portrait from one branch to the other can be viewed as a way of switching between two different states of tolerance of the particular reaction network. It is obvious that in the vicinity of the transition zone even small parameter perturbations could turn this switch. These small parameter changes might easily result from externally induced alterations of the cell internal environment (Pahlich 1990).

It should be added that network dynamics can also result in even more sophisticated types of portraits which have two or more homoeostatic branches within a given frame of parameter changes. An example which is well suited for demonstrative purposes was given by Schellenberger *et al.* (1981, 1983) with a reconstituted glycolytic system. The authors have presented data showing the energy state (ATP-level) as a function of the V_{max} of the phosphofructokinase: a high and a low homoeostatic level of ATP exists in this system under the given conditions. Switching in this bi-stable system is like a jump. At a certain parameter combination (threshold) the system jumps into a new state (it is triggered). This example shows also that it is very hard to extrapolate backwards and forwards to the development and the character of metabolic states from sets of single measurements. This is especially valid if (intermediate) pools are taken as evidence for changes. In the cited example, for instance, the phosphoenol pyruvate pool changes by a factor of nearly 25 at the borders of the high energy state. This pool change certainly would be interpreted as a massive change of the pathway state. But the superior argument developed from the energy state of the cell is to the contrary.

A study is now in progress to correlate the patterns of parameter portraits of the intermediary metabolism of barley strains with differences in drought and temperature resistance.

The proline pool

Water deficit is an external factor that changes the pools of amino acids. Glutamic acid, glutamine, aspartic acid and its amide and proline, of course, are known to be increased by witholding water from plants (Borowitzka 1981). This accumulation process deserves attention here in the context of network dynamics.

Metabolic pools can be the result of near-equilibrium or irreversible processes apart from equilibrium. The reactions of several dehydrogenases and transaminases can most likely be considered to be near-equilibrium, and reversible processes in the cytoplasm. These enzymes typically exert high activities and therefore keep substrates and products at levels which are close to their chemical equilibrium. Through the dehydrogenases these reactions are linked with the redox pairs of the respective compartment, i.e. the catabolic NAD/NADH and the anabolic NADP/NADPH redox state of the cytoplasm (Fig. 12.2). Changes in the redox state (the mole fraction of NAD(P)) in the cell will therefore quickly equilibrate the redox equivalents over a number of metabolite carriers through the action of these dehydrogenases and transaminases (malate/oxalacetate/amino acids) which in fact is equivalent to the formation of 'buffer pools'. No specific regulatory properties concerning biosynthesis or degradation of the mentioned metabolites must be met in order to change these pools. Even seemingly complex pool variations can be explained by the classical concept of equilibrium thermodynamics. The reversibility of the equilibrium reactions makes 'buffer pools', sources under changed conditions.

Proline biosynthesis is an irreversible pathway (Krüger *et al.* 1986), and its thermodynamic structure (Pahlich *et al.* 1981, 1982) identifies the proline pool as a deep sink. The pool concentration of proline could be filled to a nearly molar concentration in a single compartment of the cell because of its equilibrium constant. In contrast to equilibrium processes this pool concentration can only be regulated by independent changes of the biosynthetic flow and/or the output reactions. The pool concentration of proline is additionally sequestered in the cell in a vacuolar and an extra-vacuolar fraction (Pahlich *et al.* 1983; Fricke and Pahlich 1990). The partition of proline between vacuole and extra-vacuole has interesting dynamic properties depending on the external conditions.

It has been shown that an increasing total proline pool in the cell is accompanied by a rise of the extra-vacuolar fraction in potato cells (Fricke and Pahlich 1990). Since the bulk water fraction of the cytosol is likely to disappear under water shortage (Pahlich 1990), and since the solute capacity

(Atkinson 1969) of this compartment is drastically reduced, it must be postulated that an alternative attractor for proline in the extra-vacuole is present. The steepened gradient of proline concentration in cells towards the vacuole should favour its deposition in the latter compartment from a thermodynamic point of view. However, the opposite is observed. Obviously the vicinal or bound water fraction of macromolecules in the presumably gel-like cytoplasmic compartment is part of this sink (protective effect of proline). Further studies of the distribution patterns of proline between gel and sol water fractions, under varying conditions, are necessary in order to elucidate this particular effect of sequestering of the proline pool.

Conclusion

Changes in the external environment of plants are mirrored by respective changes of enzyme (and transport) parameters, and by the mole fraction and capacity changes of signalling pairs of the adenylates in the cell interior. Principally all parameters in a cell are reached by these perturbations and yet the response is individual. These parametric changes may or may not affect respective network fluxes depending on the respective parameter profiles of the networks: reaction systems which are located on sensitive branches of their profiles will be forced into a transient, and will approach a new flux regime (bi- or multi-homoeostasis); reaction systems on their homoeostatic branches will buffer the alterations. 'Specificity' of environmentally induced metabolic perturbations can easily result from these patterns. For the same reason it should be possible to explain qualities like susceptibility, tolerance, and even resistance to stress. The genetic background of plant varieties or cultivars and of different plant species determines the shape of the parameter portraits and hence the differences in the qualities mentioned above. System perturbations by the external environment produce system responses and adaptive capabilities of plants are system effects.

References

Argandonia, V. and Pahlich, E. (1991). Water stress on proline content and enzyme activities in barley seedlings. *Phytochemistry*, **30**, 1093–4.

Atkinson, D.E. (1969). Limitations of metabolic concentrations and the conservation of solvent capacity in the living cell. In *Current topics of cell regulation* (ed. B.L. Horecker and E.R. Stadtman), pp. 29–42, Academic Press, New York.

Atkinson, D.E. (1986). Dynamic interactions between metabolic sequences. In *Dynamics of Biochemical Systems* (ed. S. Damjanovich, T. Keleti, and L. Tron), pp. 129–143. Elsevier, Amsterdam.

Barwell, C.J. and Hess, B. (1970). The transient time of the hexokinase/pyruvate kinase/lactate dehydrogenase system *in vitro*. *Hoppe-Seyler's Zeitschrift Physiologische Chemie*, **351**, 1531–6.

Borowitzka, L. J. (1981). Solute accumulation and regulation of cell water activity. In *The physiology and biochemistry of drought resistance in plants* (ed. L.G. Paleg and D. Aspinall), pp. 97–104, Academic Press, New York.

Citri, N. (1973). Conformational adaptibility in enzymes. In *Advances in enzymology* (ed. A. Meister), pp. 398–621. Wiley, New York.

Cleland, W. W. (1963a). The kinetics of enzyme-catalyzed reactions with two or more substrates or products II. Inhibition: nomenclature and theory. *Biochimica et Biophysica Acta*, **67**, 173–87.

Cleland, W. W. (1963b). The kinetics of enzyme-catalyzed reactions with two or more substrates or products I. Nomenclature and rate equations. *Biochimica et Biophysica Acta*, **67**, 104–37.

Corcuera, L. J., Hintz, M., and Pahlich, E. (1989a). Effect of polyethylene glycol on protein extraction and enzyme activities in potato cell cultures. *Phytochemistry*, **28**, 1569–71.

Corcuera, L. J., Hintz, M., and Pahlich, E. (1989b). Proline metabolism in *solanum tuberosum* cell suspension cultures under water stress. *Journal of Plant Physiology*, **134**, 290–3.

Easterby, J. S. (1986). Application of the transient kinetic behavior of coupled enzyme reactions to the analysis of metabolic pathway dynamics. In *Dynamics of biochemical systems* (ed. S. Damjanowitch, T. Keleti, and L. Tron), pp. 145–58. Elsevier, Amsterdam.

Easterby, J. S. (1973). Coupled enzyme assays: a general expression for the transient. *Biochimica et Biophysica Acta*, **293**, 552–8.

Fricke, W. and Pahlich, E. (1990). The effect of water stress on the vacuole-extravacuole compartmentation of proline in potato cell suspension cultures. *Physiologia Plantarum*, **78**, 374–8.

Heinrich, R. and Rapoport, T. (1977). Ist der steady state eine nützliche Fiktion? In *Ergebnisse der experimentellen Medizin* (ed. S. M. Rapoport), Vol. 24, pp. 147–156. VEB Verlag Volkundgesundheit, Berlin.

Heinrich, R. and Rapoport, T. (1974). A linear steady-state treatment of enzymatic chains. General properties, control and effector strength. *European Journal of Biochemistry*, **42**, 89–95.

Hess, B. and Wurster, W. (1970). Transient time of the pyruvate kinase–lactate dehydrogenase system of rabbit muscle *in vitro*. *FEBS Letters*, **9**, 73–7.

Hess, B. (1973). Organization of glycolysis: oscillatory and stationary control. In *Rate control of biological processes XXVII* (ed. D. D. Davies), pp. 105–31. Cambridge University Press.

Kacser, H. and Burns, J. A. (1973). The control of flux. In *Rate control of biological processes XXVII* (ed. D. D. Davies), pp. 65–104. Cambridge University Press.

Krüger, R. (1987). Ph.D. Thesis, Justus-Liebig-Universität, Giessen.

Krüger, R., Jäger, H.-J., Hintz, M., and Pahlich, E. (1986). Purification to homogeneity of pyrroline–5-carboxylate reductase of barley. *Plant Physiology*, **80**, 142–4.

Pahlich, E. (1990). Switching of proline metabolism in water stressed plants. *Bulletin de la Société Botanique de France* **137**, 3–11.

Pahlich, E., Jäger, H.-J., and Kaschel, E. (1981). Thermodynamische Betrachtungen über die reversible Reaktionssequenz Glutaminsäure-Prolin. *Zeitschrift für Pflanzenphysiologie*, **101**, 137–44.

Pahlich, E., Jäger, H.-J. and Horz, M. (1982). Weitere Untersuchungen zur thermodynamischen Struktur der Biosynthese-sequenz Glutaminsäure-Prolin in wassergestreßten Buschbohnen. *Zeitschrift für Pflanzenphysiologie*, **105**, 475–8.

Pahlich, E., Kerres, R., and Jäger, H.-J. (1983). Influence of water stress on the vacuole/extravacuole distribution of proline in protoplasts of *Nicotiana rustica*. *Plant Physiology*, **72**, 590–1.

Reich, J. G. and Sel'kov, E. E. (1981). *Energy metabolism of the cell*. Academic Press, New York.

Rhodes, D., Handa, S., and Bressan, R. A. (1986). Metabolic changes associated with adaptation of plant cells to water stress. *Plant Physiology*, **82**, 890–903.

Schellenberger, W., Eschrich, K., and Hofmann, E. (1981). Self-stabilization of the energy charge in a reconstituted enzyme system containing phosphofructokinase. *European Journal of Biochemistry*, **118**, 309–14.

Schellenberger, W., Eschrich, K., and Hofmann, E. (1983). Diminution of stationary enzyme activities at increases of pyruvate kinase concentration in a reconstituted enzyme system. *Biomedica Biochimica Acta*, **42**, 57–72.

Stewart, G. R. and Lee, J. A. (1974). The role of proline accumulation in halophytes. *Planta* (Berl.), **120**, 279–89.

13. Barley seed storage proteins — structure, synthesis, and deposition

P.R. SHEWRY

Department of Agricultural Sciences, University of Bristol, AFRC Institute of Arable Crops Research, Long Ashton Research Station, Bristol BS18 9AF, UK

Introduction

Cereals are the most important crops in the world, with total annual yields approaching 2×10^9 tonnes. Proteins account for about 8 to 15 per cent of the grain dry weight, and the total protein yield exceeds by several-fold that of the more protein-rich legume seeds. They therefore form the major source of protein for the nutrition of humans and lifestock.

Wheat is the most important cereal in Europe (as it is in the world as a whole), and barley the second most important. In both cases the amount and composition of the grain storage proteins, which account for about half of the total grain proteins, have important effects on the quality of the grain for its major end uses: as feed for mono-gastric livestock, for bread-making and for malting, brewing and distilling.

This chapter discusses the structures, classical and molecular genetics, and synthesis and deposition of the storage proteins of barley, focusing on their role within the plant as a sink of applied nitrogen. Despite great variation in detail, the seed storage proteins of other cereals are similar in their biological and chemical characteristics.

Barley storage proteins

The major storage protein present in barley grain is an alcohol-soluble protein (or prolamin) called hordein. It is not, however, the only storage protein present. Hordein is only synthesized in the starchy endosperm, not in the embryo or aleurone (the outer layer of the endosperm). The major protein present in both these tissues is a storage globulin related to the 7S vicilins of legumes (Burgess and Shewry 1986; Yupsanis *et al.* 1990). In addition, a number of other proteins, such as β-amylase (Shewry *et al.* 1988*a*), accumulate in the grain during development and probably have a secondary role as storage proteins. Since these other proteins are minor components in terms of their amounts and impact on the quality of the grain they will not be considered further.

Hordein polymorphism and genetics

Hordein is a complex mixture of proteins which vary in their values of M_r (about 30 000 to 100 000 by SDS–PAGE), pI and proportions (Fig. 13.1). There is also considerable variation in the components present in different genotypes (Fig. 13.2A). These polypeptides can be classified into two major groups (B and C hordeins), and at least two minor groups (D and γ-type hordeins).

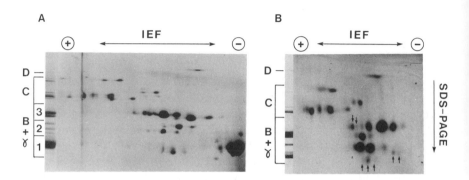

Fig. 13.1. 2-D analysis (IEF/SDS – PAGE) of total reduced and pyridylethylated hordein fractions from barley cultivars Sundance (A) and Carlsberg II (B). The groups of hordein polypeptides (B,C,D,γ) are indicated on the 1-D SDS–PAGE separations shown at the left of each 2-D gel, and the B hordeins in cv. Sundance are subdivided into B1, B2 and B3. The arrows in part B indicate γ-type hordeins. Taken from Rahman *et al.* (1982) and Kreis *et al.* (1983*b*).

The B hordeins account for about 80 per cent of the total hordein fraction (Kirkman *et al.* 1982; Shewry *et al.* 1983*b*), and consist of about 8 to 16 major components with M_r values by SDS–PAGE ranging from about 30 000 to 50 000. They are classified into two subfamilies, called class I/II and class III, on the basis of their cyanogen bromide peptide maps (Faulks *et al.* 1981), and the cross-hybridization behaviour of their cDNAs and genes (Kreis *et al.* 1983*a*). They are encoded by a multigene family at a single locus, designated *Hor2*, located on the short arm of chromosome 5 (Fig. 13.3) (See Shewry *et al.* 1990). Analysis of the *Hor2* locus by Southern blotting using a B hordein probe shows a copy number of about 20–30 genes per haploid genome, with variation in the numbers and sizes of the hybridizing fragments present in different genotypes (Fig. 13.2C). The degree of restriction-fragment polymorphism is consistent with the variation in the polypeptide composition while the gene copy number is sufficiently high for each protein to be encoded by at least one gene.

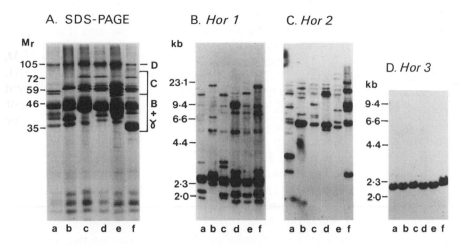

Fig. 13.2. Polymorphism in hordein polypeptides and genes. Part A shows SDS–PAGE analyses of total reduced and pyridylethylated hordein fractions from the cultivars Athos (a), Keg (b), Jupiter (c), Hoppel (d), Igri (e), and Sundance (f). Parts B,C and D show total genomic DNAs from the same six cultivars, digested with *Hin*dIII and probed with cDNA clones related to C hordein (part B), B hordein (part C) and D hordein (part D). Re-arranged from Bunce *et al.* (1986).

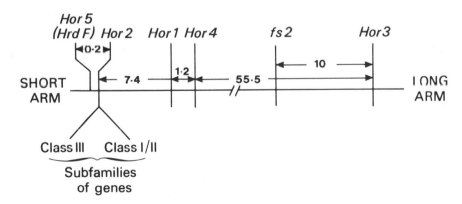

Fig. 13.3. The locations of the hordein structural loci on chromosome 5 (1H) of barley. Taken from Shewry *et al.* (1990).

Although recombinaton within the *Hor2* locus is generally considered to have contributed to the variation in hordein patterns between genotypes of barley, it had not until recently been observed in crosses. However, we recently identified a single line in which recombination within *Hor2* had occurred between the class I/II and class III subfamilies of genes (Shewry *et al*. 1990). Analysis of this line also enabled us to tentatively determine the order of the two subfamilies of genes on the chromosome arm (Fig. 3).

C hordein accounts for about 10 to 20 per cent of the total hordein fraction. It consists of a number of polypeptides with M_r values by SDS – PAGE of about 50 000–70 000, and exhibits a similar degree of polymorphism to B hordein (see Figs 13.1, 13.2A, and Shewry *et al*. 1985). It is also encoded by a single multigenic locus, designated *Hor1*, which is linked to *Hor2* on the short arm of chromosome 5 (Fig. 13.3). The degree of polymorphism at this locus again reflects that of the encoded proteins (cf. Fig. 13.2 A, B) (Shewry *et al*. 1985).

The D and γ-hordein groups each account for only a small proportion of the total fraction. D hordein gives a single band by SDS–PAGE, with an M_r of about 105 000. The true M_r is probably considerably less, (about 90 000) as SDS–PAGE has been shown to give anomalously high M_r values for the related HMW subunits of wheat glutenin (see Bunce *et al*. 1985). It is probably encoded by a single gene, located at the *Hor3* locus on the long arm of chromosome 5 (Shewry *et al*. 1983a; Bunce *et al*. 1986). There is little variation in the mobility of D hordein on SDS–PAGE (Fig. 13.2A) or in the sizes of DNA restriction fragments that hybridize to a D hordein-related cDNA clone (Fig. 13.2D) (Bunce *et al*. 1986).

Gamma hordein consists of several quantitatively minor polypeptides with similar M_r values on SDS–PAGE to B hordein (Fig. 13.1B). Little is known of the extent of polymorphism between genotypes, and RFLP analysis shows several hybridizing fragments, indicating that γ-hordein is encoded by a small multigene family (Cameron-Mills and Brandt 1988; Shewry *et al*. 1990).

Hordein structure

The prolamins of wheat, barley and rye can be classified into three groups on the basis of their amino acid sequences and the chromosomal locations of their structural genes (see Miflin *et al*. 1983). These groups, called the sulphur-poor, sulphur-rich, and high-molecular-weight prolamins, comprise in barley C hordein, B hordein and γ-hordein, and D hordein respectively. Although all have high contents of glutamine and proline, the precise amounts of these vary, as do the proportions of other amino acids such as glycine (high in D hordein), phenylalanine (high in C hordein), and sulphur amino acids (low in the S-poor C hordeins, higher in the other groups) (see Table 13.1).

Information on the detailed structures of hordein polypeptides is derived from physico-chemical analyses of purified proteins and peptides, the direct determination of amino acid sequences, and the deduction of amino acid sequences from the nucleotide sequences of cloned cDNAs and genes. In

Table 13.1 Amino acid compositions (expressed as mol%) of the groups of hordein polypepties

	B	C	D	γ
Asx	1.4	1.0	1.3	2.9
Thr	2.1	1.0	8.1	3.1
Ser	4.7	4.6	9.7	5.5
Glx	35.4	41.2	29.6	32.4
Pro	20.6	30.6	11.6	16.5
Gly	1.5	0.3	15.7	5.9
Ala	2.2	0.7	2.5	2.6
Cys	2.5	0	1.5	2.7
Val	5.6	1.0	4.5	3.7
Met	0.6	0.2	0.2	1.2
Ile	4.1	2.6	0.7	2.9
Leu	7.0	3.6	3.3	8.6
Tyr	2.5	2.3	3.9	1.7
Phe	4.8	8.8	1.4	4.7
His	2.1	1.1	3.4	2.0
Lys	0.5	0.2	1.1	1.6
Arg	2.4	0.8	1.5	2.0
Ref	1	2	3	4

1. Total B hordein from cv. Julia (Shewry *et al.* 1980).
2. Total C hordein from cv. Julia (Shewry *et al.* 1980).
3. D hordein from Risø mutant 1508 (Kreis *et al.* 1984).
4. γ-Hordein band r from Risø 56 (Kreis *et al.* 1983*b*). Other γ hordeins from the same line have compositions more like that of B hordein, with 29.9 mol% Glx and 27.1% Pro.

addition, further information can be deduced from studies of related, and in some cases more fully characterized, proteins from wheat. This chapter aims only to summarize our present state of knowledge, and the reader is referred to review articles for more detailed accounts (Shewry *et al.* 1988*b*; Kreis *et al.* 1985*a*, *b*; Tatham *et al.* 1990).

Although the groups of S-rich, S-poor, and HMW prolamins vary considerably in their amino acid sequences and conformations, they are all structurally related, and are considered to belong to a protein super-family which includes seed proteins of other types (see Kreis *et al.* 1985*a*, *b*; Kreis and Shewry 1989). They all also have two features in common. The first is that the sequence can be divided into regions, or domains, which differ in their amino acid compositions and in their secondary structures. The second is that in all cases at least one of these domains consists of repeated sequences, based on one or more short peptide motifs. These repetitive domains are responsible for the unusual amino acid compositions of prolamins (and in particular for their high contents of glutamine and certain other amino acids) and probably determine many of their characteristic properties such as solubility.

The structures of B hordein, C hordein, γ-hordein, and a HMW subunit of wheat (a homologue of D hordein) are shown diagrammatically in Fig. 13.4. Two types of HMW subunit, called x- and y-types, occur in wheat. Of these the y-types are most closely related to D hordein in their N-terminal amino acid sequences (Shewry *et al.* 1988c) and in their amino acid compositions (notably the presence of over 1 mol per cent cysteine) (Shewry *et al.* 1989). A y-type subunit encoded by the B genome of bread wheat is therefore shown in Fig. 4.

The HMW subunit in Fig. 13.4 consists of 684 residues, and has three structural domains. Non-repetitive domains of 104 and 42 residues at the N- and C-terminus, respectively, flank a repetitive domain of 538 residues. The latter consists of tandem and interspersed repeats based on two motifs: a nonapeptide (consensus gly.tyr.tyr.pro.thr.ser.leu.gln.gln.) and a hexapeptide (consensus pro.gly.gln.gly.gln.gln.). These domains also have different conformations. Whereas the non-repetitive domains appear to be globular with regions of α-helix, the repetitive domain adopts a loose spiral structure based on regularly repeated β-turns (Miles *et al.* 1991). This gives an extended rod-shaped structure to the whole molecule (Field, *et al.* 1987), and may be relevant to how the protein is packaged within the protein body. Cysteine residues are located predominantly in the non-repetitive domain, five at the N-terminus and one at the C-terminus. This is of particular interest as HMW prolamins are found only in high M_r polymers stabilized by inter-chain disulphide bonds.

C hordein also has an extensive repetitive domain, which appears to account for all of the protein (which consists of about 440 residues) with the exception of short unique sequences of 12 and 6 residues at the N- and C-terminus, respectively. The repetitive domain consists predominantly of octapeptides, based on the consensus motif pro.gln.gln.pro.phe.pro.gln.gln. Although this motif is not related to those present in the HMW subunits, the domain appears to adopt a similar spiral structure, with the whole protein forming an extended semi-rigid rod (Field *et al.* 1986). C hordeins do not contain cysteine, and are present in the grain as monomers.

The S-rich B and γ-hordeins differ from the other groups in that repetitive sequences form less than half of the total sequence, and are present at the

---

Fig. 13.4. Diagrammatical representations of the structures of B1 hordein, γ-hordein, C hordein, and a HMW subunit of wheat glutenin (1By9). SH indicates the positions of cysteine residues. Based on data reported by Cameron-Mills and Brandt (1988); Field *et al.* (1986, 1987); Forde *et al.* (1985a); Halford *et al.* (1987); Shewry *et al.* (1987a); Tatham *et al.* (1990).

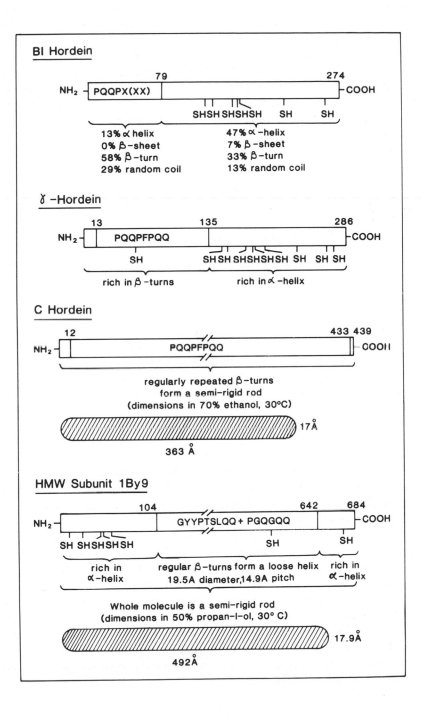

N-terminus of the protein. In γ-hordein, but not in B hordein, the repeats are preceded by a short unique N-terminal sequence. In both proteins the repeats are related to those present in C hordein.

The unique C-terminal domains contain most or all of the cysteine residues. The specificity of disulphide bond formation is not completely understood, but B hordein appears to form mainly polymers stabilized by inter-chain disulphide bonds (like D hordein), while at least some of the γ-hordeins are monomeric with intra-chain disulphide bonds.

The C-terminal domains of B and γ-hordeins appear to be globular, being rich in α-helix but also containing elements of β-sheet, β-turn, and random coil. In contrast the N-terminal domains are rich in β-turns, although it it not known whether these are organized to form a spiral structure (as in C hordein).

Hordein synthesis and deposition

Hordein is synthesized only in the starchy endosperm of the developing seed, not in the aleurone layer (which is also endosperm tissue) or in the embryo.

Synthesis occurs on microsomes bound to the rough endoplasmic reticulum (ER) (Brandt and Ingversen, 1976; Matthews and Miflin 1980). The protein synthesized *in vitro* using poly A^+ mRNA as a template migrates slightly more slowly on SDS–PAGE than that synthesized on membrane-bound polysomes, indicating the presence of a signal peptide which is cleaved co-translationally (Matthews and Miflin 1980). The newly synthesized polypeptide passes into the lumen of the ER, and protein synthesized *in vitro* on membrane-bound polysomes is resistant to digestion with chymotrypsin (Cameron-Mills *et al.* 1978; Cameron-Mills and Ingversen 1978).

The later stages of prolamin deposition are less well- understood: although it is accepted that hordeins are deposited in membrane-bound protein bodies, the precise mechanisms of deposition, and the origin of the membrane have been the subject of dispute. Early work in this area used two approaches, which gave conflicting answers.

Transmission electron microscopy of developing endosperms showed protein bodies consisting of a homogeneous matrix with more densely staining granular inclusions (see Miflin and Shewry 1979; Cameron-Mills and von Wettstein 1980) (Fig. 13.5). Ingversen (1975) suggested that these phases consisted of different protein types, the homogeneous component containing a high content of prolamins, and the granular component a high content of glutelins. This hypothesis was supported by analysis of protein bodies from Risø Mutant 1508. This line contains only a low proportion of prolamins, and the protein bodies consist of a granular component in which a few homogeneous spheres are embedded (Ingversen 1975; Miflin and Shewry

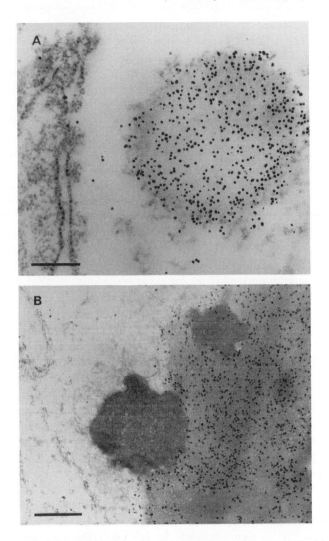

Fig. 13.5. Immuno-gold labelling of protein bodies present in subaleurone (A) and starchy endorsperm (B) cells of developing barley endosperm, using an antibody raised against C hordein (Festenstein *et al.* 1984). Note the presence of homogeneous and granular phases in the starchy endosperm protein body, of which only the former reacts with antibody. The bars are 0.25 μm (part A) and 0.5 μm (part B). Unpublished results of J. Henderson, B. J. Miflin, P. R. Shewry, and N. Harris.

1979). Recent studies of wheat indicate that these granular inclusions may in fact consist of legumin-like globulins (called triticins in wheat). Immuno-gold labelling showed that these regions did not label when probed with antisera raised against gliadins and glutenins, but were densely labelled with antiserum to triticin (Bechtel *et al.* 1989, and unpublished results). A weaker reaction of the triticin antiserum with the homogeneous matrix may have resulted from cross-reaction with prolamins. Similar studies with barley show that the homogeneous, but not the granular, phase labels with antibodies raised against hordeins (Fig. 13.5) (unpublished results of J. Henderson, P. R. Shewry, B. J. Miflin, and N. Harris).

Electron microscopy of developing endosperms has provided clear evidence that at least some protein bodies are formed by deposition within the vacuole. Cameron-Mills and von Wettstein (1980) showed the presence of protein deposits within the vacuoles of developing endosperms, and their identity as prolamins has been confirmed by immuno-gold labelling (unpublished results of J. Henderson *et al.*). The pathway of deposition may include transport in vesicles derived from the golgi apparatus, as described in wheat (Kim *et al.* 1988) and in legumes (see Chrispeels 1984).

An alternative mode of protein body formation has been proposed by Miflin and co-workers, based on the analysis of protein bodies prepared from developing endosperms by density gradient ultra-centrifugation (Miflin *et al.* 1981). They showed that protein bodies prepared from barley, wheat and maize were associated with activity of NADH-cytochrome *c* reductase, an enzyme characteristic of the endoplasmic reticulum, but not with enzymic activities characteristic of vacuoles (RN-ase, phospho-diesterase, and *N*-acetylglucosaminidase). In contrast protein bodies isolated by a similar procedure from peas were clearly associated with vacuolar enzymes. Miflin and Burgess (1982) further showed that isolated protein bodies of barley and wheat were digested by proteinase K, while those of maize and peas were protected from digestion. They interpreted these differences as indicating that the protein bodies of barley and wheat were enclosed by incomplete membranes, whereas the membranes of pea and maize protein bodies were complete. It is well-established that the protein bodies of maize are formed by direct deposition and accumulation of prolamins within the lumen of the ER, and are surrounded by a complete membrane of ER origin (Larkins and Hurkman 1978). Miflin and co-workers proposed that a similar mechanism occurs in barley, with the exception that the ER membrane does not reform around the protein deposit to give a continuous membrane. This hypothesis was supported by electron micrographs of protein bodies with incomplete membranes and associated with the rough ER (Miflin and Shewry 1979; Shewry and Miflin 1983).

Is it possible to reconcile these two different views of protein body formation? I think that it is if we bear in mind the changing nature of the developing endosperm and the limitations of the two experimental approaches.

The most convincing and successful micrographs are of fairly young tissue (up to about 20 days after anthesis), before the accumulation of large amounts of starch poses problems for sample preparation. At this stage most protein bodies are clearly of vacuolar origin. A second mechanism, direct accumulation within the ER, may become more important later in seed development, when the cells become distended with starch and protein transport may be obstructed. Protein bodies formed by this mechanism may be more resistant to damage during isolation than those of vacuolar origin, and therefore be highly represented in protein body fractions prepared by density gradient ultra-centrifugation.

Regulation of hordein synthesis

The time course of hordein accumulation in the developing endosperm of barley was first studied by Bishop (1939), and subsequent studies have essentially confirmed his observations (Brandt 1976; Shewry *et al*. 1979; Rahman *et al*. 1982). Hordein can only be detected after about 14 days after anthesis, when the endosperm has attained about 10 per cent of its final dry weight, and then accumulates rapidly to account for about 40 per cent of the nitrogen in the mature grain. (Rahman *et al*. 1982) (see Fig. 13.6*a*). Changes also occur in the composition of the hordein fraction, the proportions of C and B3 (class II) hordeins decreasing and that of B1 (class I/II) hordein increasing during development of the cultivar Sundance (Rahman *et al*. 1982) (Fig. 13.6B).

Rahman *et al*. (1984) subsequently determined the populations of mRNAs related to C, B1 and B3 hordeins by dot hybridization with specific cDNA probes. Hordein-related sequences were first detected at about 14 days after anthesis, slightly before hordein protein could be detected by SDS–PAGE, and increased in amount up to 26 days. During this period the increases in sequences related to B3 and C hordeins were one third of those in sequences related to B1 hordein, reflecting the changes in proportions of the proteins (Fig. 13.6C, D). However, the total amounts of mRNAs decreased after 26 days, whereas the amounts of protein continued to rise. This discrepancy may suggest that the hordein mRNAs are translated preferentially to other mRNA species during the later stages of development.

The total amount of hordein and the proportions of individual hordein groups and subgroups are also affected by the availabilities of nitrogen and sulphur. Kirkman *et al*. (1982) showed that the proportion of hordein increased from about 35 to 50 per cent of the total grain protein when the nitrogen content was increased from 1.27 to 2.01 per cent dry wt (equivalent to 7.2 to 11.5 per cent protein), with an increase in the proportion of C hordein from 13 to 18.5 per cent of the fraction. These effects are even more

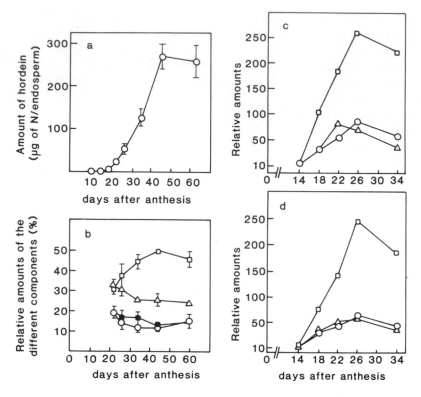

Fig. 13.6. Hordein synthesis in developing endosperms of cv. Sundance. (a) amount of total hordein; (b) proportions of hordein polypeptides; (c) proportions of membrane-bound polysomal RNAs; (d) proportions of poly (A^+) RNAs. Key to parts b,c,d: $\bigcirc-\bigcirc$, C hordein; $\square-\square$, B1 hordein; $\blacksquare-\blacksquare$, B2 hordein (part b only); $\triangle-\triangle$, B3 hordein. Taken from Rahman *et al.* (1984).

extreme at very high levels of nitrogen, with hordein accounting for 60 per cent of the total grain nitrogen and C hordein representing 40 per cent of the total (see Fig. 13.7).

Giese *et al.* (1983) also showed that hordein acted as the main nitrogen sink in liquid-cultured detached spikes grown at high levels of nitrogen availability, and that the amount of C hordein increased disproportionately. Giese and Hopp (1984) subsequently showed parallel increases in the populations of B and C hordein mRNAs.

Kirkman *et al.* (1982) speculated that the increase in the proportion of the S-poor C hordein resulted from a limitation of sulphur relative to nitrogen, and Shewry *et al.* (1983*b*) showed that C hordein accounted for 71 and 82.5 per cent of the total hordein fraction in seeds of cultivars Athos and

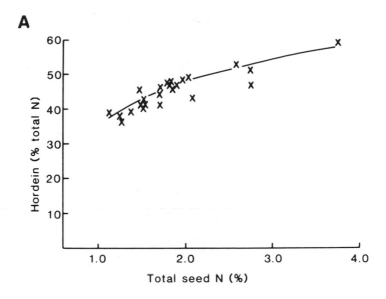

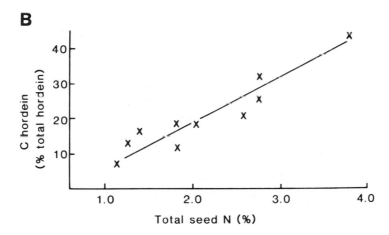

Fig. 13.7. Proportions of total hordein (A) and C hordein (B) present in barley grain containing different amounts of nitrogen. Based on data in Kirkman *et al.* (1982) and author's unpublished results.

Sundance, respectively, grown under conditions of sulphur deficiency (see Fig. 13.8). The proportion of D hordein was also reduced in both cultivars.

Rahman *et al.* (1983) reported more detailed studies of the cultivar Sundance grown under conditions of sulphur stress. C hordein accounted for about 90 per cent of the total fraction at 14 days after anthesis, and 80 per

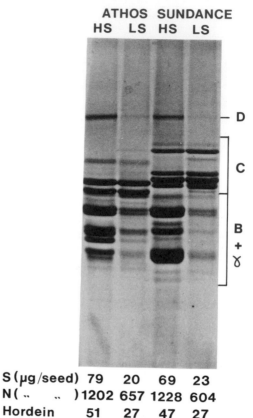

Fig. 13.8. SDS–PAGE of reduced and pyridylethylated total hordein fractions from grain of two barley cultivars, Athos and Sundance, grown at high and low levels of sulphur. Taken from Shewry *et al.* (1983*b*).

cent at 30 days after anthesis, and maturity (Fig. 13.9A). Over the same period the ratio of B1 (class I/II) to B3 (class III) hordein increased from 1:1 to 1.8:1. These differences are consistent with the changes reported by Rahman *et al.* (1982) in developing grain grown with adequate sulphur. The populations of mRNAs in the early (14 day) and late (30 day) developing endosperms were also assessed by *in vitro* translation and hybrid dot analysis (Fig. 13.9B). Although the populations of mRNAs were broadly consistent with the levels of hordein accumulation, sulphur deficiency also appeared to increase the efficiency of translation of C hordein mRNAs and decrease that of B hordein mRNAs. This could presumably result from limitation in the

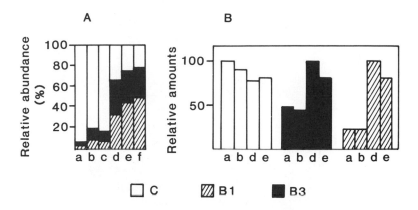

Fig. 13.9. Relative abundances of B1, B3, and C hordeins (part A) and relative amounts of corresponding membrane-bound polysomal RNAs (part B) in developing endosperms of cv. Sundance grown with adequate and limiting levels of sulphur. (a) low sulphur, early development; (b) low sulphur, late development; (c) low sulphur, mature seed; (d) high sulphur, early development; (e) high sulphur, late development; (f) high sulphur mature seed. Taken from Rahman *et al.* (1983).

availability of sulphur amino acids for protein synthesis.

The amount and composition of hordein is also affected by high lysine genes. One spontaneous gene of this type and a number of induced mutant genes have been identified, and most are recessive and 'regulatory'. An exception is Risø mutant 56, which is co-dominant and results from a deletion mutation at the *Hor2* locus (see below). In all cases the high lysine phenotype arises, at least in part, from a decrease in synthesis of the lysine-poor hordein fraction. This is accompanied by increases in the proportions of other more lysine-rich protein fractions (e.g. albumins, globulins), and in some cases, in specific 'high lysine' proteins. Although a full discussion of high lysine genes is outside the scope of this chapter (but see Shewry *et al.* 1987*b*), a brief description of the effects of one series of mutant genes on the amount and composition of hordein is relevant. This is the Risø series of mutants, which were induced in two parental cultivars (Bomi and Carlsberg II) using a range of chemical and physical mutagens.

A very dramatic change occurs in Risø 56, where deletion of the *Hor2* locus (Fig. 13.10B) results in a complete elimination of B hordein (Fig. 13.10A, tracks a and b) (Kreis *et al.* 1983*b*). Although the amount of C hordein per seed is increased about threefold and γ-type hordeins may also be increased, there is little or no change in the amount of D hordein. D hordein does, however, represent a higher proportion of the total hordein fraction than in Carlsberg II. The net effect is a reduction in total hordein of about 20–30 per cent.

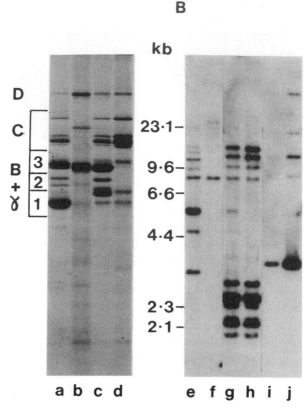

Fig. 13.10. SDS–PAGE (A) and Southern blot analysis (B) of high lysine barley mutants. A. SDS–PAGE of reduced and pyridylethylated total hordein fractions from (a) cv. Bomi; (b) Risø mutant 1508; (c) cv. Carlsberg II; (d) Risø mutant 56. B. Southern blotting of *Hin*dIII digests of total genomic DNA from cv. Carlsberg II (tracks e and g) and Risø mutant 56 (tracks f and h) probed with cDNA clones related to B hordein (tracks e and f) and C hordein (tracks g and h). Tracks i and j are reconstructions to represent one copy, and ten copies respectively, of the C hordein sequence per haploid genome. Taken from Forde *et al.* (1985*b*), Kreis *et al.* (1983*b*, 1984).

The regulatory *lys3a* gene of Risø 1508 has an even more dramatic effect on hordein synthesis, with the total fraction accounting for only about 15 per cent of the grain proteins (Kreis *et al.* 1984). The C hordeins appear to be completely eliminated and the B1 (class I/II) hordeins affected more than the B3 (class III) (see Fig 13.10A, tracks c and d). In contrast, the content of D hordein is increased about fourfold.

Other mutants of the Risø series also have altered proportions of B hordein polypeptides, as well as decreased total amounts of hordein (see

Table 13.2) (Shewry *et al.* 1987*b*). Risø mutants 7, 527, and 13 are all in the cultivar Bomi, and have differential effects on class I/II and class III polypeptides: whereas mutants 7 and 527 have higher proportions of B1 (class I/II) hordein, the proportion of B3 (class III) is greater in Risø 13. Risø 527 also has less D hordein. Mutants 29 and 86 are in Carlsberg II, and it is less easy to interpret the changes in B hordein pattern in relation to the specific subfamilies.

Molecular analyses of Risø 56 and Risø 1508 have essentially confirmed the biochemical studies. Developing endosperms of Risø 56 did not contain mRNA for B hordeins, and the deletion of most, if not all, of the B hordein (*Hor2*) genes was confirmed by Southern blot analysis (Kreis *et al.* 1983*b*). Similarly, analysis of Risø 1508 showed only traces of mRNAs for C hordein, while the abundances of mRNAs for I/II and class III hordeins were reduced to about 5 and 40 per cent, respectively, of those present in Bomi (Kreis *et al.* 1984). In contrast, the proportion of D hordein mRNAs was increased by more than twofold.

Taken together, the studies of grain development and of the effects of mineral nutrition and mutant high lysine genes (summarized in Table 13.2) enable us to draw some conclusions about the regulation of hordein synthesis.

Table 13.2 Summary of the effects of grain maturation, mineral nutrition, and mutant high lysine genes on the proportions of hordein polypeptides

	Amount of total hordein	Proportions of groups				
		D	C	B	γ	Ratio of class I/II to class III B hordein
a. Grain maturation						
Increasing age	↑		↓	↑		↑
b. Mineral nutrition						
High N	↑	↓	↑	↓		−
Low S	↓	↓	↑	↓		↑
c. Mutant genes						
Risø 56 (*Hor2* ca)	↓	↑	↑	0	↑	−
Risø 1508 (*lys* 3a)	↓	↑	0	↓	↓	
Risø 527 (*lys* 6i)	↓	↓				↑
Risø 7 −	↓					↑
Risø (*lys* 5f)	↓					↓

↑ increase in proportion
↓ decrease in proportion
0 group absent
− no apparent change

1. The rates of accumulation of the different groups of hordein polypeptides are closely related to the populations of their respective mRNAs, which could be determined by effects on their rates of transcription, their stability, or a combination of these.

2. Some control of hordein synthesis may also be exerted at the level of translation, for example under conditions of sulphur deficiency.

3. Although B, C, and D hordeins are expressed in a co-ordinated fashion, their relative rates of accumulation (and the populations of their mRNAs) vary during grain development, under different conditions of nutrient availability, and under the influence of mutant genes.

4. Differences are also observed in the synthesis of the two major sub-families of B hordeins, indicating differential regulation of the two sub-families of genes.

These early studies of hordein gene expression have provided a basis for more recent analyses of the structure and expression of isolated genes.

Hordein gene structure and expression

Five hordein genes have to date been isolated and characterized. These include three B hordein genes of the class I/II sub-family (Forde *et al.* 1985*a*; Brandt *et al.* 1985; Chernyshev *et al.* 1989), and one gene each for γ-type hordein (Cameron-Mills and Brandt 1988) and C hordein (Entwistle, 1988). The latter is probably a pseudo-gene, as the encoded protein is atypical of C hordeins in its N-terminal and C-terminal amino acid sequences and also contains an in frame stop colon. Although genes encoding D hordein have not been isolated, information on their structure can be inferred by reference to the well-characterized family of genes encoding the related HMW subunits of wheat glutenin (see Shewry *et al.* 1989). No genes encoding class III (B3) hordein polypeptides have been characterized.

The structures of the genes are compared in Fig. 13.11 and regions of conserved sequence are aligned in Table 13.3. The genes all have TATA boxes about 80 to 100 bp upstream of the ATG initiation codons, and one or more putative polyadenylation signals (AATAAA) between about 50 and 150 bp downstream of the stop codon. The B hordein and HMW subunit genes also have 'CATC' boxes about 30 and 60 bp, respectively, upstream of the TATA boxes, but these do not appear to be present in the γ-hordein and C hordein genes. The five barley genes also have a conserved region of about 30 bp approximately 300 bp upstream of the ATG initiation codon. Related sequences are present in other prolamin genes from wheat (α- and γ-gliadins) and maize (the M_r 19 000 and M_r 22 000 α-zeins) (see Forde *et al.* 1985*a*; Kreis *et al.* 1986), and Forde *et al.* (1985*a*) speculated

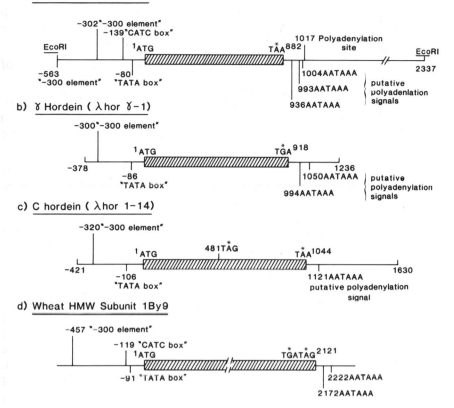

Fig. 13.11. Summary of the structures of genes for B1 hordein, γ-hordein, C hordein, and a HMW subunit of wheat glutenin. Based on data in Forde *et al.* (1985*a*), Halford *et al.* (1987), Entwistle (1988), and Cameron-Mills and Brandt (1988).

that these '−300 elements' play a role in regulating prolamin gene expression, similar to the *cis*-regulating elements identified in the 5′ upstream regions of other multigene families.

A conserved '−300 element' does not appear to be present in the HMW subunit genes of wheat, but a shorter sequence (24 bp) with limited homology is present further upstream (about 450 bp). This region is deleted in two silent (i.e. unexpressed) genes from cultivars Cheyenne and Chinese Spring, but does not appear to be responsible for their lack of expression (Halford *et al.* 1989) (see below).

Little is known about the total size of the loci encoding B, C, and γ-hordeins, or the organization of the individual genes within them. Kreis

Table 13.3 Conserved sequences in the 5′ upstream regions of hordein genes and in a HMW subunit gene from wheat. Based on data in Brandt *et al.* (1985), Forde *et al.* (1985*a*), Halford *et al.* (1987), Cameron-Mills and Brandt (1988), Entwistle (1988), and Chernyshev *et al.* (1989)

A. Putative CATC and TATA boxes

Gene product	Clone	CATC box		TATA box	
B1 hordein	pBHR184	−139	ACATCCAAACA	−80	CTATAAATA
"	λhor 2 – 4	−140	GCATCCAAACA	−80	CTATAAATA
"	λ r9	−139	ACATCCAAACA	−80	CTATAAATA
γ hordein	λhor γ – 1		not present	−86	CTATAAAGA
C hordein	λhor 1 – 14		not present	−106	CTATAAATA
HMW subunit	λHMW 47	−119	CCATGGTCCTG	−91	CTATAAAAG

B. ‘−300 elements’

Gene product	Clone	−300 element	
B1 hordein	pBHR184	−300	ACATG..TAAAGTGAATAAGG.TGAGTCATG
"	λhor 2 – 4	−303	ACATG..TAAAGTGAATAAGG.TGAGTCATG
"	λ r9	−300	ACATG..TAAAGTGAATAAGG.TGAGTCATG
γ – hordein	λhor γ – 1	−300	AGATG..TAAAGTGAATAAGA.TGAGTCAGC
C hordein	λhor 1 – 14	−320	TAGTG..TAAAGTAAAAAAAA.TGACTCATC
HMW Subunit	λHMW 47	−457	GACATGCTTAGAA.GCTTTT.AGTGA

et al. (1983*b*) estimated a minimum size of 85 kb for the *Hor2* locus, based on Southern blot analysis of the deletion mutant Risø 56. However, the gene copy number (11–13) was considerably less than that indicated by later studies (20–30) (Bunce *et al.* 1986), indicating that this estimate of minimum size should probably be doubled (i.e. about 170 kb). These estimates do not allow for the presence of either non-coding DNA or other genes interspersed with the B hordein genes.

A maximum size for the *Hor2* locus may be indicated by the recent work of Sørensen (1989) in which genomic DNA was digested with rare-cutting enzymes and the fragments separated by pulsed field gel electrophoresis. Southern blotting using a B hordein probe indicated that two fragments of DNA, of approximately 220 and 300 kb, respectively, could together account for the entire *Hor2* locus.

Two approaches have been used to study the regulation of prolamin gene expression: expression of chimaeric genes in transgenic tobacco, and the identification of putative regulatory proteins that bind to conserved sequences in the 5′ flanking region.

Marris *et al.* (1988) linked 549 bp of the 5′ upstream region of the B1 hordein gene of Forde *et al.* (1985*a*) to the reporter gene encoding chloramphenicol acetyl transferase (CAT), and transferred the chimaeric gene into tobacco plants, using *Agrobacterium tumefaciens*. CAT enzyme activity was detected only in the developing endosperm from about 15 days after pollination, not in other parts of the seed (the embryo and testa) or in the leaves. Kreis and co-workers also transferred the entire 2.9 kb *Eco*R1 fragment containing the B1 hordein gene into tobacco, using a similar transformation procedure (Shewry *et al.* 1988*b*). B1 hordein transcripts were detected in the developing seeds (but not the leaves) of the transgenic plants, but it was not possible to detect hordein protein. These experiments demonstrate conclusively that all the information required for seed and developmental specificity of expression of the B1 hordein gene is located in the 549 bp of 5′ flanking sequence, but do not provide conclusive information on the role of the '−300 element'. Experiments are currently in progress to analyse this region in more detail, using chimaeric constructs with the β-glucuronidase (GUS) reporter gene (C. Marris and M. Kreis, personal communication).

Although D hordein gene expression has not been studied, it is again possible to extrapolate from studies of the HMW subunits of wheat glutenin. Colot *et al.* (1987) demonstrated that 433 bp of 5′ upstream sequence was sufficient to confer endosperm-specific CAT activity, and Halford *et al.* (1989) made more detailed studies using the GUS reporter system. They demonstrated that only 280 bp of 5′ sequence from an expressed HMW subunit gene was sufficient to confer tissue and development-specific expression. The results clearly demonstrated that the '−300 element'-like sequence present about 450 bp upstream of the ATG initiation codon does not have a role in determining the specificity of HMW subunit gene expression.

There is little upstream homology between the 5′ upstream sequences of the HMW subunits of wheat and those of the S-rich and S-poor prolamins of barley and wheat. This, taken together with the studies of hordein and glutenin gene expression in transgenic tobacco, indicates that the HMW prolamins may differ fundamentally from the other prolamin groups in their control of gene expression. This is consistent with the effects of the *lys3a* gene of Risø 1508, which results in almost total loss of B and C hordein but an increase in D hordein (see p. 216).

It is also of interest that the HMW prolamins are encoded by a small multigene family in wheat (six genes per haploid genome in hexaploid bread wheat) and probably by single genes in barley (Bunce *et al.* 1986). In contrast the S-rich and S-poor groups of prolamins are encoded by larger multigene families. This difference appears to be associated with a difference in promoter strength, as the levels of expression of the chimaeric HMW subunit gene constructs reported by Halford *et al.* (1989) were much higher than those observed using the B1 hordein promoter in similar constructs (C. Marris and M. Kreis, personal communication).

The identification and characterization of DNA binding proteins is technically very demanding, and has been less successful when applied to hordein genes. Kreis *et al.* (1986) reported the use of a modified Western blotting procedure, in which proteins from transcriptionally active nuclei are separated by SDS–PAGE, transferred to a nitrocellulose filter, and incubated with ^{32}P-labelled DNA. Analysis of fractions from shoots and endosperms showed six major protein bands (M_r values 37 000 to 200 000) which bound a 600 bp upstream fragment from the B1 hordein gene. Some of these were detected only in the nuclei from endosperms, including one protein of M_r about 120 000 which also bound to a synthetic oligonucleotide representing the '−300 element' (see Shewry *et al.* 1988*b*). Further studies aimed at defining the roles of these proteins are in progress, using a range of approaches.

Conclusions

Eleven years ago we presented a paper entitled *The synthesis of proteins in normal and high lysine barley seeds* at a previous Phytochemical Society Symposium. Reference to the published account (Miflin and Shewry 1979) shows the enormous advances that have been made in the intervening period. The work reviewed then was mainly descriptive, with electrophoretic and other analyses of crudely separated protein fractions. There was, however, reference also to the application of molecular biology (*in vitro* protein synthesis), which has since revolutionized our understanding of the structure and regulation of hordein genes, and has facilitated (by provision of complete amino acid sequences) the use of physico-chemical methods to study protein structure. The remaining problems centre on gene regulation, in relation to the development of the cereal seed and the perception and response to external stimuli (such as availability of mineral nutrients).

Progress in this field is slow, but the results should eventually enable us to manipulate the structure and expression of barley seed protein genes so that the plants make better use of mineral nutrients, and to produce grain more suited for its end use, whether this is as animal feed or for industrial processing.

References

Bechtel, D. B., Wilson, J. D., and Shewry, P. R. (1989). Identification of legumin-like proteins in thin sections of developing wheat endosperms by immunocyto-chemical procedures. *Cereal Foods World*, **34**, 784.

Bishop, L. R. (1939). The proteins of barley grain, with special reference to hordein. M. Sc. Thesis, University of Birmingham, UK.

Brandt, A. (1976). Endosperm protein formation during kernel development of wild type and a high-lysine barley mutant. *Cereal Chemistry*, **53**, 890–901.

Brandt, A. and Ingversen, J. (1976). *In vitro* synthesis of barley endosperm proteins on wild type and mutant templates. *Carlsberg Research Communications*, **41**, 312–20.

Brandt, A., Montembault, A., Cameron-Mills, V., and Rasmussen, S. K. (1985). Primary structure of a B1 hordein gene from barley. *Carlsberg Research Communications*, **50**, 333–45.

Bunce, N., White, R. P., and Shewry, P. R. (1985). Variation in estimates of molecular weights of cereal prolamins by SDS–PAGE. *Journal of Cereal Science*, **3**, 131–42.

Bunce, N. A. C., Forde, B. G., Kreis, M., and Shewry, P. R. (1986). DNA restriction fragment length polymorphism at hordein loci: application to identifying and fingerprinting barley cultivars. *Seed Science Technology*, **14**, 419–29.

Burgess, S. R. and Shewry, P. R. (1986). Identification of homologous globulins from embryos of wheat, barley, rye and oats. *Journal of Experimental Botany*, **37**, 1863–71.

Cameron-Mills, V. and Brandt, A. (1988). A γ-hordein gene. *Plant Molecular Biology*, **11**, 449–61.

Cameron-Mills, V. and Ingversen, J. (1978). *In vitro* synthesis and transport of barley endosperm proteins: reconstitution of functional rough microsomes from polyribosomes and stripped microsomes. *Carlsberg Research Communications*, **43**, 471–89.

Cameron-Mills, V. and von Wettstein, D. (1980). Endosperm morphology and protein body formation in developing wheat grain. *Carlsberg Research Communications*, **45**, 577–94.

Cameron-Mills, V., Ingversen, J., and Brandt, A. (1978). Transfer of *in vitro* synthesised barley endosperm proteins into the lumen of the endoplasmic reticulum. *Carlsberg Research Communications*, **43**, 91–102.

Chernyshev, A. K., Davletova, Sh. K., Bashkirov, V. I., Shakhmanov, N. B., Mekhedov, S. L., and Ananiev, E. V. (1989). Nucleotide sequence of BI hordein gene of barley *Hordeum vulgare* L. *Genetica*, **25**, 1349–1355.

Chrispeels, M. J. (1984). Biosynthesis, processing and transport of storage proteins and lectins in cotyledons of developing legume seeds. *Philosophical Transactions of the Royal Society of London B.*, **304**, 309–22.

Colot, V., Robert, L.S., Kavanagh, T. A., Goldsbrough, A. P., Bevan, M. W., and Thompson, R. D. (1987). Localisation of sequences in wheat endosperm protein genes which confer tissue- specific expression in tobacco. *EMBO Journal*, **6**, 3559–64.

Entwistle, J. (1988). Primary structure of a C-hordein gene from barley. *Carlsberg Research Communications*, **53**, 247–58.

Faulks, A. J., Shewry, P. R., and Miflin, B. J. (1981). The polymorphism and structural homology of storage polypepties (hordein) coded by the *Hor 2* locus in barley (*Hordeum vulgare* L.). *Biochemical Genetics*, **19**, 841–58.

Festenstein, G. N., Hay, F. C., Miflin, B. J., and Shewry, P. R. (1984). Immunochemical studies on barley seed storage proteins. The specifity of an antibody to 'C' hordein and its reaction with prolamins from other cereals. *Planta*, **162**, 524–31.

Field, J. M., Tatham, A. S., Baker, A., and Shewry, P. R. (1986). The structure of C hordein. *FEBS Letters*, **200**, 76–80.

Field, J. M., Tatham, A. S., and Shewry, P. R. (1987). The structure of a high molecular weight subunit of wheat gluten. *Biochemistry Journal*, **247**, 215–21.

Forde, B. G., Heyworth, A., Pywell, J., and Kreis, M. (1985a). Nucleotide sequence of a B1 hordein gene and the identification of possible upstream regulatory elements in endosperm storage protein genes from barley, wheat and maize. *Nucleic Acids Research*, **13**, 7327–39.

Forde, B. G., Kreis, M., Williamson, M., Fry, R., Pywell, J., Shewry, P. R., Bunce, N., and Miflin, B. J. (1985b). Short tandem repeats shared by B- and C-hordein cDNAs suggest a common evolutionary origin for two groups of cereal storage protein genes. *EMBO Journal*, **4**, 9–15.

Giese, H. and Hopp, H. E. (1984). Influence of nitrogen nutrition on the amount of hordein, protein Z and β-amylase messenger RNA in developing endosperms of barley. *Carlsberg Research Communications*, **49**, 365–83.

Giese, H., Andersen, B., and Doll, H. (1983). Synthesis of the major storage protein, hordein, in barley. Pulse labelling study of grain filling in liquid cultured detached spikes. *Planta*, **159**, 60–65.

Halford, N. C., Forde, J., Anderson, O. D., Greene, F. C., and Shewry, P. R. (1987). The nucleotide and deduced amino acid sequences of an HMW glutenin subunit gene from chromosome 1B of bread wheat (*Triticum aestivum* L.), and comparison with those of genes from chromosomes 1A and 1D. *Theoretical and Applied Genetics*, **75**, 117–26.

Halford, N. G., Forde, J., Shewry, P. R., and Kreis, M. (1989). Functional analysis of the upstream regions of a silent and an expressed member of a family of wheat seed protein genes in transgenic tobacco. *Plant Science*, **62**, 207-16.

Ingversen, J. (1975). Structure and composition of protein bodies from wild-type and high-lysine barley endosperm. *Hereditas*, **81**, 69-76.

Kim, W.T., Franceschi, V.R., Krishnan, H., and Okita, T.W. (1988). Formation of wheat protein bodies: involvement of the Golgi apparatus in gliadin transport. *Planta*, **176**, 173-82.

Kirkman, M.A., Shewry, P.R., and Miflin, B.J. (1982). The effect of nitrogen nutrition on the lysine content and protein composition of barley seeds. *Journal of the Science of Food and Agriculture*, **33**, 115-27.

Kreis, M. and Shewry, P.R. (1989). Unusual features of seed protein structure and evolution. *Bio-Essays*, **10**, 201-7.

Kreis, M., Rahman, S., Forde, B.G., Pywell, J., Shewry, P.R., and Miflin, B.J. (1983a). Sub-families of hordein mRNA encoded at the *Hor2* locus of barley. *Molecular and General Genetics*, **191**, 194-200.

Kreis, M., Shewry, P.R., Forde, B.G., Rahman, S., and Miflin, B.J. (1983b). Molecular analysis of a mutation conferring the high lysine phenotype on the grain of barley (*Hordeum vulgare*). *Cell*, **34**, 161-7.

Kreis, M., Shewry, P.R., Forde, B.G., Rahman, S., Bahramian, M.B., and Miflin, B.J. (1984). Molecular analysis of the effects of the mutant *lys 3a* gene on the expression of *Hor* loci in developing endosperms of barley (*Hordeum vulgare*). *Biochemical Genetics*, **22**, 231-55.

Kreis, M., Forde, B.G., Rahman, S., Miflin, B.J., and Shewry, P.R. (1985a). Molecular evolution of the seed storage proteins of barley, rye and wheat. *Journal of Molecular Biology*, **183**, 499-502.

Kreis, M., Shewry, P.R., Forde, B.G., Forde, J., and Miflin, B.J. (1985b). Structure and evolution of seed storage proteins and their genes, with particular reference to those of wheat, barley and rye. In *Oxford surveys of plant cell and molecular biology*, Vol. 2 (ed. B.J. Miflin), pp. 253-317. Oxford University Press, Oxford.

Kreis, M., Williamson, M.S., Forde, J., Schmutz, D., Clark, J., Buxton, B., Pywell, J., Marris, C., Henderson, J., Harris, N., Shewry, P.R., Forde, B.G., and Miflin B.J. (1986). Differential gene expression in the developing barley endosperm. *Philosphical Transactions of the Royal Society of London B*, **314**, 355-65.

Larkins, B.A. and Hurkman, W.J. (1978). Synthesis and deposition of zein in protein bodies of maize endosperm. *Plant Physiology, Lancaster*, **62**, 256-63.

Marris, C., Gallois, P., Copley, J., and Kreis, M. (1988). The 5'-flanking region of a barley B hordein gene controls tissue and development specific CAT expression in tobacco plants. *Plant Molecular Biology*, **10**, 359-66.

Matthews, J.A. and Miflin, B.J. (1980). *In vitro* synthesis of barely storage proteins. *Planta*, **149**, 262-8.

Miflin, B.J. and Burgess, S.R. (1982). Protein bodies from developing seeds of barley, maize, wheat and peas: the effects of protease treatment. *Journal of Experimental Botany*, **33**, 251-60.

Miflin, B.J. and Shewry, P.R. (1979). The synthesis of proteins in normal and high lysine barley seeds. In *Recent advances in the biochemistry of cereals* (ed. D. Laidman and R. G. Wyn Jones), pp. 239–73, Academic Press, New York.

Miflin, B. J., Burgess, S. R., and Shewry, P. R. (1981). The development of protein bodies in the storage tissues of seeds. *Journal of Experimental Botany*, **32**, 119–219.

Miflin, B. J., Field, J. M., and Shewry, P. R. (1983). Cereal storage proteins and their effects on technological properties. In *Seed proteins* (ed. J. Daussant, J. Mosse, and J. Vaughan), pp. 255–319. Academic Press, New York.

Miles, M. J., Carr, H. J., McMaster, T., Belton, P. S., Morris, V. J., Field, J. M., Shewry, P. R., and Tatham, A. S. (1991). Scanning tunnelling microscopy of a wheat gluten protein reveals details of a spiral supersecondary structure. *Proceedings of the National Academy of Sciences, USA*, **88**, 68–71.

Rahman, S., Shewry, P. R., and Miflin, B. J. (1982). Differential protein accumulation during barley grain development. *Journal of Experimental Botany*, **33**, 717–28.

Rahman, S., Shewry, P. R., Forde, B. G. Kreis, M., and Miflin, B. J. (1983). Nutritional control of storage protein synthesis in developing grain of barley (*Hordeum vulgare* L.). *Planta*, **159**, 366–72.

Rahman, S., Kreis, M., Forde, B. G., Shewry, P. R., and Miflin, B. J. (1984). Hordein gene expression during development of the barley (*Hordeum vulgare*) endosperm. *Biochemical Journal*, **223**, 315–22.

Shewry, P. R., Pratt, H. M., Leggatt, M. M., and Miflin, B. J. (1979). Protein metabolism in developing endosperms of high-lysine and normal barley. *Cereal Chemistry*, **56**, 110–17.

Shewry, P. R., Field, J. M., Kirkman, M. A., Faulks, A. J. and Miflin, B. J. (1980). The extraction, solubility and characterisation of two groups of barley storage polypeptides. *Journal of Experimental Botany*, **31**, 393–407.

Shewry, P. R. and Miflin, B. J. (1983). Characterisation and synthesis of barley seed proteins. In *Seed proteins: genetics, chemistry and nutritive value* (ed. W. Gottschalk and H. Muller), pp. 143–205. Martinus Nijoff, The Hague.

Shewry, P. R., Finch, R., Parmar, S., Franklin, J., and Miflin, B. J. (1983*a*). Chromosomal location of *Hor3*, a new locus governing storage proteins in barley. *Heredity*, **50**, 179–89.

Shewry, P. R., Franklin, J., Parmar, S., Smith, S. J., and Miflin, B. J. (1983*b*). The effects of sulphur starvation on the amino acid and protein compositions of barley grain. *Journal of Cereal Science*, **1**, 21–31.

Shewry, P. R., Bunce, N. A. C., Kreis, M., and Forde, B. G. (1985). Polymorphism at the *Hor 1* locus of barley (*Hordeum vulgare* L.). *Biochemical Genetics*, **23**, 389–402.

Shewry, P. R., Field, J. M., and Tatham, A. S. (1987*a*). The structures of cereal seed storage proteins. In *Cereals in a European context* (ed. I. D. Morton), pp. 421–37. Ellis Horwood, Chichester.

Shewry, P. R., Williamson, M. S., and Kreis, M. (1987*b*). Effects of mutant genes on the synthesis of storage components in developing barley endosperms. In *Mutant genes that affect plant development* (ed. H. Thomas and D. Grierson), pp. 95–118. Cambridge University Press.

Shewry, P. R., Parmar, S., Buxton, B., Gale, M. D., Liu, C. J., Hejgaard, J. and Kreis, M. (1988*a*). Multiple molecular forms of β-amylase in seeds and vegetative tissues of barley. *Planta*, **176**, 127–34.

Shewry, P. R., Tatham, A. S., Field, J. M., Forde, B. G., Clark, J., Gallois, P., Marris, C., Halford, N., Forde, J., and Kreis, M. (1988*b*). The structure of barley and wheat prolamins and their genes. *Biochemie und Physiologie der Pflanzen*, **183**, 117–27.

Shewry, P. R., Tatham, A. S., Pappin, D. J., and Keen, J. (1988*c*). N-terminal amino acid sequences show that D hordein of barley and HMW secalins of rye are homologous with HMW glutenin subunits of wheat. *Cereal Chemistry*, **65**, 510–11.

Shewry, P. R., Halford, N. G., and Tatham, A. S. (1989). The high molecular weight subunits of wheat, barley and rye: genetics, molecular biology, chemistry and role in wheat gluten structure and functionality. In *Oxford surveys of plant molecular and cell biology*, Vol. 7 (ed. B. J. Miflin), pp. 163–219. Oxford University Press.

Shewry, P. R., Parmar, S., Franklin J., and Burgess, S. R. (1990). Analysis of a rare recombination event within the multigenic *Hor2* locus of barley (*Hordeum vulgare* L.), *Genetical Research* (Cambridge), **55**, 171–6.

Sørensen, M. B. (1989). Mapping of the *Hor2* locus in barley by pulsed field gel electrophoresis. *Carlsberg Research Communications*, **54**, 109–20.

Tatham, A. S., Shewry, P. R., and Belton, P. S. (1990). Structural studies of cereal prolamins, including wheat gluten. In *Advances in cereal science and technology*, Vol. 10 (ed. Pomeranz, Y.), pp. 1–78. AACC St Paul, Minnesota.

Yupsanis, T., Burgess, S. R., Jackson, P. J., and Shewry, P. R. (1990). Characterisation of the major protein component from aleurone cells of barley (*Hordeum vulgare* L.). *Journal of Experimental Botany*, **41**, 385–92.

14. Polyamine metabolism and compartmentation in plant cells

NELLO BAGNI and ROSSELLA PISTOCCHI

Department of Biology, Institute of Botany, Via Irnerio 42, 40126 Bologna, Italy

Introduction

Polyamines are ubiquitous nitrogenous compounds classified as plant growth substances (Bagni 1989). They act mainly in processes based on cell division, with a broad spectrum of effects on various plant tissues, but very little is known about the basic mechanisms and events of this action. Studies on the role of polyamines are complicated by the lack of plant mutants deficient in polyamines. Additional knowledge of their action can be obtained by studying in greater detail their biosynthetic pathways, as well as by blocking them at various steps by using the various inhibitors now available. The study of their transport and compartmentation could also help to explain some effects, such as the different sensitivity of a plant tissue or organ to applied polyamines and the different rate of polyamine synthesis within the same plant, besides providing information on the nature of polyamine action.

This chapter will mainly review these two aspects of polyamines, namely their metabolism and compartmentation.

Polyamine metabolism

Aliphatic polyamines are synthesized in both prokaryotic and eukaryotic organisms. The most common ones are a diamine, putrescine (1,4 diamino-butane), a triamine, spermidine (1,8-diamino-4-azaoctane), and a tetramine, spermine (1,12-diamino-4,9-diazododecane). Many other di- and polyamines are present in plants and micro-organisms; for example, the diamines 1,3-diaminopropane, and cadaverine (1,5-diaminopentane). In particular many unusual polyamines have been detected in thermophilic bacteria, which are fascinating organisms for both biochemists and molecular biologists. In the extreme thermophile, *Thermus thermophilus* at least 14 polyamines have been isolated, among these polyamines longer than tetramines. So far two straight-chained pentamines and two hexamines have been detected (Table 14.1). Caldopentamine is present in considerable quantities and, in cells grown at extremely high temperatures such as 80°C or higher,

Table 14.1 Polyamines in the cells of extreme thermophilic bacteria

Trivial name	Systematic	Chemical structure
1,3-Diaminopropane	1,3-Diaminopropane	$NH_2(CH_2)_3NH_2$
Putrescine	1,4-Diaminobutane	$NH_2(CH_2)_4NH_2$
Norspermidine (= caldine)	1,7-Diamino-4-azaheptane	$NH_2(CH_2)_3NH(CH_2)_3NH_2$
Spermidine	1,8-Diamino-4-azaoctane	$NH_2(CH_2)_3NH(CH_2)_4NH_2$
sym-Homospermidine	1,9-Diamino-5-azanonane	$NH_2(CH_2)_4NH(CH_2)_4NH_2$
Thermine	1,11-Diamino-4.8-diazaundecane	$NH_2(CH_2)_3NH(CH_2)_3NH$-$(CH_2)_3NH_2$
Spermine	1,12-Diamino-4,9-diazadodecane	$NH_2(CH_2)_3NH(CH_2)_4NH$-$(CH_2)_3NH_2$
Thermospermine	1,12-Diamino-4,8-diazadodecane	$NH_2(CH_2)_3NH(CH_2)_3NH$-$(CH_2)_4NH_2$
Homospermine	1,13-Diamino-4,9-diazatridecane	$NH_2(CH_2)_3NH(CH_2)_4NH$-$(CH_2)_4NH_2$
Caldopentamine	1,15-Diamino-4,8,12-triazapentadecane	$NH_2(CH_2)_3NH(CH_2)_3NH$-$(CH_2)_3NH(CH_2)_3NH_2$
Homocaldopentamine	1,16-Diamino-4,8,12-triazahexadecane	$NH_2(CH_2)_3NH(CH_2)_3NH$-$(CH_2)_3NH(CH_2)_4NH_2$
Caldohexamine	1,19-Diamino-4,8,12,16-tetraazanonadecane	$NH_2(CH_2)_3NH(CH_2)_3NH$-$(CH_2)_3NH(CH_2)_3NH_2$
Homocaldohexamine	1,20-Diamino-4,8,12,16-tetraazaeicosane	$NH_2(CH_2)_3NH(CH_2)_3NH$-$(CH_2)_3NH(CH_2)_4NH_2$
	Tris(3-aminopropyl)amine	$[NH_2(CH_2)_3]_3N$
	Tetrakis(3-aminopropyl) ammonium	$[NH_2(CH_2)_3]_4N^+$

this pentamine was one of the major polyamines (Oshima 1989). Even though these unusual polyamines are typical of thermophilic bacteria, some of them also occur in higher plants, for example a spermidine homologue, *sym*-homospermidine, which is present in leaves of the sandalwood tree (*Santalum album*) (Radhakrishanan 1974). Spermine is normally present in plants in lower amounts or in traces in respect to putrescine and spermidine. Much of the older data on the occurrence of this amine are over-estimated due to an insufficient separation by thin layer chromatography from other amines.

Polyamines arise from amino acids through decarboxylation. The basic amino acids arginine, ornithine, and lysine furnish the main part of the carbon skeleton, while methionine contributes the propylamino group to putrescine to form spermidine and to spermidine to form spermine (Fig. 14.1). Putrescine normally arises from one of two major pathways: via

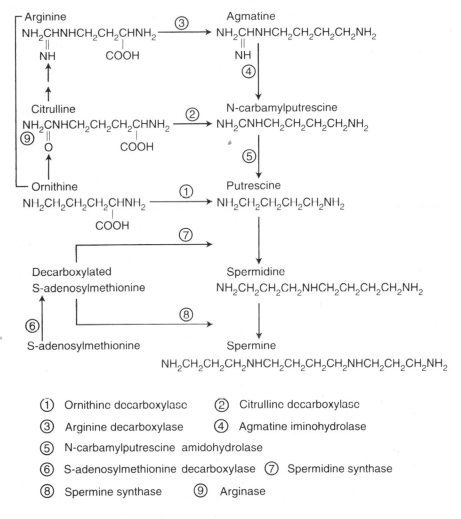

Fig. 14.1. Biosynthetic pathways of polyamines in plants.

ornithine decarboxylase (ODC) directly or via arginine decarboxylase (ADC) through agmatine and *N*-carbamylputrescine (see Fig. 14.1). While in animals and some phyto-pathogenic fungi the decarboxylation of ornithine is the only known route for the synthesis of putrescine, in bacteria, other fungi, and higher plants, there are at least two pathways involved, ODC and ADC. Recently it was shown that, in the presence of pyridoxal phosphate, extracts of wheat, barley, sugar beet, and rape leaves are able to catalyse the release of CO_2 from the carboxyl group of ornithine without the formation of putrescine (Smith and Marshall 1988). This oxygen-dependent reaction is:

$$\underset{\text{ornithine}}{\underset{|}{\underset{\text{COOH}}{H_2N-CHCH_2CH_2CH_2NH_2}}} + O_2 \longrightarrow CO_2 + H_2O + \underset{\text{4-aminobutanamide}}{\underset{\overset{\|}{O}}{H_2NCCH_2CH_2CH_2NH_2}}$$

The determination of real ODC activity should therefore be established by relating CO_2 release and putrescine formation. In addition, since arginase is very active in plants, the results for ADC activity, using $1 - {}^{14}C$ arginine to measure ${}^{14}CO_2$ release, are often over-estimated if this enzyme has not been inhibited during the ODC assay. Ornithine can also be converted back to arginine through the ornithine cycle.

In some cases, such as in *Sesamum* leaves and *Helianthus tuberosus* tuber explants (Crocomo and Basso 1974; Speranza and Bagni 1978), a citrulline decarboxylase activity, which forms *N*-carbamylputrescine, was detected; in *Sesamum*, during potassium deficiency, this enzyme was solely responsible for putrescine synthesis.

Cadaverine (1,5-diaminopentane) is formed by lysine decarboxylase. This enzyme is present in some bacteria, but also in higher plants, in particular in some Gramineae, Leguminosae, and Solanaceae. In the latter family, cadaverine is related to alkaloid synthesis in *Nicotiana*; in fact, cadaverine is a precursor of anabasine, the main alkaloid of *Nicotiana glauca*, whose biosynthesis takes place mainly in the roots (Bagni *et al.* 1986). Also, other diamines such as putrescine can be utilized as precursors of nicotine in tobacco (Leete 1982). To date no relationship has been found between cadaverine and other major polyamines, such as putrescine, spermidine, and spermine. In the case of animal tumour cells exposed to α-difluoromethylornithine, an irreversible inhibitor of ODC, cadaverine, which is not normally produced by animal cells, was formed as a compensatory mechanism to the blockage of putrescine and spermidine synthesis, and was rapidly converted to analogues of spermidine and spermine, namely aminopropylcadaverine and bis(aminopropyl)cadaverine (Jänne *et al.* 1981). We cannot exclude the possibility that some plants, such as legumes in which cadaverine is present in abnormal amounts (Federico and Angelini 1988), are also able to convert cadaverine to the above analogues.

S-adenosylmethionine is not only the precursor for spermidine and spermine synthesis via *S*-adenosylmethionine decarboxylase (Fig. 14.1), but it also forms 1-aminocyclopropane-1-carboxylic acid, a precursor of ethylene. A number of physiological effects of ethylene in plants are known to be antagonized by treatment with polyamines, leading to the suggestion that the two compounds may interact with each other in their effect on growth and senescence (Galston and Kaur-Sawhney 1987). This interaction may derive from the fact that ethylene and polyamines share, and probably compete for, the same intermediate, *S*-adenosylmethionine.

Polyamines in plants are known to occur as free molecules but are also conjugated to small molecules, such as amides of hydroxycinnamic acid, or to proteins. Di- and polyamines conjugated with cinnamic acids are found in many higher plants, the Solanaceae in particular (Smith *et al.* 1983). The main putrescine conjugates which have been identified are *p*-coumaroyl-caffeoyl- and feruloyl-putrescine, which are also known as cinnamoyl-putrescines (Fig. 14.2). The relative proportions of free and conjugated polyamines are variable, but it has been reported that up to 90 per cent of the polyamine pool in tobacco can occur in the conjugated form (Torrigiani

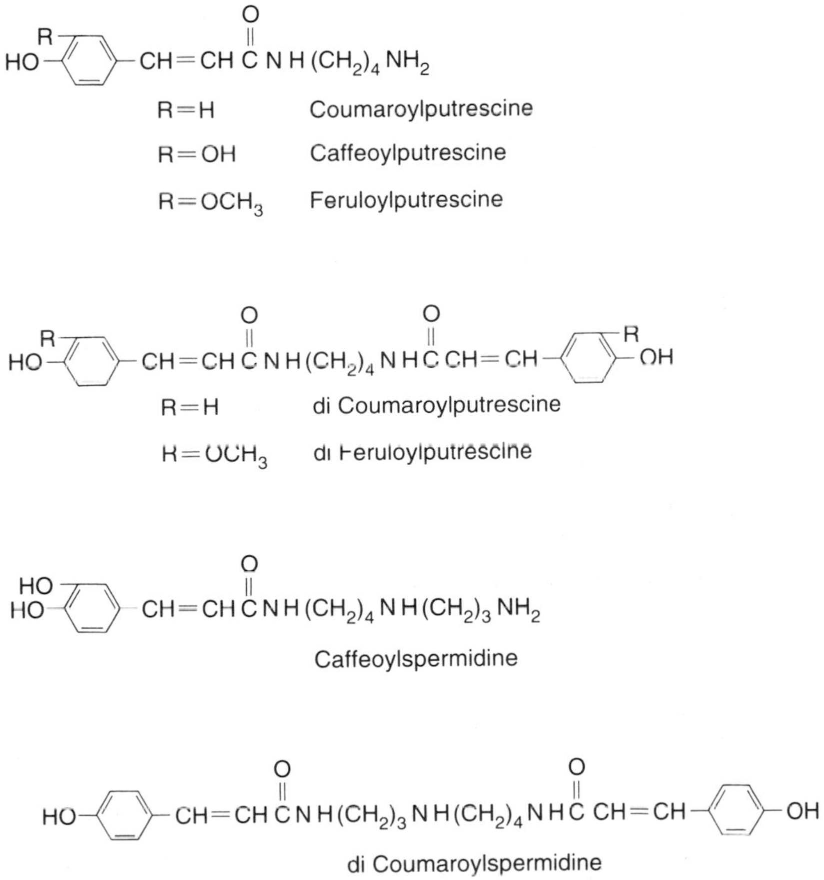

$$HO-\overset{R}{\underset{}{\bigcirc}}-CH=CH\ \overset{O}{\underset{\|}{C}}\ NH(CH_2)_4\ NH_2$$

R=H	Coumaroylputrescine
R=OH	Caffeoylputrescine
R=OCH$_3$	Feruloylputrescine

$$HO-\overset{R}{\underset{}{\bigcirc}}-CH=CH\ \overset{O}{\underset{\|}{C}}\ NH(CH_2)_4\ NH\overset{O}{\underset{\|}{C}}\ CH=CH-\overset{R}{\underset{}{\bigcirc}}-OH$$

| R=H | di Coumaroylputrescine |
| R=OCH$_3$ | di Feruloylputrescine |

$$\overset{HO}{\underset{HO}{\bigcirc}}-CH=CH\ \overset{O}{\underset{\|}{C}}\ NH(CH_2)_4\ NH(CH_2)_3\ NH_2$$

Caffeoylspermidine

$$HO-\bigcirc-CH=CH\ \overset{O}{\underset{\|}{C}}\ NH(CH_2)_3\ NH(CH_2)_4\ NH\overset{O}{\underset{\|}{C}}\ CH=CH-\bigcirc-OH$$

di Coumaroylspermidine

Fig. 14.2. Amides formed between hydroxycinnamic and di- and polyamines found in tobacco plants.

et al. 1987). Their occurrence in the tobacco plant has been extensively studied and more recently an enzyme, caffeoyl-CoA N-caffeoyl transferase, has also been characterized (Negrel 1989). The function of polyamine conjugates is still unclear. It was suggested that they represent a storage form of polyamines and (or) that they regulate the free polyamine pool in the cell. A transglutaminase or transglutaminase-like activity, an enzyme capable of catalysing the covalent binding of polyamines to proteins, was recently found in plants (Serafini-Fracassini *et al.* 1989); this finding has provided a possible mechanism for the formation of polyamine-protein complexes.

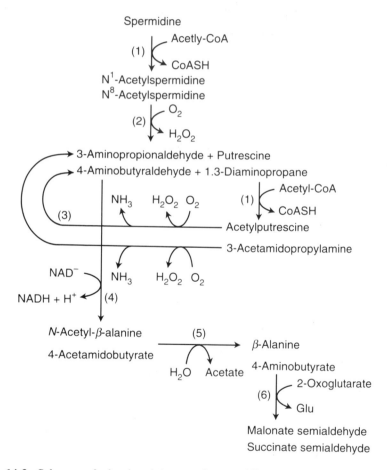

Fig. 14.3. Scheme of the breakdown of spermidine and putrescine in fungi. The enzymes involved are: (1) diamine acetyltransferase; (2) polyamine oxidase; (3) acetylputrescine oxidase; (4) acetamidoaldehyde dehydrogenase; (5) acetamidoalkanoate deacetylase; (6) ω-amino acid aminotransferase.

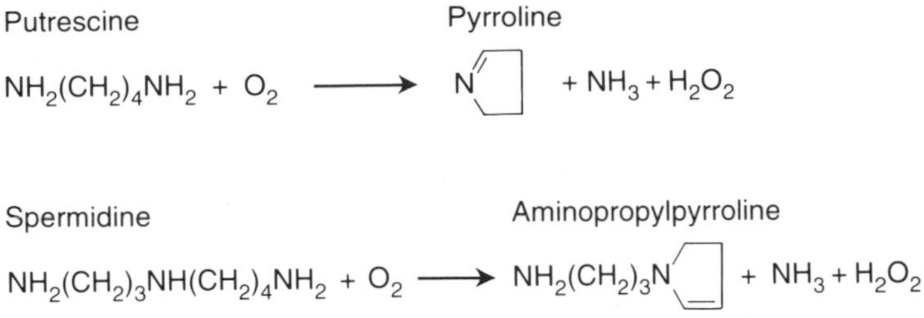

Putrescine   Pyrroline

$$NH_2(CH_2)_4NH_2 + O_2 \longrightarrow \quad N \qquad + NH_3 + H_2O_2$$

Spermidine   Aminopropylpyrroline

$$NH_2(CH_2)_3NH(CH_2)_4NH_2 + O_2 \longrightarrow NH_2(CH_2)_3N \qquad + NH_3 + H_2O_2$$

Fig. 14.4. Diamine oxidase in plants.

Polyamines covalently bound to peptides (molecular weight about 5000) or proteins were previously found in different plants such as apple pollen (Bagni *et al.* 1981) and *Helianthus tuberosus* tubers (Serafini-Fracassini *et al.* 1989).

Acetylated polyamines were discovered for the first time in urine and were related to animal excretion (Seiler *et al.* 1983). More recently, however, it has been established that for vertebrates and yeasts, N^1-acetylation is the first step in the degradative transformation of one polyamine into another (Fig. 14.3) (Gillyon *et al.* 1987), and that this inter-conversion could regulate intracellular polyamine metabolism. Also in plants a compound, tentatively identified as acetylputrescine, has been detected in different parts of sugar beet seedlings (Christ *et al.* 1989). If this finding is confirmed in other plant materials, the acetylation of polyamines could be considered a general phenomenon.

Although the di- and polyamine oxidases of plants have an apparently limited distribution, in some species they are remarkably active. Diamine oxidase (DAO), which is found especially in the Leguminosae, has a broad specificity and oxidizes putrescine and other diamines, but also spermidine through the mechanism shown in Fig. 14.4. Polyamine oxidases (PAO), found only in the monocotyledons, mainly the Gramineae, are quite distinct from diamine oxidase (Fig. 14.5). Aminopropylpyrroline derived from

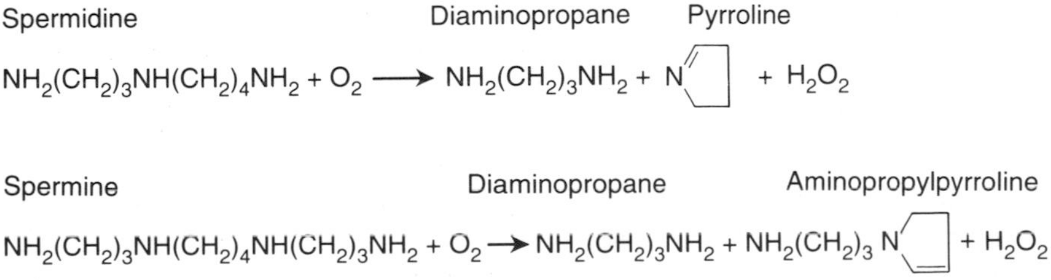

Spermidine   Diaminopropane   Pyrroline

$$NH_2(CH_2)_3NH(CH_2)_4NH_2 + O_2 \longrightarrow NH_2(CH_2)_3NH_2 + N \qquad + H_2O_2$$

Spermine   Diaminopropane   Aminopropylpyrroline

$$NH_2(CH_2)_3NH(CH_2)_4NH(CH_2)_3NH_2 + O_2 \longrightarrow NH_2(CH_2)_3NH_2 + NH_2(CH_2)_3 N \qquad + H_2O_2$$

Fig. 14.5. Polyamine oxidase in plants.

di- and polyamine oxidases can be also cycled to diazobicyclonanane according to Smith and Barker (1988). The degradative enzymes of polyamines in bacteria and fungi are different. Many fungi catabolize polyamines through acetyltransferase (see Fig. 14.3), furthermore the oxidation of spermidine produces 3-aminopropanaldehyde, unlike plant PAOs, but like certain animal enzymes.

Polyamine transport and subcellular compartmentation in different plant systems

The uptake of some monoamines (methyl- or ethylamine) has been extensively studied since these molecules are used as ammonia analogues because of the lack of a convenient radioisotope for ammonia. These studies were undertaken mostly in algae such as *Chara corallina*, *Hydrodictium africanum*, or *Nitella clavata* which have giant cells and can be studied using micro-electrodes. Methylamine uptake has been reported to be inhibited by ammonia, to follow saturation kinetics and to consist in a $\Delta\psi$-driven uniport through specific channels (for a review see Kleiner 1981). In contrast, scarce information is available on monoamine accumulation in higher plants. In some case the uptake of these molecules has been used for estimating the ΔpH across biological membranes, especially if these are not accessible to standard electro-physiological techniques (Kleiner 1981); however some researchers believe that this is not a reliable method for eukaryotic systems (Bertl *et al.* 1984).

The uptake of di- and polyamines was first investigated in bacteria and animals with the aim of understanding the basic mechanisms of this event (Tabor and Tabor 1966; Pohjanpelto 1976). In plants this subject was studied later and initially not extensively. On the other hand the enzymes for polyamine biosynthesis have been found in all cellular types investigated, thus a regulation of endogenous content by means of transport/efflux phenomena was not regarded as a necessary event. Furthermore the first studies on polyamine transport did not encourage further investigations: Joshi *et al.* (1983) found that a lag period of about 20 h was necessary before the onset of rapid polyamine uptake in protoplasts of *Vigna unguiculata*; Young and Galston (1983) reported that the injection of radiolabelled putrescine or spermidine into cotyledons of pea seedlings did not result in recovery of labelled polyamines in the shoot or root.

Investigations made in our laboratory revealed that by spraying polyamines on apple trees, at the time of fertilization, fruit-set and yield per tree as well as fruit growth increased, suggesting that these substances were readily absorbed. In fact, in isolated apple corymbs, the fruitlets and the young leaves were found to absorb exogenous putrescine and to synthesize spermidine and spermine from it. Putrescine and probably spermidine and spermine, were translocated via the peduncle from leaves to fruitlets and *vice versa* (Bagni *et al.* 1984).

These observation stimulated us to study the characteristics and mechanisms of both short- and long-distance polyamine transport in greater detail. The importance of these studies lies in the need to confirm the fact that polyamines are plant hormones. In fact, cell-to-cell and long-distance transport of plant hormones from zones of higher production rate (for example, meristems) can be regarded as a regular feature of their action. Polyamines are ubiquitous, have a high endogenous titre and have a cationic nature which facilitates their binding to membranes and cell walls; in addition, the scarce number of studies on their interaction with proteins and on their uptake as well as translocability favoured the interpretation of their role as that of 'second messengers' rather than that of growth regulators. The uptake studies were extended to an analysis of their compartmentation in order to identify the cellular localization of the absorbed molecules. The subcellular compartmentation of polyamines could overcome another difficulty in the classification of these substances as growth hormones, which is the presence of high concentrations of polyamines within cells with respect to other plant hormones.

Cultured carrot cells

The first investigation on the mechanism of polyamine uptake was carried out on petals of *Saintpaulia ionantha*. Uptake was linear with time for up to 2 h, concentration-dependence followed saturation kinetics and pH-dependence showed two maxima, one at pH 4–5 and another at pH 8 (Bagni and Pistocchi 1985; Pistocchi *et al.* 1986). A similar dependence on external pH was observed also in the lichen *Evernia prunastri* (Escribano and Legaz 1985).

Our studies were later extended to cells from carrot root phloem parenchyma cultured *in vitro*; this represents an active, rapidly dividing system which is more responsive to exogenously applied substances. Carrot cells displayed a more rapid uptake than *Saintpaulia* petals, with saturation after 1–2 min (Fig. 14.6). The uptake dependence on external polyamine concentration showed a biphasic pattern, with saturation kinetics in the concentration range of 1 μM to 5 mM (system I). K_m values in this system were 41.9 and 27.3 μM respectively for putrescine and spermidine. Higher polyamine concentrations of up to 100 mM revealed the existence of a second system which consisted of a saturable component for putrescine with a K_m value of 29.2 mM and a linear component for spermidine. Spermine uptake was characterized by a much lower affinity in both systems ($K_m = 7.7$ and 15.9 mM respectively), probably due to the lack of endogenous spermine in our carrot cell line and therefore to the lack of an activated transport system (Pistocchi *et al.* 1987).

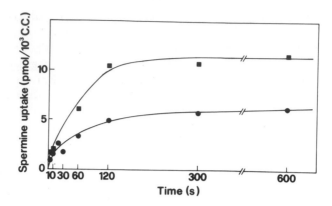

Fig. 14.6. Time course of spermine uptake in carrot cell cultures in the presence (■) or absence (●) of 1 mM Ca^{2+}.

The absorbed putrescine and spermidine were released from the cell in large amount (80 per cent) when the corresponding unlabelled molecules were present in the external solution at 2.5 mM concentration. This efflux therefore seemed to occur by means of an exchange mechanism between internal and external molecules. The same exchange occurred between putrescine and calcium, while spermidine exchange with this cation was limited. In contrast to the other two polyamines, spermine release from the cells did not exceed 20 per cent.

The localization of the absorbed polyamines, determined by differential centrifugation, revealed that most of the putrescine was in the cytoplasmic soluble fraction (Table 14.2), while spermidine and spermine were mostly bound to the cell wall (73 and 77.5 per cent respectively). This structure, by sequestering the supplied polyamines, therefore represents an obstacle to the study of transport across the plasmalemma. For this reason a comparative experiment with whole cells and isolated cell walls was performed. Spermine uptake in the latter did not increase with time from 10 s to 10 min in contrast to that in whole cells. At 10 s the uptake value was the same both for cell walls and cells, so that it may represent the external binding.

Experiments designed to investigate the driving force for polyamine uptake demonstrated that it was only partially (about 30 per cent) inhibited by metabolic inhibitors (Pistocchi *et al.* 1987).

The effect of calcium on polyamine uptake was investigated as this cation is considered necessary for transport phenomena. We found that putrescine uptake was stimulated at low calcium concentration (10 μM) and inhibited at higher levels; spermidine and spermine uptake was increasingly stimulated by Ca^{2+} in the 10 μM to 1 mM range. La^{3+} mimicked this stimulatory

Table 14.2 Distribution of supplied [^{14}C] polyamines in fractions of cultured carrot cells obtained by differential centrifugation

Fraction	pmol/10^3 cell clusters	%
Putrescine		
Cell walls	0.84	24.2
Particulate	0.24	6.9
Cytoplasmic	2.39	68.9
Total pmol taken up	3.47	
Spermidine		
Cell walls	11.0	73.0
Particulate	2.9	19.3
Cytoplasmic	1.2	7.7
Total pmol taken up	15.1	

effect but higher concentrations were strongly inhibitory even in the presence of calcium. Mg^{2+} at concentration ranging from 10 μM to 10 mM neither stimulated nor inhibited spermine uptake. The effect of Ca^{2+} was investigated in greater detail and we observed that the presence of this cation (1 mM) did not affect either the K_m or the V_{max} of the first transport system. Both Ca^{2+}/spermine co-transport and antiport were ruled out by the observation that spermine did not stimulate the uptake of Ca^{2+}, and that the efflux of the polyamine was not enhanced by the presence of the cation. Washing out the endogenous Ca^{2+} present in the cell walls with EGTA (0.5 mM) did not inhibit spermine uptake as expected; on the contrary it was enhanced, probably because the Ca^{2+} washed out by the treatment was replaced by the polyamine (Pistocchi and Bagni 1990). These results suggest therefore a non-specific effect of Ca^{2+} and the presence of interference with the cell walls.

Carrot protoplasts and vacuoles

Subsequent studies were carried out using protoplasts owing to the presence of interference with the cell wall as well as a large amount of polyamines bound to this fraction; the compartmentation analysis was later extended to vacuoles isolated from them. Protoplasts and vacuoles were isolated from phloem parenchyma of carrot tap-roots using the method described by Keller (1988). Owing to the presence of negative charges on its surface, the plasma membrane could also represent a binding site for polyamines. This was evidenced by the addition of a 100 mM solution of unlabelled polyamine during the filtration procedure of protoplasts from the incubation medium.

The solution was inserted between two layers of silicon oil through which the protoplasts had to pass during the centrifugation performed at the end of the incubation period. The presence of this 'washing layer' reduced the amount of polyamines absorbed by both protoplasts and vacuoles, thus confirming the occurrence of binding of the exogenous molecules to the membranes (plasmalemma and tonoplast). The percentage of bound molecules was higher when the incubation was performed with low external polyamine concentrations (Pistocchi *et al.* 1988).

The uptake of all three polyamines into protoplasts and vacuoles was very rapid, reaching a maximum after 1–2 min. It was increasingly enhanced by Ca^{2+} at concentrations up to 1 mM. Ca^{2+} also enhanced the uptake of spermine throughout the concentration range of 10 μM to 1 mM. Thus, in contrast to the data obtained with intact cells, it affected the V_{max}, which rose from 45 to 363 nmol/mg protein/30 s, as well as the K_m, which shifted from 292 to 122 μM. Spermine uptake in the second system (from 1 to 50 mM external spermine) displayed a linear pattern and the effect of Ca^{2+} was not investigated. La^{3+} also enhanced the uptake of spermine about fourfold.

Ca^{2+}-stimulated uptake was unaffected by the presence of the ionophore A23187 (2 μM) while it was inhibited by about 50 per cent in the presence of 10 μM carbonyl cyanide *p*-trifluoromethoxy-phenylhydrazone (FCCP). Thus Ca^{2+} seems to act on external membrane sites and to have a specific effect on transport activation. In fact La^{3+}, which does not enter the cell, also stimulated polyamine transport; however La^{3+}-stimulated transport and uptake in the absence of either Ca^{2+} or La^{3+} was not inhibited by FCCP (Pistocchi and Bagni 1990).

Polyamine uptake into vacuoles displayed a biphasic concentration dependence: a saturable component below 1 mM ($K_m = 61.8$ μM) was followed by a linear component up to 50 mM concentration. pH-Dependence in vacuoles showed a distinct optimum at pH 7.0 in contrast to that in protoplasts which increased linearly from pH 5.5 to 7.0.

Compartmentation analysis revealed that 24 and 28 per cent of the endogenous putrescine and spermidine, respectively, were vacuolar in location (Table 14.3). The compartmentation of the exogenously supplied molecules in the vacuoles was studied in two different ways:

1. Protoplasts were incubated with 100 mM unlabelled spermidine, and vacuoles were then isolated from them; results showed that the spermidine content of vacuoles increased sixfold with respect to the endogenous content.

2. Protoplasts were incubated with a lower concentration (6.6 μM) of [^{14}C] spermidine; after isolating the vacuoles, 27 per cent of it was found to be localized in these organelles.

Table 14.3 Endogenous polyamine content of carrot tap-root cells, protoplasts and vacuoles and subcellular localization of exogenously applied spermidine

Endogenous content in	Putrescine	Spermidine
	nmol/10^6 cells or protoplasts or vacuoles	
Cells	3.57 ± 0.54	10.67 ± 1.44
Protoplasts	3.23 ± 0.15 (100%)	3.74 ± 0.38 (100%)
Vacuoles	1.36 ± 0.23 (42%)	1.06 ± 0.16 (28%)
Exogenous content in protoplasts preloaded with 100 mM spermidine		
Protoplasts	2.98 ± 0.38 (100%)	73.70 ± 2.87 (100%)
Vacuoles	1.28 ± 0.33 (43%)	6.04 ± 0.10 (8%)
Exogenous content in protoplasts preloaded with 6.6 µM spermidine		
Protoplasts	n.d.[a]	0.41 ± 0.06 (100%)
Vacuoles	n.d.	0.11 ± 0.01 (27%)

[a] Not determined.

It should be pointed out that all the values reported here concerning vacuolar content could be affected by an observed leakage of polyamines during the isolation procedure. Anyway it appears clearly that the vacuole is likely to represent a storage site for polyamines (Pistocchi *et al.* 1988).

Helianthus tuberosus mitochondria

The study of polyamine transport and compartmentation was also extended to mitochondria isolated from tubers of *Helianthus tuberosus*. These organelles, as well as pine chloroplasts, were previously found to contain endogenous polyamines (Torrigiani and Serafini-Fracassini 1980; Torrigiani *et al.* 1986) and the polyamine biosynthetic enzymes (Torrigiani *et al.* 1986); thus the presence of endogenous polyamines could be attributed to a biosynthetic event. We have shown that the presence of polyamines in mitochondria can also be attributed to transport from the cytoplasm. In fact, isolated mitochondria have been shown to take up exogenous labelled spermidine very rapidly; this uptake is dependent on respiration, as antimycin A completely blocked it (Fig. 14.7). The existence of interactions between polyamines and membranes was evidenced by the presence of a consistent amount of label also in mitochondria incubated without the respiratory

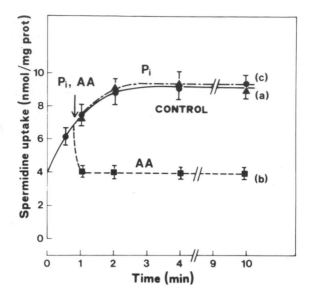

Fig. 14.7. Effect of antimycin A (b) or inorganic phosphate (c) on the time course (a) of spermidine uptake.

substrate or in the presence of respiratory chain inhibitors (Fig. 14.7, trace b) or uncouplers. This amount was therefore subtracted in further experiments. The concentration dependence showed saturation kinetics with an apparent K_m of 89 μM; Mg^{2+} (0.2–1 mM) and K^+ (20–100 mM) reduced the amount of spermidine absorbed while Ca^{2+} (20–100 μM) did not affect it. The driving force for the transport of spermidine was identified in the membrane potential ($\Delta\psi$) as FCCP (3 μM) and valinomycin (10 nM), which abolish the total electrochemical gradient and $\Delta\psi$, respectively, completely blocked it, while nigericin (100 nM) which suppresses ΔpH, only slightly affected spermidine accumulation (Fig. 14.8) (Pistocchi *et al.* 1990).

With regard to the physiological role of polyamine transport within the plant, long-distance transport studies have revealed that tomato and maize seedlings take up exogenous putrescine via roots and translocate it and its metabolites to the upper parts. The translocation is temperature- and relative humidity-dependent and occurs mainly through xylem vessels. Basipetal transport also occurs, but to a lesser extent (Rabiti *et al.* 1989). The existence of a physiological transport of polyamines was confirmed by the findings of Friedman *et al.* (1986) who reported the presence of endogenous polyamines in xylem and phloem exudates of various plant species. Christ *et al.* (1989) also observed that, after treatment of one sugar

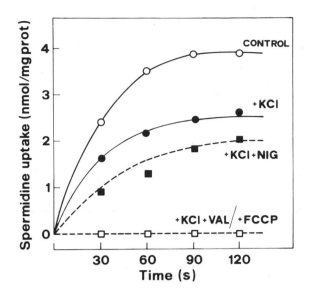

Fig. 14.8. Effect of K^+, valinomycin, nigericin, and FCCP on spermidine uptake.

beet cotyledon with labelled putrescine, small amounts of putrescine and acetylputrescine were found in other parts of the seedling. Massé *et al.* (1989) followed the uptake, translocation and metabolism of a synthetic analogue of spermine (*N,N'*-bis[3-aminopropylamino]-ethane) during the growth of potato and corn plantlets. This exogenous polyamine was found in roots and shoots in appreciable amounts, while a degradation product was detected in all parts of the plantlets. From the evidence reported above polyamine transport within plants can be regarded as part of their normal physiological behaviour.

Conclusion

Studies on polyamine compartmentation are partly limited by redistribution occurring during the experimental procedure for organelle isolation as well as by efflux phenomena, especially when particular methods or controls are not applied, and so it might be easier to study the compartmentation of the enzymes responsible for polyamine biosynthesis or degradation. At present it is established that *S*-adenosylmethionine decarboxylase is an enzyme located in the cytoplasmic soluble fraction (Torrigiani *et al.* 1986), while ODC and ADC are most probably particulate enzymes, distributed in nuclei, chloroplasts, and mitochondria (Panagiotidis *et al.* 1982; Torrigiani *et al.* 1986). The function and significance of these two enzymes is debatable.

The simplest hypothesis could be that they have a different function, and thus a different localization within cellular compartments. As for their function it was suggested that ODC is particularly involved in cell division while ADC is related to stress phenomena (pH, water and salt stress) and to cell extension (Bagni 1989) but no conclusive data have been obtained; in fact neither a preferential localization nor the existence of different isoenzymes has been found.

Additional help in understanding the significance of the localization of the enzymes involved in putrescine synthesis could derive from studies on compartmentation and metabolic regulation of the enzymes involved in ornithine and arginine biosynthesis in plant cells. The results thus far obtained suggest that arginine degradation to ornithine via arginase, occurring in mitochondria, appears to be distinctly separate from arginine synthesis occurring partly in the cytoplasm (Shargool *et al.* 1988). The mitochondrial and plastid membranes would therefore represent a barrier separating ornithine as a catabolic product from ornithine destined to arginine biosynthesis.

Apparently, the localization of polyamine oxidative enzymes is clearer. DAO and PAO occur in the cell walls and there are at least two possible reasons for this location:

1. The peroxide which is formed during the oxidative process may be utilized in lignification (Angelini and Federico 1986). However this would not explain the high activity also present in cotyledons and other non-lignified tissues.

2. These enzymes may regulate the transport of amines between the cells.

However, it is important to note that DAO and PAO activities are not only restricted to the Leguminosae and Gramineae, respectively, where their activity is abnormally high; either of the two enzymes may be present in many other plants, with an activity comparable to that of the biosynthetic enzymes; furthermore their localization is not only restricted to the cell walls because they are also present in protoplasts and mitochondria (Torrigiani, personal communication). This lends support to the hypothesis of regulation by the oxidative enzymes of the intracellular polyamine pool as well as of polyamine transport between cells. The data reported above on the localization of the enzymes catalysing polyamine breakdown cannot be generalized as extensive studies on the subcellular localization of these enzymes for putrescine breakdown, made in the yeast *Candida boidinii*, showed that acetylputrescine oxidase, a particular diamine oxidase, was localized in peroxisomes, while the acetylation of putrescine occurred in mitochondria (Fig. 14.9) (Gillyon *et al.* 1987). No information is available on the different types of di- or polyamine oxidases localized in fungal cell walls.

If we consider the compartmentation of endogenous polyamines in plants,

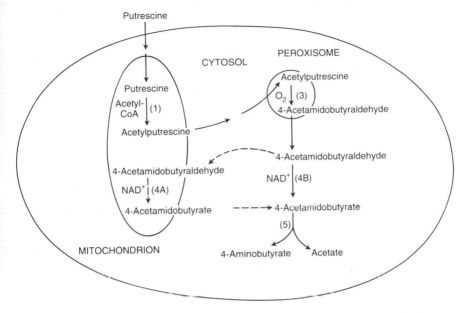

Fig. 14.9. Compartmentation of the enzymes involved in putrescine metabolism in *Candida boidinii*. Numbers denote the enzymic steps of Fig. 14.3. 4A and 4B are the two acetamidoaldehyde dehydrogenases (from Gillyon *et al.* 1987).

with the exclusion of fungi and algae, we see that spermidine and spermine are mostly localized in the cell walls (Goldberg and Perdrizet 1984), while putrescine is mostly present in the cytoplasmic soluble fraction, further identified as vacuolar sap (Pistocchi *et al.* 1988). This distribution confirms our results, obtained with the study of polyamine uptake and transport, concerning their preferential localization in cell walls and vacuoles (Pistocchi *et al.* 1987, 1988).

The presence of polyamines stored in the vacuole, their localization in other compartments such as mitochondria, their interaction with cell wall constituents, the existence of conjugated polyamines and of di- and polyamine oxidase activities suggest that all these events may participate in the regulation of the cytoplasmic levels of polyamines. In fact, despite their high internal concentration (in the mM range, with a maximum in plant tumour cells where it reaches 30 mM) there is requirement for exogenous polyamines (at concentrations ranging from 10 to 100 μM) for cell division to occur when endogenous polyamines are lacking or below 1 μM. Thus the presence of subcellular compartmentation, similar to that suggested for Ca^{2+}, and the rapid inter-conversion between free and conjugated forms, are a suitable explanation for the modulation of the plant response to polyamines.

References

Angelini, R. and Federico, R. (1986). In *Biochemical studies of natural polyamines* (ed. C.M. Caldarera, C. Cló and C. Guarnieri), p. 183. CLUEB, Bologna.

Bagni, N. (1989). In *The physiology of polyamines*, Vol. II (ed. U. Bachrach and Y.M. Heimer), p. 107. CRC Press, Boca Raton, Florida.

Bagni, N. and Pistocchi, R. (1985). Putrescine uptake in *Saintpaulia ionantha* petals. *Plant Physiology*, 77, 398–402.

Bagni, N., Baraldi, R., and Costa, G. (1984). Uptake, translocation and metabolism of aliphatic polyamines in leaves and fruitlets of *Malus domestica* (cv. 'Ruby spur'). *Acta Horticulturae*, 149, 173–8.

Bagni, N., Creus, J.A., and Pistocchi, R. (1986). Distribution of cadaverine and lysine decarboxylase activity in *Nicotiana glauca* plants. *Journal of Plant Physiology*, 125, 9–15.

Bagni, N., Adamo, P., Serafini-Fracassini, D., and Villanueva, V.R. (1981). RNA, proteins and polyamines during tube growth in germinating pollen. *Plant Physiology*, 68, 727–30.

Bertl, A., Felle, H., and Bentrup, F.-W. (1984). Amine transport in *Riccia fluitans*. Cytoplasmic and vacuolar pH recorded by a pH-sensitive micro-electrode. *Plant Physiology*, 76, 75–8.

Christ, M., Harr, J., and Felix, H. (1989). Transport of polyamines in sugar beet seedlings. *Zeitschrift für Naturforschung*, 44c, 59–63.

Crocomo, O.J. and Basso, L.C. (1974). Accumulation of putrescine and related amino acids in potassium deficient *Sesamum*. *Phytochemistry*, 13, 2659–605.

Escribano, M.I. and Legaz, M.E. (1985). Putrescine accumulation does not affect RNA metabolism in the lichen *Evernia prunastri*. *Endocytology Cell Research*, 2, 239–248.

Federico, R. and Angelini, R. (1988). Distribution of polyamines and their related catabolic enzymes in etiolated and light-grown leguminosae seedlings. *Planta,.*173, 317–21.

Friedman, R., Levin, N., and Altman, A. (1986) Presence and identification of polyamines in xylem and phloem exudates of plants. *Plant Physiology*, 82, 1154–7.

Galston, A.W. and Kaur-Sawhney, R. (1987). In *Plant hormones and their role in plant growth and development* (ed. P.J. Davies), p. 280. Martinus Nijhoff, Dordrecht.

Gillyon, C., Haywood, G.W., Large, P.J., Nellen, B., and Robertson, A. (1987). Putrescine breakdown in the yeast *Candida boidinii*: subcellular location of some of the enzymes involved and properties of two acetamidoaldehyde dehydrogenases. *Journal of General Microbiology*, 133, 2477–85.

Goldberg, R. and Perdrizet, E. (1984). Ratio of free and bound polyamines during maturation in mung-bean hypocotyl cells. *Planta*, 161, 531–5.

Jänne, J., Alhonen-Hongisto, L., Seppänen, P., and Höltta, E. (1981). In *Advances in polyamine research*, Vol. 3 (ed. C.M. Caldarera, V. Zappia, and U. Bachrach), p. 85. Raven Press, New York.

Joshi, S., Pleij, C.W.A., Haenni, A.-L., and Bosch, L. (1983). Age dependence of cowpea protoplasts for uptake of spermidine and infectibility by alfalfa mosaic virus. *Plant Molecular Biology*, **2**, 89–94.

Keller, F. (1988). A large-scale isolation of vacuoles from protoplasts of mature carrot taproots. *Journal of Plant Physiology*, **132**, 199–203.

Kleiner, D. (1981). The transport of NH_3 and NH_4^+ across biological membranes. *Biochimica et Biophysica Acta*, **639**, 41–52.

Leete, E. (1982). In *Secondary plant products. Encyclopedia of plant physiology NV*, Vol. 8 (ed. E.A. Bell and B.W. Charlwood), pp. 65–91. Springer-Verlag, Berlin.

Massé, J., Laberche, J.-C., and Jeanty, G. (1989). Translocation of polyamines in plants growing on media containing N,N'-bis(3-aminopropylamino)ethane. *Plant Physiology and Biochemistry*, **27**, 489–93.

Negrel, J. (1989). The biosynthesis of cinnamoylputrescines in callus tissue cultures of *Nicotiana tabacum*. *Phytochemistry*, **28**, 477–81.

Oshima, T. (1989). In *The physiology of polyamines*, Vol. II (ed. U. Bachrach and Y.M. Heimer), p. 35. CRC Press, Boca Raton, Florida.

Panagiotidis, C.A., Georgatsos, J.G., and Kyriakidis, A. (1982). Superinduction of cytosolic and chromatin-bound ornithine decarboxylase activities of germinating barley seeds by actinomycin D. *FEBS Letters*, **146**, 193–6.

Pistocchi, R. and Bagni, N. (1990). Effect of calcium on spermine uptake in carrot cell cultures and protoplasts. *Journal of Plant Physiology*, **136**, 728–33.

Pistocchi, R., Bagni, N., and Creus, J.A. (1986). Polyamine uptake, kinetics, and competition among polyamines and between polyamines and inorganic cations. *Plant Physiology*, **80**, 556–60.

Pistocchi, R., Bagni, N., and Creus, J.A. (1987). Polyamine uptake in carrot cell cultures. *Plant Physiology*, **84**, 374–80.

Pistocchi, R., Keller, F., Bagni, N., and Matile, Ph. (1988). Transport and subcellular localization of polyamines in carrot protoplasts and vacuoles. *Plant Physiology*, **87**, 514–18.

Pistocchi, R., Antognoni, F., Bagni, N., and Zannoni, D. (1990). Spermidine uptake by mitochondria of *Helianthus tuberosus*. *Plant Physiology*, **92**, 690–5.

Pohjanpelto, P. (1976). Putrescine transport is greatly increased in human fibroblasts initiated to proliferate. *Journal of Cell Biology*, **68**, 512–20.

Rabiti, A.L., Pistocchi, R., and Bagni, N. (1989). Putrescine uptake and translocation in higher plants. *Physiologia Plantarum*, **77**, 225–30.

Radhakrishanan, A.N. (1974). Some unusual amino compounds from sandal, *Santalum album* L. *Journal of Scientific Indian Research*, **33**, 461–6.

Seiler, N., Knodgen, B., Bink, G., Sarhan, S., and Bolkenius, F. (1983). In *Advances in polyamine research*, Vol. 4 (ed. U. Bachrach, A. Kaye, and R. Chayen), p. 135. Raven Press, New York.

Serafini-Fracassini, D., Del Duca, S., and Torrigiani, P. (1989). Polyamine conjugation during the cell cycle of *Helianthus tuberosus*: non-enzymatic and transglutaminase-like binding activity. *Plant Physiology and Biochemistry*, **27**, 659–68.

Shargool, P.D., Jain, J.C., and McKay, G. (1988). Ornithine biosynthesis and arginine biosynthesis and degradation in plant cells. *Phytochemistry*, **27**, 1571–4.

Smith, T.A. and Barker, J.H.A. (1988). In *Advances in polyamine research*, Vol. 4 (ed. U. Bachrach, A. Kaye, and R. Chayen), p. 573. Raven Press, New York.

Smith, T.A. and Marshall, J.H.A. (1988). The oxidative decarboxylation of ornithine by extracts of higher plants. *Phytochemistry*, **27**, 703–10.

Smith, T.A., Negrel, J., and Bird, C.R. (1983). In *Advances in polyamine research*, Vol.4 (ed. U. Bachrach, A. Kaye, and R. Chayen), p. 347. Raven Press, New York.

Speranza, A. and Bagni, N. (1978). Products of L-[^{14}C-carbamoyl] citrulline metabolism in *Helianthus tuberosus* activated tissue. *Zeitschrift fur Pflänzenphysiologie*, **88**, 163–8.

Tabor, C.W. and Tabor, H. (1966). Transport system for 1,4-diaminobutane, spermidine, and spermine in *Escherichia coli*. *Journal of Biological Chemistry*, **241**, 3714–23.

Torrigiani, P. and Serafini-Fracassini, D. (1980). Early DNA synthesis and polyamines in mitochondria from activated parenchyma of *Helianthus tuberosus*. *Zeitschrift fur Pflanzenphysiologie*, **97**, 353–359.

Torrigiani, P., Serafini-Fracassini, D., Biondi, S., and Bagni, N. (1986). Evidence for the subcellular localization of polyamines and their biosynthetic enzymes in plant cells. *Journal of Plant Physiology*, **124**, 23–9.

Torrigiani, P., Altamura, M.M., Pasqua, G., Monacelli, B., Serafini-Fracassini, D., and Bagni, N. (1987). Free and conjugated polyamines during de novo floral and vegetative bud formation in thin cell layers of tobacco. *Physiologia Plantarum*, **70**, 453–60.

Young, N.D. and Galston, A.W. (1983). Are polyamines transported in etiolated peas? *Plant Physiology*, **73**, 912–14.

15. The biology of the cyanogenic glycosides: new developments

ADOLF NAHRSTEDT

Institut für Pharmazeutische Biologie und Phytochemie der Westf. Wilhelms-Universität, Hittorfstr. 56, D-4400 Münster, FRG

Introduction

Several reviews of the cyanogenic compounds have been presented during the past years, particularly of the cyanogenic glycosides. Among them are reviews on their occurrence (Hegnauer 1986; Nahrstedt 1987), their function (Nahrstedt 1985; Hegnauer 1986), toxicity (Poulton 1983), biosynthesis (Conn 1988), and their occurrence and physiology in insects (Nahrstedt 1988). A comprehensive volume, *Cyanide compounds in biology*, was published in 1988 by the Ciba Foundation. All these activities indicate that the cyanogenic compounds are an important subject for biological research. This chapter aims to summarize recent progress in this area.

Metabolism

The term cyanogenesis is used for the production of hydrogen cyanide (HCN) under physiological conditions, excluding the so-called pseudo-cyanogenic glycosides (Seigler 1975). Figure 15.1 shows cyanogenic pathways that have been more or less established for higher plants. In small amounts (about 1 μM and less) cyanide is produced:

(1) during the formation of ethylene from 1-amino-cyclopropane-1-carboxylic acid (ACC) (Yang and Hoffmann 1984);

(2) by the action of horseradish peroxidase (HRP) on L-amino acids (Porter and Bright 1987); and

(3) from glyoxylate and hydroxylamine via an enzymatically catalysed reaction (Hucklesby *et al.* 1982).

The last pathway may also be used by some micro-algae which also produce cyanide from D-amino acids (Vennesland *et al.* 1981) (not shown in Fig. 15.1).

Another pathway is able to produce large amounts of cyanide, up to approximately 1 per cent in plants (Poulton 1983) and 0.2 per cent in insects (Davis and Nahrstedt 1985). It too originates from amino acids such as valine, isoleucine, leucine, phenylalanine, and tyrosine as well as the

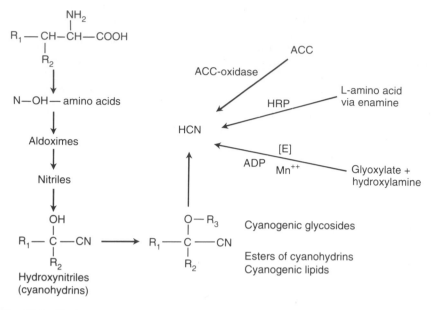

Fig. 15.1. Hydrogen cyanide (HCN) in living organisms generally arises from the hydrolysis of cyanogenic glycosides biogenetically derived from amino acids (left column); in minor quantities, HCN is formed during the biosynthesis of ethylene from 1-amino-cyclopropane-1-carboxylic acid (ACC), by the action of horseradish peroxidase (HRP) on amino acids, and by enzymatic oxidation of glyoxylate and hydroxyl amine in the presence of manganese ions.

non-proteinogenic amino acids cyclopentenylglycine and, probably, nicotinic acid (Nahrstedt *et al.* 1982*a*) as precursors. Products accumulated are cyanohydrins, which occur freely in some plants and arthropods, or which (usually) are stabilized either by glycosylation to form the cyanogenic glycosides, or (more seldom) by esterification as in mandelonitrile benzoate found in some millipedes (Duffey 1981); esterification with fatty acids leads to the special group of the cyanolipids that occur exclusively in the seeds of many sapindaceous plants (Mikolajczak 1977). This pathway has been thoroughly investigated by Professor E. E. Conn using mainly his laboratory plant *Sorghum bicolor* Moench (Poaceae), and was found to be bound to the microsomal fraction as a highly channeled system (Conn 1983). Halkier and Moller [1989] have recently purified the dhurrin-synthesizing system from etiolated seedlings of *S. bicolor* on a Sephacryl S-1000 column and succeeded in a tenfold increase of specific activity. Further purification using different methods was not successful as the activity was reduced drastically, probably by dissociation of the essential components from the enzyme system. (See *note added in proof*, p. 269.)

An open question was the identity of the precursor of mandelonitrile glycosides bearing a *m*-hydroxyl on their aromatic ring such as holocalin (Fig. 15.10) or xeranthin (Fig. 15.11). Schütte (1973) has argued that m-tyrosine might be the precursor, while others suggest tyrosine (van Valen 1978) or phenylalanine (Nahrstedt 1976). The fruits of *Xeranthemum cylindraceum* contain both *m*-hydroxylated and non-hydroxylated mandelonitrile glycosides (Schwind *et al.* 1990) (see below). Feeding the ripening inflorescences with 0.5 mM solutions of L-tyr, D,L-*meta*-tyr, and L-phe affected the concentration of the cyanogenic glucoside zierin (2-β-D-glucopyranosyloxy-2S-2-(3-hydroxy) phenyl acetonitrile) at a rate of zero, 46 per cent decrease and 28 per cent increase respectively. L-[U-^{14}C]phe was incorporated into zierin at a rate of approximately 0.5 per cent, whereas L-[U-^{14}C]tyr was not. When glyphosate, an inhibitor of enolpyruvylshikimate phosphate synthase (Amrhein 1986) was fed (0.6 mM), a decrease of approximately 50 per cent zierin was observed that was clearly antagonized by 2 mM L-phe but not by L-tyr or D,L-*m*-tyr (Fig. 15.2). These results show that phenylalanine is the precursor and that hydroxylation occurs at a later step during biosynthesis of the 3-hydroxymandelonitrile glycosides, at least in *X. cylindraceum* (Schwind 1990).

The cyanogenic glycosides are readily hydrolysed by more or less specific

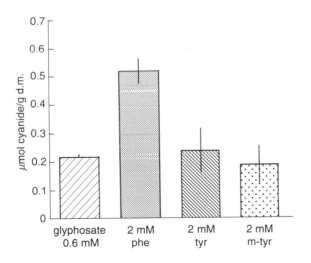

Fig. 15.2. The inflorescences of the asteracean plant *Xeranthemum xylindraceum* show a decreased accumulation of the meta-hydroxylated aromatic cyanogenic glucoside zierin at an amount of *c*.0.2 μmol g^{-1}d.m. when incubated in 0.6 mM glyphosate[R]. This effect is antagonized by phenylalanine (phe) but not by tyrosine (tyr) and *meta*-tyrosine (m-tyr) indicating that the non-hydroxylated amino acid phenylalanine is the precursor.

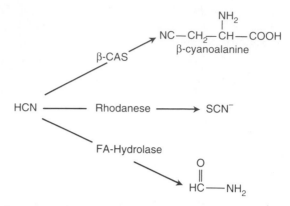

Fig. 15.3. Hydrogen cyanide (HCN) is metabolized (detoxified) by living organisms (i) to the non-proteinogenic amino acid β-cyanoalanine, catalysed by the enzyme β-cyanoalanine synthase (β-CAS); (ii) to rhodanide, catalysed by rhodanese; (iii) to formamide, catalysed by formamide (FA)-hydrolase.

β-glucosidases (Hösel and Conn 1982; Poulton 1988) to give sugars and cyanohydrins (for example Fig. 15.14). The latter decompose spontaneously depending on pH, and/or catalysed by hydroxynitrile lyases which accelerate the HCN production up to 20-fold under normal physiological conditions, as has recently been shown for the *Hevea* hydroxynitrile lyase (Selmar *et al.* 1989*b*). Thus, a hydroxynitrile lyase that is usually present in cyanogenic plants effectively enhances, and thereby optimizes, the cyanogenic system indicating a biological significance of a fast and concentrated HCN production.

Living organisms can usually handle HCN to a certain extent in that they are able to metabolize it to less toxic products (Fig. 15.3). The main pathway for plants and insects seems to be the transfer of cyanide to serine or cysteine to form β-cyanoalanine (Miller and Conn 1980; Witthohn and Naumann 1984; Nahrstedt 1988). The ubiquitous occurrence of β-cyanoalanine synthase (β-CAS) even in plants which do not accumulate cyanogenic glycosides is obviously connected with the metabolism of cyanide evolving from the production of ethylene (Manning 1986; Yip and Yang 1988). The main pathway in higher animals, but also observed in plants and insects, is the transfer of sulphur from thiosulphate to cyanide to form rhodanide (Chew 1973; Miller and Conn 1980; Frankenberg 1983; Long and Brattsten 1984; Beesley *et al.* 1985). The hydrolysis to formamide and further products by formamide hydrolase is questionable in higher plants (Miller and Conn 1980), but is realized in fungi (Knowles 1988).

In order to avoid mistakes occurring in the literature, it should be empha-

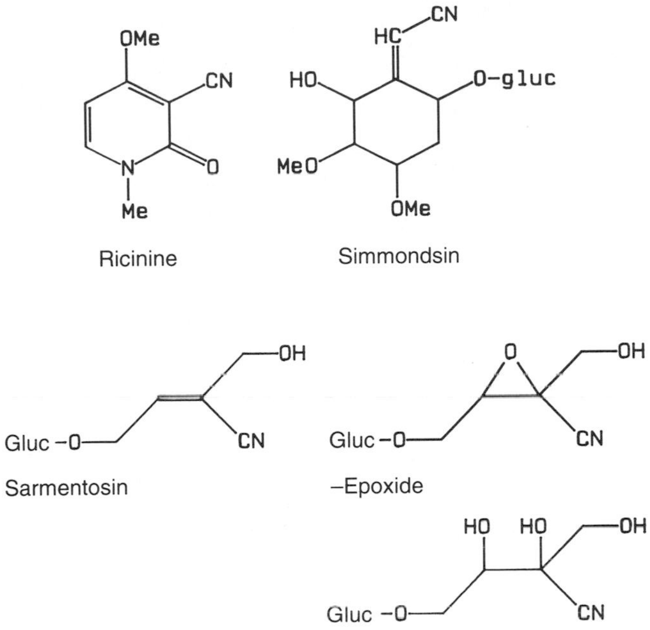

Fig. 15.4. Ricinine, simmondsine, and sarmentosin are examples of non-cyanogenic nitriles. The epoxide of sarmentosin, however, is hydrolysed to give a vicinal diol (α-hydroxynitrile) that decomposes to give HCN.

sized that some compounds bearing a nitrile group are not cyanogenic; for example, ricinine of the 3-cyanopyridone series, simmondsin of the cyclohexylcyanomethylene glucosides, or sarmentosin (Fig. 15.4). Sarmentosin-epoxide, however, is cyanogenic upon hydrolysis of the oxiran group due to the formation of a cyanogenic cyanohydrin (Fig. 15.4) (Nahrstedt *et al.* 1982*b*).

New structures

Regarding structural development, clarifying work has been done by Jaroszewski and co-workers on the somewhat confusing situation (Nahrstedt 1987) in the series of cyclopentenyl cyanohydrin glycosides typical of families of the order Passiflorales (Fig. 15.5). Thus the correct stereochemistry of the monohydroxylated (Jaroszewski and Jensen 1985), the dihydroxylated (Olafsdottir *et al.* 1989*a*), and the trihydroxylated (Kim *et al.* 1970) cyclopentenyl cyanohydrins is now known. Barterin (Spencer and Seigler 1984) was found to be identical with tetraphyllin B (Olafsdottir *et al.*

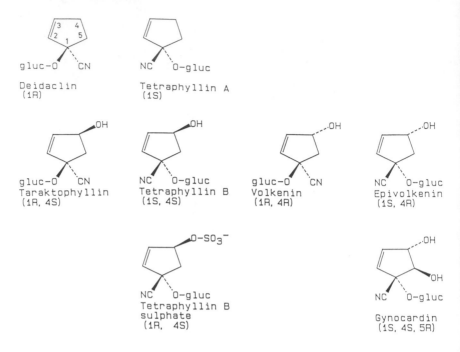

Fig. 15.5. Several cyclopentenyl cyanohydrin glycosides have been structurally established concerning their sterochemistry at C_1, C_4, and C_5 of the aglykone.

1989*a*). The configuration at C-1 and C-4 of tetraphyllin B sulphate was established (Jaroszewski and Fog 1989). New bidesmosides of this series isolated from *Passiflora capsularis* L. and *P. biflora* Lam. are derivatives of epivolkenin (Fig. 15.6), the second sugar of passicapsin (Fischer *et al.* 1982; Olafsdottir *et al.* 1989*b*) is boivinose, a member of the 2,6-dideoxyhexoses which are usually associated with pregnane and cardiac glycosides; passibiflorin (Olafsdottir *et al.* 1989*b*) carries the extremely rare antiarose, a 6-deoxyhexose.

A dihydromandelonitrile glucoside was recently claimed to be isolated from the fruits of *Ilex aquifolium* L. (Aquifoliaceae) (Willems 1988) (Fig. 15.7). The structure would have been of biogenetic interest as it can be considered to be derived from tyrosine, and, probably, a precursor of the cyclohexylcyanomethylene glucosides. However, a careful spectroscopic investigation has led to the revision of the proposed structure now identified as menisdaurin (Nahrstedt and Wray 1990) (Fig. 15.7).

The herb and the seeds of *Merremia dissecta* (Jacq.) Hallier (Convolvulaceae) are cyanogenic (Hegnauer 1989). The cyanogenic compounds of the leaves have been established to be 6'-malonyl prunasin as well as

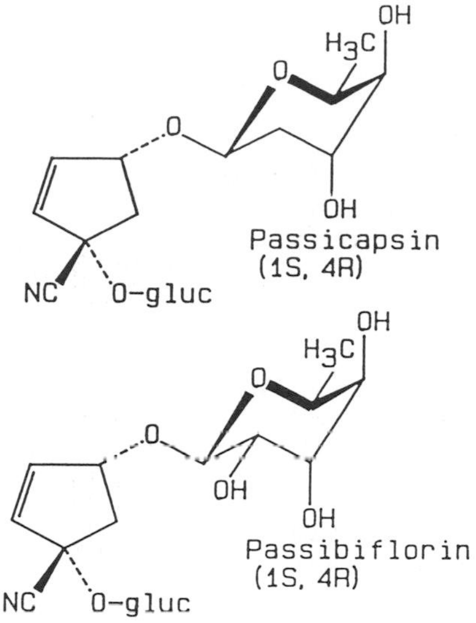

Fig. 15.6. Two bidesmosides of the cyclopentenyl cyanohydrin glycosides show unusual 6-deoxy sugars; boivinose in the case of passicapsin, and antiarose in the case of passibiflorin.

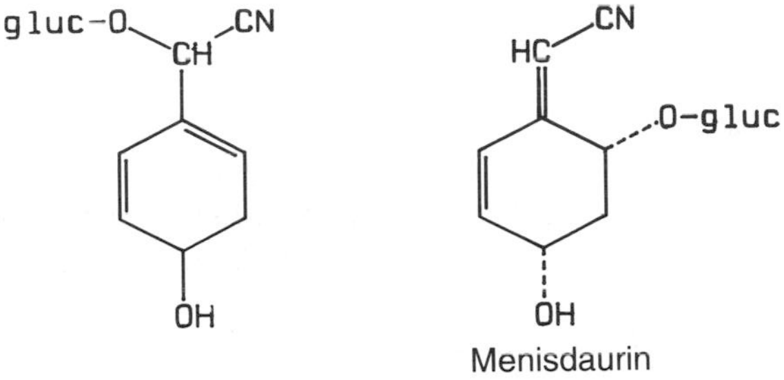

Fig. 15.7. The fruits of *Ilex aquifolium* (Aquifoliaceae) contain not cyanogenic dihydromandelonitrile glucoside (left), but the isomeric non-cyanogenic menisdaurin (right).

major amounts of prunasin (Fig. 15.10) (Nahrstedt *et al*. 1989), whereas the seeds contain as well as amygdalin its 6″-(4-hydroxy) benzoate and its 6″-(4-hydroxy)-E-cinnamate (Nahrstedt *et al*. 1990) (Fig. 15.8). Oxyanthin and its 5″-benzoate resemble new phenylalanine-derived cyanogenic glycosides obtained from *Oxyanthus* and *Psydrax* (*Canthium*) species of the

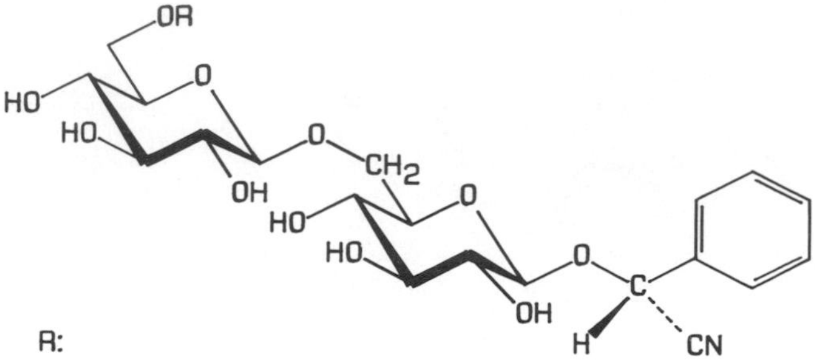

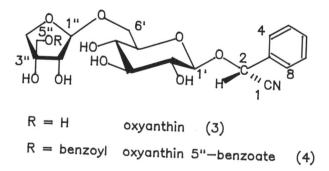

R:

H	amygdalin
4-OH-benzoyl	6''-p-hydroxybenzoyl amygdalin
4-OH-E-cinnamoyl	6''-p-coumaroyl amygdalin

Fig. 15.8. Novel acylated amygdalin derivatives obtained from the seeds of *Merremia dissecta* (Convolvulaceae).

Rubiaceae (Rockenbach *et al.* 1992) (Fig. 15.9); they are accompanied by prunasin, amygdalin and/or holocalin. Another new prunasin acyl derivative, grayanine, was isolated from the bark of *Prunus grayana* Maxim. (Shimomura *et al.* 1987), and *P. buergeriana* Maxim. (Shimomura and Sashida 1988) (Rosaceae) as its 6′-E-caffeoyl ester (Fig. 15.10). The achenes of *Xeranthemum cylindraceum* Sibth. et Sm. (Asteraceae) contain a complex set of mandelonitrile glycosides and their meta-hydroxylated derivatives, among them sambunigrin, zierin, epilucumin, and zierinxyloside. The first trisaccharide of the cyanogenic glycosides is xeranthin (Fig. 15.11) obtained in a yield of 0.006 per cent from the achenes (Schwind *et al.* 1990). This compound resembles the so-called anthemisglycoside A (Fig. 15.11) that

R = H   oxyanthin   (3)

R = benzoyl   oxyanthin 5''—benzoate   (4)

Fig. 15.9. Novel cyanogenic diglycosides bearing an apiose moiety obtained from members of the Rubiaceae.

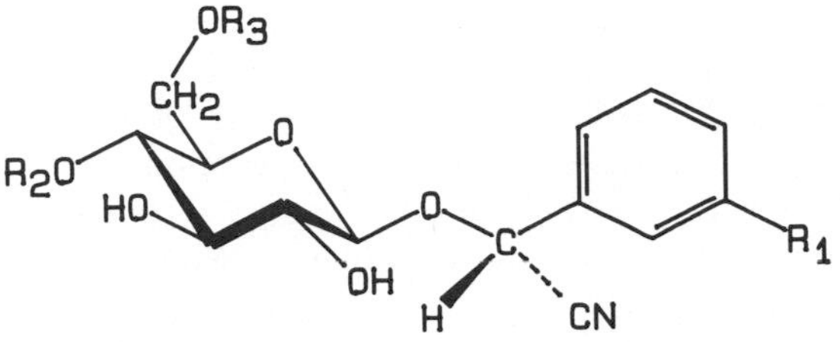

R_1	R_2	R_3	
H	H	H	prunasin
H	H	malonyl	6'-malonyl prunasin
H	H	caffeoyl	grayanin
OH	H	H	holocalin

Fig. 15.10. 6'-malonyl and 6'-caffeoyl derivatives of prunasin have been isolated from the leaves of *Merremia dissecta* and the bark of *Prunus grayana* respectively.

was isolated from the achenes of the asteraceous plant *Anthemis cairica* (Nahrstedt *et al.* 1983). Xeranthin and oxyanthin are the first examples of the cyanogenic glycosides containing an apiose in the sugar side chain; however there is still no exception to the rule that glucose is the first sugar attached to the cyanohydrin-OH in the entire set of known cyanogenic glycosides (approximately 60).

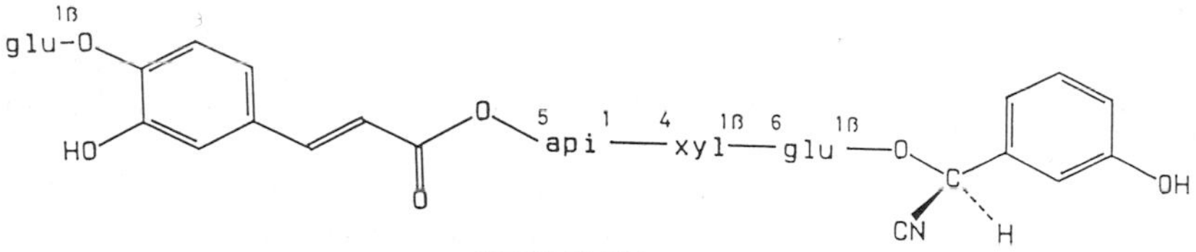

XERANTHIN

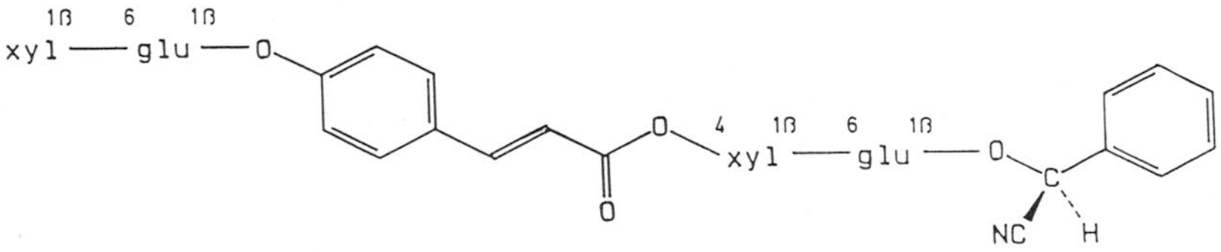

ANTHEMISGLYCOSIDE A

Fig. 15.11. Xeranthin is an acylated cyanogenic trisaccharide obtained from the fruits of *Xeranthemum cylindraceum* (Asteraceae). It shows structural similarities to anthemisglycoside A, recently isolated from the akenes of *Anthemis cairica* (Asteraceae).

Function

Although Hruska (1988) has criticised many classical papers concerning the role of cyanogenic glycosides as defensive compounds, there is good evidence from many sources that these compounds are in fact part of the chemical defence system generally used by living organisms (Nahrstedt 1985, 1988). However, working with biological systems one should be careful in generalizing on this point. Recent results give some examples.

The rate of release of HCN rather than the concentration of cyanogenic compounds is important for repelling aphids, as was indicated recently for seedlings of *Sorghum bicolor* (Hsieh 1989). This points to the need for an effective cyanogenic system consisting of the cyanogenic glycosides, an active β-glucosidase and a hydroxynitrile lyase; organisms that have optimized this system almost certainly use it as a defence mechanism (Nahrstedt 1985). However, the active principle need not necessarily be HCN: predatory ants (*Myrmica ameriana*) are clearly more repelled by benzaldehyde than by HCN, both of which are formed from mandelonitrile that is secreted by some polydesmid millipedes upon attack of predators (Peterson 1986).

Cyanogenesis might also influence the separation of species into different environments as argued by Reilly and co-workers (1987). The peach tree borer (PTB, *Synanthedon exitiosa* Say) feeds on the healthy cambium and inner bark of *Prunus persica* (L.) Batsch (Rosaceae) with relatively high concentrations of prunasin just below the soil level; the lesser peach tree borer (LPTB, *S. pictipes* Grote and Robinson) attacks the trunk and branches at injured areas where prunasin concentration is much lower. Both insects possess a β-glucosidase able to hydrolyse prunasin, but the PTB only contains a β-CAS to detoxify HCN. Furthermore, the PTB gained weight with increasing amounts of amygdalin in artificial diet, whereas the LPTB did not (Reilly *et al.* 1987). Thus each species has settled an ecological niche dependent on its sensitivity to the cyanogenic glucoside prunasin, and one of them, the PTB, has even adapted to prunasin in its food.

Regarding fungal infection, high cyanogenesis correlates with a high susceptibility of *Hevea brasiliensis* Muell.-Arg. (Euphorbiaceae) to its pest *Microcyclus ulei* (Lieberei 1988). In a recent paper Lieberei *et al.* (1989) have presented evidence that HCN, produced during the fungal attack, supresses the production of scopoletin that acts as a phytoalexin for the rubber plant (Giesemann *et al.* 1986). Thus HCN in this case represents a susceptibility factor in the plant during its competition with a fungal parasite. Very recently (Pourmohseni 1989), it has been shown that the cyanogenic glucoside epiheterodendrin (Erb *et al.* 1979), and further structurally related (non-cyanogenic) nitriles (Fig. 15.12) are responsible for the susceptibility of *Hordeum vulgare* L. to its pest *Erysiphe graminis* DC f. sp. *hordei*. These compounds account for 90 per cent of the glycosidic fraction in the epidermis

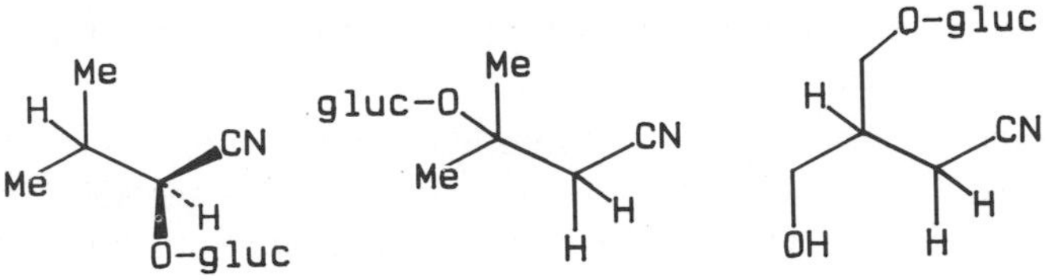

epiheterodendrin epidermin nn

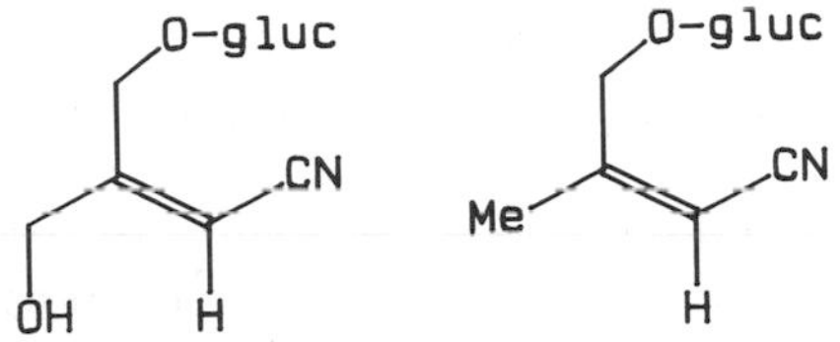

sutherlandin epihordenin

Fig. 15.12. The cyanogenic glucoside epiheterodendrin and further non-cyanogenic nitriles (epidermin, sutherlandin, epihordenin, and a no-name compound nn) are located in the epidermal cells of barley leaves. They are responsible for the susceptibility of certain *H. vulgare* cv. to its pest *Erysiphe graminis*. Incompatible cultivars contain much less of these compounds in the epidermis.

cells of the compatible cultivars, whereas incompatible ones contain much less (Pourmohseni 1989). These results, and earlier observations from flax (Lüdtke and Hahn 1953), show that cyanogenic glycosides probably provide no chemical defence against adapted fungi. Of course, such plants are not defenceless: for example *Hordeum* species contain the fungitoxic hordatines A and B, regarded to be effective natural anti-fungal compounds (Gross 1989).

Formerly plant physiologists favoured the hypothesis that cyanogenic glycosides contribute to the formation of proteins using the nitrogen of HCN (for example, Paech 1950). The nitrile group arising from cyanogenic glycosides, ethylene biosynthesis or fed directly as HCN is indeed incorporated into asparagine via β-cyanoalanine (Tschiersch 1964; Blumenthal *et al.* 1968; Dziewanowska and Lewak 1982; Peiser *et al.* 1984). Starting from the observation that the cyanogenic glucoside linamarin disappears from the endosperm of the seeds of *Hevea brasiliensis* during germination without HCN release, Selmar *et al.* have formulated the hypothesis of the 'linustatin pathway' (Selmar and Lieberei 1988) that is favoured by several experimental results:

1. A β-glucosidase capable of hydrolysing linamarin (linamarase) but not

the corresponding diglucosides, is apoplastic in the endosperm of *Hevea* seeds (Selmar *et al*. 1987*a*, 1989*a*).

2. Linamarin (2-β-D-glucosyloxy-2-methylpropionitrile) is obviously glucosylated in the seeds to give linustatin (2-β-D-gentiobiosyloxy-2-methylpropionitrile) that is found in the exudation liquor of the endosperm (Selmar *et al*. 1987*b*) thus passing the apoplastic space.

3. Expanding leaves contain a diglucosidase (linustatinase) activity and β-CAS which may introduce the cyanide nitrogen into protein metabolism (Selmar and Lieberei 1988).

The enzyme activities in the leaves clearly correspond in time with the exudation phase of linustatin from the seeds, thus strongly indicating that linustatin functions as a transport form for linamarin protected against apoplastic linamarase and used, probably, for protein synthesis in the growing leaves.

A similar situation may exist with the cyanogenic lipids in *Ungnadia speciosa* (Sapindaceae) (Selmar *et al*. 1990). During germination of the seeds a drastic decrease of the lipids (type I) is observed without releasing HCN. The plantlets gain a much smaller amount (approximately 25 per cent) of the corresponding glucoside proacacipetalin (Fig. 15.13). Selmar *et al*. conclude that the cyanolipids serve as storage products for reduced nitrogen.

The latter example may well indicate the multi-functional properties of secondary plant constituents. Cyanolipids are known to be toxic to insects (Janzen *et al*. 1977; Mikolajczak *et al*. 1984), and thus may protect the seeds as long as they are able to germinate. During germination the stored nitrogen, as well as the fatty acids, may be used for *de novo* syntheses. Both models, *Hevea* and *Ungnadia*, need further investigation but they underline the ideas of former generations of plant physiologists and prove that the cyanogenic glycosides have both primary and secondary functions as shown, for example, for the pyrrolizidine alkaloids (Boppré 1990).

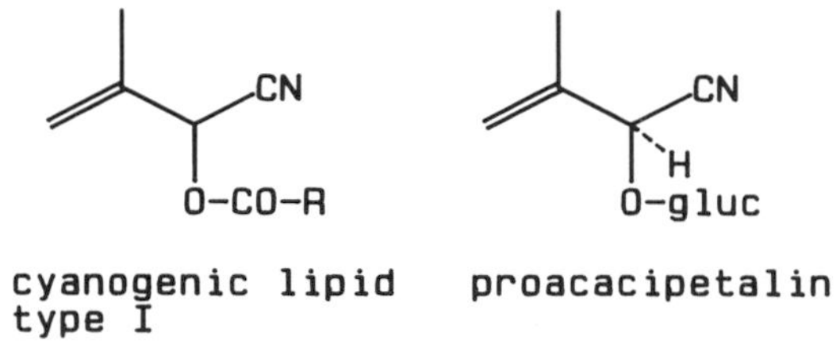

cyanogenic lipid   proacacipetalin
type I

Fig. 15.13. The type 1 cyanogenic lipids ($-CO-R$ = unsaturated fatty acids of 20 and 22 carbons chain length) in seeds of some Sapindaceae may serve as storage products for nitrogen, and are probably precursors of cyanogenic glucosides such as proacacipetalin during germination.

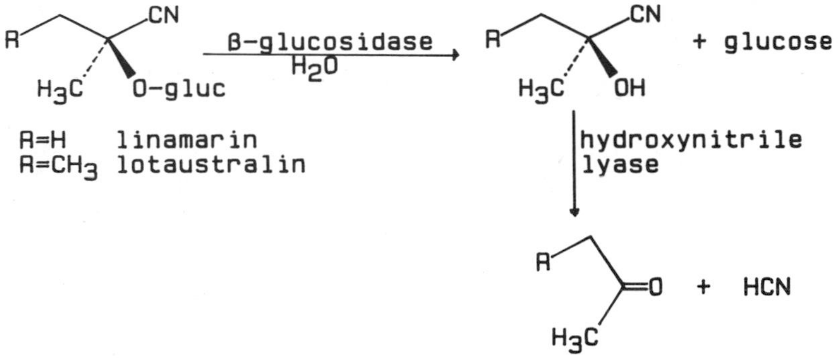

Fig. 15.14. Species of the moth family Zygaenidae contain the cyanogenic glucosides linamarin and lotaustralin which are hydrolysed by a specific β-glucosidase to give the corresponding cyanohydrins; the latter decompose catalysed by a hydroxynitrile lyase to give a carbonyl (R = H: acetone, R = CH₃: methylethyl ketone) and hydrogen cyanide.

The cyanogenic system in Lepidoptera

Cyanogenesis also occurs in the arthropods. Chilopoda and Diplopoda usually produce mandelonitrile and/or its derivatives (Duffey 1981), while several species of butterflies and moths (Insecta, Lepidoptera) contain the cyanogenic glucosides linamarin and lotaustralin (Fig. 15.14) (Davis and Nahrstedt 1985; Nahrstedt 1988). Most subfamilies of the moth family Zygaenidae are characterized by cyanogenic species. Experiments mainly with *Zygaena trifolii* Esper have shown that these insects obtain linamarin and lotaustralin by *de novo* synthesis and by uptake from their food plant *Lotus corniculatus* (Fabaceae), indicating a hitherto unique situation of the relationship between plants and insects (Nahrstedt 1988, 1989). The biosynthetic pattern in *Zygaena*, as far as investigated, resembles that of the plants, using the same amino acids as precursors (valine, isoleucine) via the corresponding nitriles (Fig. 15.15) (for summary see Nahrstedt 1988). First results obtained on the localization of the biosynthesis indicated that the glucosides are produced by an organ within the haemolymph (Franzl *et al.* 1988). Meanwhile one step of the biosynthetic sequence could be localized in the fat body of the larvae of *Z. trifolii* (Table 15.1). When incorporating the isolated fat body with valine and isoleucine, only 1/20 of the amount of cyanide (released from the cyanohydrins produced) was obtained in comparison to 140–150 nmol when incubated with the corresponding nitriles, 2-methylpropane nitrile and 2-methylbutane nitrile. When homogenized the activity was found in the major particles (15 000 g pellet) of the fat body; the

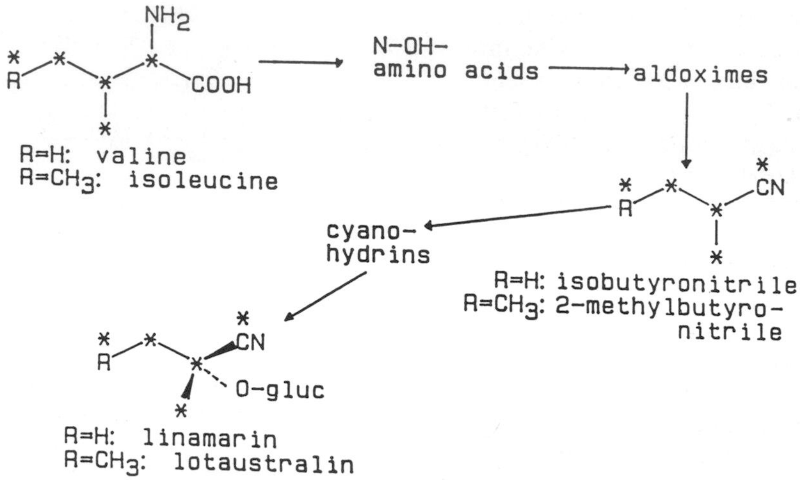

Fig. 15.15. The biosynthetic pathway leading to linamarin and lotaustralin seems to be very similar in plants and lepidopterans. Valine and isoleucine have been established as precursors, isobutyronitrile and 2-methylbutyronitrile as intermediates in larvae of *Zygaena* sp. Carbons of the precursors marked (*) appear in the corresponding position (*) in the glucosides.

latter were significantly (five times) more active when NADPH was added to the incubation mixture (Franzl *et al.*, in prep). Thus, at least for this step, similarities exist to the plant system (Conn 1983) which is membrane bound and needs NADPH.

Similarities also exist in the enzymes involved in cyanogenesis (Fig. 15.14) of all stages of *Zygaena* specimens when injured. From the haemolymph a β-glucosidase was isolated and characterized that fairly specifically cleaves the natural substrates linamarin and lotaustralin with preference for mono-glucosides and the β-configuration of the glycosidic linkage (Table 15.2) (Franzl *et al.* 1989). The dimeric enzyme shares some properties with the plant glucosidases (Poulton 1988) such as the acidic pH optimum (4–4.5) and the molecular weight of the subunit (66 kDa), but in contrast exhibits a temperature optimum (40 °C) below that of plant glucosidases hydrolysing cyanogenic glycosides (>50 °C), and a fairly narrow substrate specificity when compared, for example, to the unspecific linamarase of *Trifolium repens* (Fabaceae) that has been intensively studied recently (Pocsi *et al.* 1989).

A hydroxynitrile lyase (HNL) was also isolated from the haemolymph of the larvae of *Z. trifolii* with a molecular weight of 145 kDa consisting of two

Table 15.1 The isolated intact fat body of larvae of *Zygaena trifolii* produces hydrogen cyanide (arising from non glucosylated cyanohydrins) when incubated with isobutyronitrile (= 2-methylpropanenitrile) and 2-methylbutyronitrile (= 2-methylbutanenitrile); the activity is located with the larger particles (15 000 g pellet) of the homogenized fat body and needs NADPH as a cofactor

Incubation mixture	CN^-/larva and 17 h [nmol]
Fat body	2
Fat body + val/ileu (4 mM)	7
Fat body + 2-Mc-propanenitrile	149
Fat body + 2-Me-butanenitrile	139
Homogeneneous fat body 15 000 g pellet + 2-Me-propanenitrile	15
Homogeneneous fat body 15 000 g pellet + 2-Me-propanenitrile + NADPH	80

Table 15.2 The endogenous glucosides linamarin and lotaustralin are the preferred substrates of the β-glucosidase isolated from the haemolymph of *Zygaena trifolii* when compared to other cyanogenic glycosides, salicin, and storage (trehalose) or food (cellobiose) carbohydrates. The *p*-nitrophenolglycosides (4-NP-glycosides) indicate a specificity for the β-configuration of the glycosidic linkage and for glucose as the sugar moiety

Substrate	K_m mmol/l^{-1}	V_{max} U/mg	Relative activity %
Linamarin	7.8	71.8	50
Lotaustralin	2.5	65.0	105
Prunasin	0.02	1.2	7.1
Acalyphin			≪1
Linustatin			0
Amygdalin			≪1
4-NP-β-glucoside	0.27	20.9	100
4-NP-α-glucoside			≪1
4-NP-β-galactoside			0
Salicin			≪1
Trehalose			0
Cellobiose			0

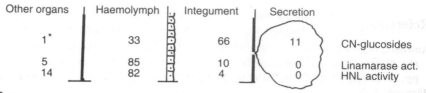

Other organs	Haemolymph	Integument	Secretion	
1*	33	66	11	CN-glucosides
5	85	10	0	Linamarase act.
14	82	4	0	HNL activity

*all measures are approximate percentages

Fig. 15.16. Linamarase activity and hydroxynitrile lyase (HNL) activity are mainly contained in the haemolymph, but to a small extend also in the integument of larvae of *Zygaena trifolii*; the substrates linamarin and lotaustralin, however, account for 1/3 in the haemolymph and 2/3 in the integument. Upon irritation, the larvae produce a secretion that contains the glucosides but no activity of catabolizing enzymes.

dimers of 71 kDa (Müller and Nahrstedt, in prep). The preferred substrates are acetone cyanohydrin (V_{max} 22 μmol/min $\times$ mg protein) and methylethylketone cyanohydrin (8 μmol) resembling the kinetic properties of the HNL partially purified from *Hevea brasiliensis* (Selmar *et al.* 1989*b*). The enzyme is a flavoprotein, similar to some, but not all, hydroxynitrile lyases from plants (Bärwald and Jänicke 1978; Poulton 1988). Interestingly the activities of the β-glucosidase and the HNL correspond to storage in the larval body (Fig. 15.16) with 80–90% in the haemolymph and 4 and 10% respectively in the integument whereas no activity is found in the sticky secretion of the larvae secreted upon irritation, although all three compartments contain cyanogenic glucosides.

Thus the larvae of *Z. trifolii* have disposed of enzymes which effectively liberate HCN from their endogenous substrates linamarin and lotaustralin. It is still an open question how this cyanogenic system, entirely (substrates and enzymes) located in the haemolymph, is compartmentalized. However, this optimized system surely has an important defence function for these insects (Nahrstedt 1988).

Acknowledgement

Thanks are due to co-workers and students engaged in the recent work on cyanogenic compounds in plants and insects: Dr Sylvia Franzl, Elisabeth Müller, Dr Philipp Schwind, Jürgen Rockenbach, and Dr Victor Wray. The technical assistance of Friedel Schratz and Bettina Quandt and the linguistic advice of Victor Wray are gratefully acknowledged.

References

Amrhein, N. (1986). The shikimic acid pathway. In *Specific inhibitors as probes into the biosynthesis and metabolism of aromatic amino acids* (ed. E.E. Conn), pp. 83–117. Plenum Press, New York.

Bärwald, K.-R. and Jaenicke, L. (1978). *D*-Hydroxnitrile lyase: Involvement of the prosthetic flavine adenine dinucleotide in enzyme activity. *FEBS Letters*, **90**, 255.

Beesley, S.G., Compton, S., and Jones, D.A. (1985). Rhodanese in insects. *Journal of Chemical Ecology*, **11**, 45–50.

Blumenthal, S.G., Hendrickson, H.R., Abrol, Y.P., and Conn, E.E. (1968). Cyanide metabolism of higher plants. III. The Biosynthesis of β-cyanoalanine. *Journal of Biological Chemistry*, **243**, 5302–7.

Boppre, M. (1990). Lepidoptera and pyrrolizidine alkaloids. Exemplification of complexity in chemical ecology. *Journal of Chemical Ecology*, **16**, 165–85.

Chew, M.Y. (1973). Rhodanese in higher plants. *Phytochemistry*, **12**, 2365.

CIBA Foundation (1988). *Cyanide compounds in biology*. John Wiley and Sons, Chichester.

Conn, E.E. (1983). Cyanogenic glycosides: a possible model for the biosynthesis of natural products. In *New front. Plant biochemistry* (ed. T. Akazawa, T. Asahi, and H. Imaseki), pp. 11–22. Japanese Science Society Press, Tokyo.

Conn, E.E. (1988). Biosynthetic relationship among cyanogenic glycosides, glucosinolates and nitro compounds. In *Biologically active natural products: potential use in agriculture*, ACS Symposium Series No. 380 (ed. H.G. Cutler), pp. 144–53. American Chemical Society, New York.

Davis, R.H. and Nahrstedt, A. (1985). Cyanogenesis in insects. In *Comprehensive insect physiology, biochemistry and pharmacology* (ed. G.A. Kerkut and L.I. Gilbert), Vol. 11, pp. 635–54. Pergamon Press, Oxford.

Duffey, S.S. (1981). Cyanide and arthropods. In *Cyanide in biology* (ed. B. Vennesland, E.E. Conn, C.J. Knowles, J. Westley, and F. Wissing), pp. 385–414. Academic Press, London.

Dziewanowska, K. and Lewak, S. (1982). Hydrogen cyanide and cyanogenic compounds in seeds. IV. Metabolism of hydrogen cyanide in apple seeds under conditions of stratification. *Physiologia Vegetarum*, **20**, 165–70.

Erb, N., Zinsmeister, H.D., Lehmann, G., and Nahrstedt, A. (1979). Epiheterodendrin a new cyanogenic glucoside from *Hordeum vulgare*. *Phytochemistry*, **18**, 1515.

Fischer, F.C., Fung, S.Y., and Lankhorst, P.P. (1982). Cyanogenesis in Passifloraceae II. Cyanogenic compounds from *Passiflora capsularis*, *P. warmingii* and *P. perfoliata*. *Planta Medica*, **45**, 425.

Frankenberg, L. (1983). Studies on cyanide detoxication. *Acta Pharmaceutica Suecica*, **20**, 72.

Franzl, S., Naumann, C.M., and Nahrstedt, A. (1988). Cyanoglucoside storing cuticle of *Zygaena larvae* (Lepidoptera, Zygaenidae). Morphological, ultrastructural and cyanoglucoside changes during the moult. *Zoomorphology*, **108**, 183–90.

Franzl, S., Ackermann, I., and Nahrstedt, A. (1989). Purification and characterization of a β-glucosidase (linamarase) from the haemolymph of *Zygaena trifolii* Esper, 1783 (Insecta, Lepidoptera). *Experientia*, **45**, 712–8.

Giesemann, A., Biehl, B., and Lieberei, R. (1986). Identification of scopoletin as a phytoalexin in the rubber tree *Hevea brasiliensis*. *Journal of Phytopathology*, **117**, 373–6.

Gross, D. (1989). Antimikrobielle Abwehrstoffe in Gramineen. *Journal of Plant Diseases and Protection*, **96**, 535–53.

Halkier, B. A. and Moller, B. L. (1989). Biosynthesis of the cyanogenic glucoside dhurrin in seedlings of *Sorghum bicolor* (L.) Moench and partial purification of the enzyme system involved. *Plant Physiology*, **90**, 1552–9.

Halkier, B. A., Olsen, C. E., and Moller, B. L. (1990). The biosynthesis of cyanogenic glucosides in higher plants. The (*E*)- and (*Z*)-isomers of *p*-hydroxyphenyl-acetaldehyde oxime as intermediates in the biosynthesis of dhurrin in *Sorghum bicolor* (L.) Moench. *Journal of Biological Chemistry*, **264**, 19487–94.

Halkier, B. A., Lykkesfeldt, J., and Moller, B. L. (1991). 2-Nitro-3-(*p*-hydroxy-phenyl)propionate and *aci*-1-nitro-2-(*p*-hydroxy-phenyl)ethane, two intermediates in the biosynthesis of the cyanogenic glucoside dhurrin in *Sorghum bicolor* (L.) Moench. *Proceedings of the National Academy of Sciences, USA*, **88**, 487–91.

Hegnauer, R. (1986). *Chemotaxonomie der Pflanzen*, Vol. VII, p. 345 ff. Birkhäuser Verlag, Basle.

Hegnauer, R. (1989). *Chemotaxonomie der Pflanzen*, Vol. VIII, p. 323. Birkhäuser Verlag, Basle.

Hösel, W. and Conn, E. E. (1982). The aglykone specificity of plant β-glycosidases. *Trends in Biochemical Sciences*, **7**, 219–21.

Hruska, A. J. (1988). Cyanogenic glycosides as defense compounds. A review of the evidence. *Journal of Chemical Ecology*, **14**, 2213–17.

Hsieh, J. S. (1989). Cyanogenesis and aphid resistance in sorghum. *Chemical Abstracts*, **111**, No. 74934.

Hucklesby, D. P., Dowling, M. J., and Hewitt, E. J. (1982). Cyanide formation from glyoxylate and hydroxylamine catalyzed by extracts of higher-plant leaves, *Planta*, **156**, 487–91.

Janzen, D. H., Juster, H. B., and Bell, E. A. (1977). Toxicity of secondary compounds of the seed-eating larvae of the Bruchid beetle *Collosobruchus maculatus*. *Phytochemistry*, **16**, 223–7.

Jaroszewski, J. and Jensen, B. (1985). Deidaclin and tetraphyllin A, epimeric glucosides of 2-cyclopentenone cyanohydrin in *Adenia globosa* ssp. globosa Eng. (Passifloraceae). Crystal structure of deidaclin tetraacetate. *Acta Chimica Scandinavica*, **39**, 867.

Jaroszewski, J. W. and Fog, E. (1987). Sulphate esters of cyclopentenoid cyanohydrin glycosides. *Phytochemistry*, **28**, 1527–8.

Kim, H. S., Jeffrey, G. A., Panke, D., Clapp, R. C., Coburn, R. A., and Long Jr., L. (1970). The X-ray crystallographic determination of the structure of gynocardin. *Chemical Communications*, **1970**, 381.

Knowles, C. J. (1988). Cyanide utilization and degradation by micro-organisms. In *Cyanide compounds in biology* (ed. CIBA Foundation), pp. 3–15. Wiley, Chichester.

Lieberei, R. (1988). Ralationship of cyanogenic capacity (HCN-c) of the rubber tree *Hevea brasiliensis* to susceptibility of *Microcyclus ulei*, the agent causing South American leaf blight. *Journal of Phytopathology*, **122**, 54–67.

Lieberei, R., Biehl, B., Giesemann, A., and Junqueira, N.T.V. (1989). Cyano-genesis inhibits active defense reactions in plants. *Plant Physiology*, **90**, 33-6.

Long, K.Y. and Brattsten, L.B. (1984). Is rhodanase important in the detoxification of dietary cyanide in southern armyworm (*Spodoptera eridania* Cramer) larvae? *Insect Biochemistry*, **12**, 367-75.

Lüdtke, M. and Hahn, H. (1953). Über den Linamaringehalt gesunder und von Colletotrichum befallener junger Leinpflanzen. *Biochemische Zeitung*, **324**, 433-42.

Manning, K. (1986). Ethylene production and β-cyanoalanine synthase activity in carnation flowers. *Planta*, **168**, 61-6.

Mikolajczak, K.L. (1977). Cyanolipids. *Progress in Chemistry of Fats and Lipids*, **15**, 97.

Mikolajczak, K.L., Madrigal, R.V., Smith Jr., C.R., and Reed, D.K. (1984). Insecticidal effects of cyanolipids on three species of stored product insects, European corn borer (Lepidoptera:Phyralidae) larvae, and striped cucumber beetle (Coleoptera:Chrysomelidae). *Journal of Economic Entomology*, **77**, 1144-8.

Miller, J.M. and Conn, E.E. (1980). Metabolism of hydrogen cyanide by higher plants. *Plant Physiology*, **65**, 1199-202.

Nahrstedt, A. (1976). Prunasin in *Holocalyx balansae*. *Phytochemistry*, **15**, 1983.

Nahrstedt, A. (1985). Cyanogenic compounds as protecting agents for organisms. *Plant Systematics and Evolution*, **150**, 35-47.

Nahrstedt, A. (1987). Recent developments in chemistry, distribution and biology of the cyanogenic glycosides. In *Biologically active natural products* (ed. K. Hostettmann and P.J. Lea), pp. 213-34. Clarendon Press, Oxford.

Nahrstedt, A. (1988). Cyanogenesis and the role of cyanogenic compounds in insects. In *Cyanide compounds in biology* (ed. CIBA Foundation), Vol. 140, pp. 131-45. Wiley, Chichester.

Nahrstedt, A. (1989). The significance of secondary metabolites for interactions between plants and insects. *Planta Medica*, **55**, 333.

Nahrstedt, A. and Wray, V. (1990). Structural revision of a cyanogenic glucoside from *Ilex aquifolium*. *Phytochemistry*, **29**, 3934.

Nahrstedt, A., Kant, J.-D., and Wray, V. (1982a) Acalyphin, a new cyanogenic glucoside from *Acalypha indica* (Euphorbiaceae). *Phytochemistry*, **21**, 101.

Nahrstedt, A., Walther, A., and Wray, V. (1982b). Sarmentosin epoxide, a cyanogenic compound from *Sedum cepaea* (Crassulaceae). *Phytochemistry*, **21**, 107.

Nahrstedt, A., Wray, V., Grotjahn, L., Fikenscher, L.H., and Hegnauer, R. (1983). New acylated cyanogenic diglycosides from fruits of *Anthemis cairica* and *A. altissima*. *Planta Medica*, **49**, 143.

Nahrstedt, A., Jensen, P.S., and Wray, V. (1989). Prunasin-6'-malonate, a cyanogenic glucoside from *Merremia dissecta*. *Phytochemistry*, **28**, 623-4.

Nahrstedt, A., Abdel Sattar, E., and El-zalabani, S.H.M. (1990). Amygdalin acyl derivatives, cyanogenic glycosides from the seeds of *Merremia dissecta*. *Phytochemistry*, **29**, 1179-81.

Olafsdottir, E.S., Andersen, J.V., and Jaroszewski, J.W. (1989a). Cyclopentenoid cyanohydrin glycosides. Part 9. Cyanohydrin glycosides of Passifloraceae. *Phytochemistry*, **28**, 127-32.

268   Adolf Nahrstedt

Olafsdottir, E. S., Cornett, C., and Jaroszewski, J. W. (1989b) Natural cyclopentenoid cyanohydrin glycosides. 8. Cyclopentenoid cyanohydrin glucosides with unusual sugar residues. Acta Chimica Scandinavica, 43, 51-5.

Paech, K. (1950). Die Blausäureverbindungen. In Biochemie und Physiologie der sekundären Pflanzenstoffe, p. 250. Springer Verlag, Berlin.

Peiser, G. D., Wang, T. T., Hofman, N. E., Yang, S. F., and Liu, H. (1984). Formation of cyanide from carbon 1 of aminocyclopropane-1-carboxylic acid during its conversion to ethylene. Proceedings of the National Academy of Science, USA, 81, 3059-63.

Peterson, S. C. (1986). Breakdown products of cyanogenesis. Repellency and toxicity to predatory ants. Naturwissenschaften, 73, 627-8.

Pocsi, I., Kiss, L., Hughes, M. A., and Nanasi, P. (1989). Kinetic investigations of the substrate specificity of the cyanogenic β-D-glucosidase (linamarase) of white clover. Archives of Biochemistry and Biophysics, 272, 496-506.

Porter, D. J. T. and Bright, H. J. (1987). The cyanogenic substrate for horseradish peroxidase is a conjugated enamine. Journal of Biological Chemistry, 262, 9154-9.

Poulton, J. E. (1983). Cyanogenic compounds in plants and their toxic effects. In Handbook of natural toxins. Plant and fungal toxins (ed. R. F. Keeler and A. T. Tu), Vol. 1, p. 117. Dekker Inc., New York.

Poulton, J. E. (1988). Localization and catabolism of cyanogenic glycosides. In Cyanide compounds in biology (ed. CIBA Foundation), p. 67. Wiley, Chichester.

Pourmohseni, H. (1989). Cyanoglycoside in der Epidermis von Sommergerste und ihre Bedeutung für die quantitative Resistenz bezw. Anfälligkeit gegenüber dem Echten Mehltau (Erysiphe graminis DC f. sp. hordei Marchal). Unpublished Ph.D. Thesis, Göttingen, 1989.

Reilly, C. C., Gentry, C. R., and McVay, J. R. (1987). Biochemical evidence for resistance of rootstocks to the peachtree borer and species separation of peachtree borer and lesser peachtree borer (Lepidoptera: Sesiidae) on peach trees. Journal of Economic Entomology, 80, 338.

Rockenbach, J., Nahrstedt, A., and Wray, V. (1992). Cyanogenic glycosides from Psydrax and Oxyanthus species (Rubiaceae). Phytochemistry. (In press.)

Schütte, H. R. (1973). Biosynthese von cyanogenen Glykosiden und Senfölglykosiden. Fortschritte der Botanik, 35, 103.

Schwind, P. (1990). Die cyanogenen Glykoside von Chlorophytum comosum (Anthericaceae) und Xeranthemum cylindraceum (Asteraceae) sowie Untersuchungen zur biogenetischen Vorstufe m-hydroxylierter aromatischer Cyanglykoside. Ph.D. Thesis, Münster.

Schwind, P., Wray, V., and Nahrstedt, A. (1990). Structure elucidation of an acylated cyanogenic triglycoside, and further cyanogenic constituents from Xeranthemum cylindraceum. Phytochemistry, 29, 1903-11.

Seigler, D. S. (1975). Isolation and characterization of naturally occurring cyanogenic compounds. Phytochemistry, 14, 9.

Selmar, D. and Lieberei, R. (1988). Mobilization and utilization of cyanogenic glycosides - The linustatin pathway. Plant Physiology, 86, 711.

Selmar, D., Lieberei, R., and Biehl, B. (1987a). Hevea linamarase - a nonspecific β-glycosidase. Plant Physiology, 83, 557.

Selmar, D., Lieberei, R., Biehl, B., and Nahrstedt, A. (1987). Occurrence of linustatin in *Hevea brasiliensis*. *Phytochemistry*, **26**, 2400–1.

Selmar, D., Frehner, M., and Conn, E.E. (1989*a*). Purification and properties of endosperm protoplasts of *Hevea brasiliensis*. *Journal of Plant Physiology*, **135**, 105–9.

Selmar, D., Lieberei, R., Biehl, B., and Conn, E.E. (1989*b*). α-Hydroxynitrile lyase in *Hevea brasiliensis* and its significance for rapid cyanogenesis. *Physiology of Plants*, **75**, 97.

Selmar, D., Grocholewski, S., and Seigler, D.S. (1990). Cyanogenic lipids: utilization during seedling development of *Ungnadia speciosa*. *Plant Physiology*, **93**, 631.

Shimomura, H. and Sashida, Y. (1988). Phenylpropanoid glucose esters from *Prunus Buergeriana*. *Phytochemistry*, **27**, 641–4.

Shimomura, H., Sashida, Y., and Adachi, T. (1987), Cyanogenic and phenylpropanoid glucosides from *Prunus grayana*. *Phytochemistry*, **26**, 2363–6.

Spencer, K.C. and Seigler, D.S. (1984). The cyanogenic glycoside barterin from *Barteria fistulosa* is epitetraphyllin B. *Phytochemistry*, **23**, 2365.

Tschiersch, B. (1964). Über den Stoffwechsel der Blausäure. *Flora*, **154**, 445.

Van Valen, F. (1978). Contribution to the knowledge of cyanogenesis in Angiosperms. 3. Communication. Cyanogenesis in Liliaceae. *Proceedings of the Koninklijke Nederlanse Akademie van Wetenschappen, Amsterdam*, **81c**, 132–40.

Vennesland, B., Pistorius, E.K. and Gewitz, H.-S. (1981). HCN production by microalgae. In *Cyanide in biology* (ed. B. Vennesland, E.E. Conn, C.J. Knowles, J. Westley, and F. Wissing), pp. 349–61. Academic Press, London.

Willems, M. (1988). A cyanogenic glucoside from *Ilex aquifolium*. *Phytochemistry*, **27**, 1852–3.

Witthohn, K. and Naumann, C.M. (1984). Die Verbreitung des β-Cyanoalanin bei cyanogenen Lepidopteren. *Zeitschrift für Naturforschung*, **39c**, 837.

Yang, S.F. and Hoffmann, N.E. (1984). Ethylene biosynthesis and its regulation in higher plants. *Annual Review of Plant Physiology*, **35**, 155.

Yip, W.-K. and Yang, S.F. (1988). Cyanide metabolism in relation to ethylene production in plant tissues. *Plant Physiology*, **88**, 473–6.

Note added in proof

Recent results of Birger Moller's group add new intermediates to the pathway of dhurrin biosynthesis in the *Sorghum* microsomal system: *N*-hydroxytyrosine is oxidized to form 2-nitro-(*p*-hydroxyphenyl)propionic acid that is decarboxylated to give 1-*aci*-nitro-2-(*p*-hydroxyphenyl)ethane (Halkier *et al.* 1991); the latter is converted to (*E*)-*p*-hydroxyphenylacetaldehyde oxime that isomerizes to its (*Z*)-isomer, which is the direct precursor of the nitrile (cf. Fig. 15.1) (Halkier *et al.* 1990).

16. Distribution and biological activity of alkaloidal glycosidase inhibitors from plants

LINDA E. FELLOWS, GEOFFREY C. KITE,
ROBERT J. NASH, MONIQUE S.J. SIMMONDS,
and ANTHONY M. SCOFIELD*

*Jodrell Laboratory, Royal Botanic Gardens, Kew, TW9 3DS UK,
and *University of London, Wye College, Wye, Ashford, Kent
TN25 5AH, UK*

Introduction

Alkaloidal glycosidase inhibitors (AGIs), which have been reported to occur in many species of higher plants, bacteria and fungi, are probably widespread in nature. They belong to five different chemical structural types, namely polyhydroxylated derivatives of piperidine, pyrrolidine, pyrroline, octahydroindolizine, and pyrrolizidine. Certain of the simpler piperidine and pyrrolidine derivatives bear an obvious structural resemblance to 1-deoxy monosaccharides, with the ring oxygen replaced by nitrogen. The orientation of hydroxy groups on the more complex bicyclic structures also suggests a structural resemblance to sugars. AGIs exhibit a range of different types of biological activity. The structures of all AGIs described to date are shown in Figs 16.1–16.3 For simplicity compounds will be referred to by their common names or abbreviations, if any, in this chapter.

Natural distribution

AGIs have been positively identified in five families of higher plants (Leguminosae, Moraceae, Euphorbiaceae, Polygonaceae, and Aspidiaceae). DNJ and DMJ occur in both higher plants and bacteria (*Streptomyces* and *Bacillus* spp.), and swainsonine in both higher plants and fungi (*Rhizoctonia* and *Metarrhizium* spp.) Simple analogues of glucose, mannose and galactose with the anomeric hydroxy group at C1 retained (namely nojirimycin, nojirimycin B, and galactostatin) have only been found in bacteria, (*Streptomyces* spp.), and the one pyrroline known was found in a fungus (*Nectria lucida* Hohnel) (reviewed in Fellows *et al.* 1989a). All other AGIs have so far only been isolated from higher plants, with the exception of HNJ (homonojirimycin) and DMDP, recently found to accumulate in

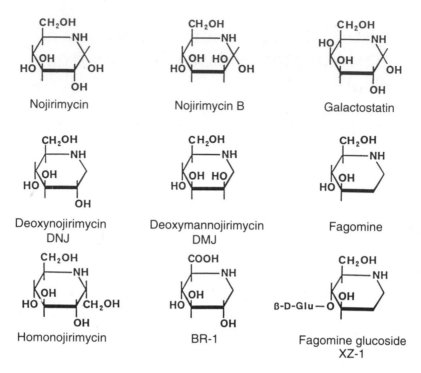

Fig. 16.1. Alkaloidal glycosidase inhibitors incorporating a 6-sided ring. Nojirimycin, 5-amino-5-deoxy-D-glucopyranose; nojirimycin B, 5-amino-5-deoxy-D-mannopyranose; galactostatin, 5-amino-5-deoxy-D-galactopyranose; deoxynojirimycin, DNJ, 1,5-dideoxy-1,5-imino-D-glucitol; deoxymannojirimycin, DMJ, 1,5-dideoxy-1, 5-imino-D-mannitol; fagomine, 1, 2, 5-trideoxy-1, 5-imino-D-arabinitol; homonojirimycin, 2,6-dideoxy-2,6-imino-D-glycero-L-gulo-heptitol; BR1, 2S-carboxy-3R, 4R, 5S-trihydroxypiperidine; fagomine glucoside, XZ-1, 1,2,5-trideoxy-4-O-(β-D-glucopyranosyl)-1,5-imino-D-arabinohexitol.

Fig. 16.3. Alkaloidal glycosidase inhibitors with a bicyclic ring. Castanospermine, (1S,6S,7R,8R,8aR)-1,6,7,8-tetrahydroxyoctahydroindolizine; 6-*epi*castanospermine,(1S,6R,7R,8R,8aR)-1,6,7,8-tetrahydroxyoctahydroindolizine; swainsonine, (1S,2R,8R,8aR)-1,2,8-trihydroxyoctahydroindolizine; alexine, (1R,2R,3R,7S,7aS)-3-hydroxymethyl-1,2,7- trihydroxypyrrolizidine, 3,7a-*diepi*alexine, (1R,2R,3S,7S,7aR)-3-hydroxymethyl-1,2,7-trihydroxypyrrolizidine; 7a-*epi*alexine, (australine) (1R, 2R, 3R, 7S, 7aR)-3-hydroxymethyl-1, 2, 7-trihydroxypyrrolizidine; 1,7a-*diepi*alexine, (1S,2R,3R,7R,7aS)-3-hydroxymethyl-1,2,7-trihydroxypyrrolizidine; 7, 7a-*diepi*alexine, (1R, 2R, 3R, 7R, 7aR)-3-hydroxymethyl-1, 2, 7-trihydroxypyrrolizidine.

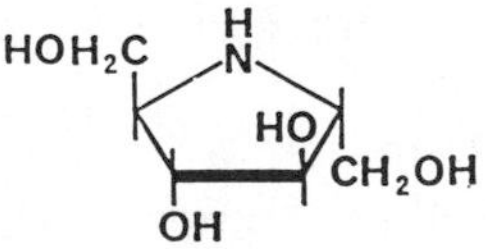

DMDP

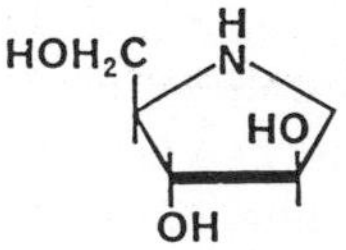

DAB-1

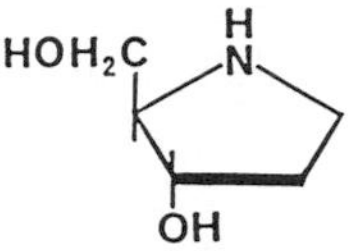

CYB-3

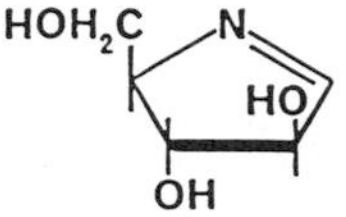

FR-900483

Fig. 16.2. Alkaloidal glycosidase inhibitors incorporating a 5-sided ring. DMDP, 2R, 5R-dihydroxymethyl-3R, 4R-dihydroxypyrrolidine; DAB-1, 1, 4-dideoxy-1, 4-imino-D-arabinitol; CYB-3, 2R-hydroxymethy-3S-hydroxypyrrolidine; FR-900483, (3R,4R,5R)-3,4-dihydroxy-5-hydroxymethyl-1-pyrroline.

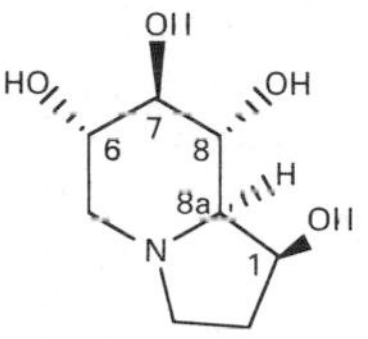

Castanospermine

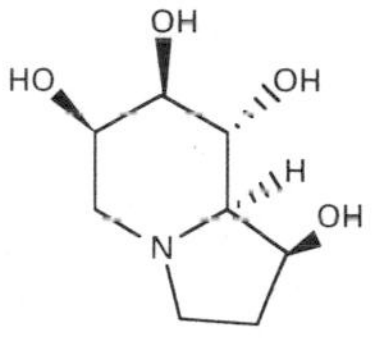

6-*epi* Castanospermine

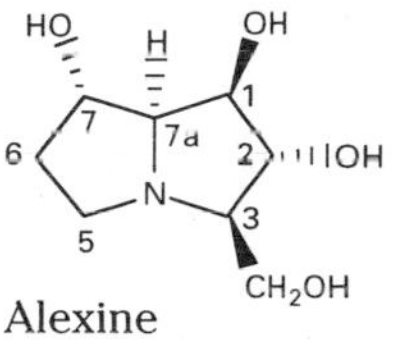

Alexine

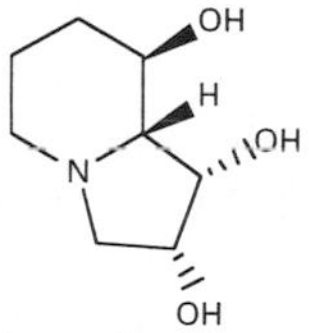

Swainsonine

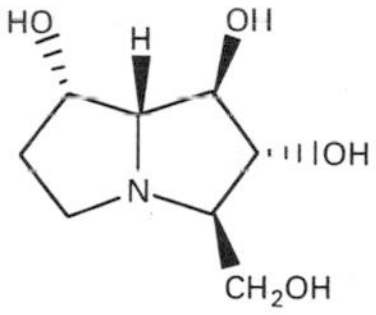

7a-*epi* Alexine

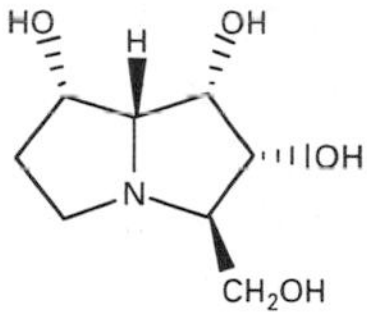

1,7a-*diepi* Alexine

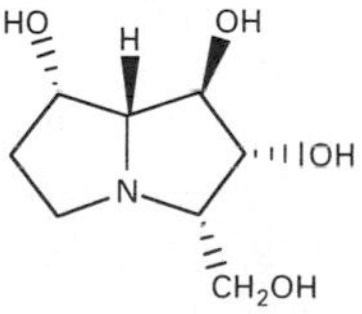

3,7a-*diepi* Alexine

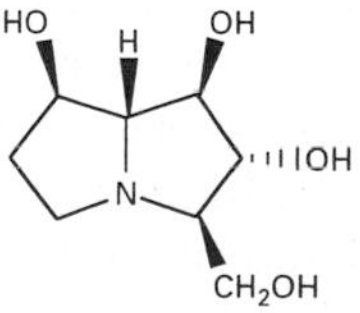

7,7a-*diepi* Alexine

the moth *Urania fulgens* whose host plant *Omphalea diandra* (Euphorbiaceae) contains both (Kite *et al.* 1990).

AB1 is the only AGI so far found in a fern. Castanospermine, 6-epicastanospermine, swainsonine, and the alexine series of pyrrolizidines have not yet been found outside the Leguminosae. In contrast the simpler AGIs, DNJ, DMDP, and DMJ have been detected in this laboratory in many different plant families. (Fellows and Fleet 1988; Fellows *et al.* 1989*a*; R. J. Nash, unpublished).

Biological activity

Certain AGIs exhibit anti-viral, anti-insect, anti-cancer, and hypoglycaemic effects, as well as immune modulatory properties. Most of these effects can be shown to result from the direct or indirect inhibition of glycosidase enzymes. The inhibitory effect of a given AGI on a particular enzyme cannot yet be predicted in advance: there is wide variation in the susceptibility of glycosidases with different substrate specificities and from different species to inhibition by these compounds.

Inhibition of digestive glycosidases

DNJ and several semi-synthetic derivatives inhibit the rise in blood glucose which follows a meal, and are of potential use as antidiabetic drugs (Rhinehart *et al.* 1987). We have shown that DNJ and DAB-1 inhibit α- and DMDP β-glucosidase activity in mouse gut and human amniotic fluid. In contrast DMDP proved far more powerful than DNJ as an inhibitor of α-glucosidase in insects (*Callosobruchus maculatus* and *Spodoptera littoralis*). DNJ is more active than DMDP against trehalase in all animals studied (mouse, human, and several insects) (Fellows 1986; Scofield *et al.* 1986; Fellows *et al.* 1989*b*; Scofield, unpublished.) Some other species differences are shown in Table 16.1. Note that while DNJ is inactive against β-glucosidase in mouse and *Spodoptera* digestive tract, it inhibits the enzyme of *Penicillium*. It has no effect on yeast sucrase (a β-fructofuranosidase), despite its action on the enzyme of mouse gut (an α-glucosidase). Castanospermine is so far reported to be more powerful against β- than α-glucosidase in insects although the reverse is true in mammals. It may be possible to exploit such differences in the formulation of selective drugs or pesticides. Campbell *et al.* (1987) reported considerable species differences in the susceptibility of insects to castanospermine. In this laboratory we have confirmed some, but not all, of their findings. For example we were unable to detect castanospermine-sensitive sucrase activity in any insect tested (*S. littoralis, Agrotis segetum, Heliconius melpomone, Aphis fabae, Myzus persicae*, and *Brevicoryne brassicae*). This discrepancy might reflect dif-

Table 16.1 Species differences in the susceptibility of alpha- and beta-glucosidase activity to inhibition by AGIs

	DMDP	DAB1	DNJ	HNJ	CAST	6-EPI CAST
Inhibition of alpha-glucosidase activity						
Mouse gut	3.0×10^{-4}	4.7×10^{-5}	8.3×10^{-7}	2.2×10^{-7}	2.8×10^{-6}	1.8×10^{-5}
Spodoptera littoralis gut	4.8×10^{-7}	2.8×10^{-6}	8.0×10^{-6}	1.3×10^{-5}	NI	NI
Saccharomyces cerevisiae	1.5×10^{-5}	5.0×10^{-6}	NI	7.5×10^{-5}	NI	NI
Inhibition of beta-glucosidase activity						
Mouse gut	1.0×10^{-5}	NI	NI	1.4×10^{-4}	1.7×10^{-5}	NI
S. littoralis gut	2.0×10^{-5}	NI	NI	3.0×10^{-4}	NI	NI
Penicillium expansum	2.0×10^{-5}	2.3×10^{-4}	5.0×10^{-6}	5.0×10^{-5}	3.2×10^{-5}	NI

Concentration of AGI giving 50% inhibition under standard conditions. Substrate: synthetic p-nitrophenyl-α or β-D-glucopyranoside. NI = < 50 per cent inhibition at 3.3×10^{-4} M. For details of assay see Scofield *et al*. 1986. Data from Fellows *et al*. 1989*a*, and Scofield, unpublished.

ferences in insect strains, or differences in culturing technique, in particular in diet (Scofield, unpublished).

The recently discovered pyrrolizidines of the alexine series are poor inhibitors of digestive enzymes, but strong inhibitors of fungal (*Aspergillus*) amylo-glucosidase, an exo-1, 4-glucosidase acting on oligosaccharides (Nash *et al*. 1990).

Insect feeding deterrence

Several AGIs have been shown to deter insects from feeding. *Locusta migratoria* L. was deterred from feeding on food treated with DMDP and the aphid *Acyrthosiphon pisum* Harris rejected a diet containing castano-spermine (reviewed in Fellows *et al*. 1989*a*). In recent studies we have shown that several AGIs reduce feeding in caterpillars. Studies with the African leafworm *Spodoptera littoralis* have shown that AGIs interact in different ways with the chemoreceptors used by caterpillars in food selection. Taste sensilla on the caterpillar's mouthparts contain neurons that, when activated by the sugars glucose, fructose, and sucrose stimulate food intake. However, when the sensilla are exposed to DMDP their sensitivity to the sugars, especially to fructose, is reduced. This effect lasts for up to 2 h and is associated with reduced feeding. Castanospermine and DAB-1 have similar, thought less profound effects on the sensitivity of the sensilla to glucose and sucrose respectively (Simmonds *et al*. 1990). It was suggested that the role of these compounds in plants might be to render insects temporarily unable to detect sugars (Fellows *et al*. 1989*a*).

By studying the amplitude and shape of the action potentials associated with the neurons in the taste sensilla we can distinguish which neuron is stimulated by a particular compound. The AGIs that deter feeding in *S. littoralis* stimulate a neuron different from that which responds to sugars. Stimulating a sensillum with a mixture of an alkaloid and a sugar results, however, in a lower rate of firing than when the chemicals are used singly. For example, in a recent study it was shown that when the lateral styloconica sensillum is stimulated with solutions of increasing concentrations of fructose (1–100 mM), firing of the sugar-sensitive neuron increases in a dose-dependent manner; if the sugar concentration is then held constant while the concentration of DMDP is steadily increased (1–10 mM), the firing rate of the alkaloid-sensitive neuron increases while that of the sugar-sensitive neuron declines (Simmonds *et al*. 1990).

Exactly how DMDP affects the response of the sugar-sensitive neuron is not as yet known. One possibility is that DMDP competes with fructose for a binding site on the sugar-sensitive neuron without triggering a response. Alternatively, it has been suggested that a glucosidase is present at or near the sugar receptor site and plays a part in its function (Shimada *et al*. 1974).

DMDP and other AGIs could act as inhibitors of such activity (Simmonds *et al.* 1989, 1990).

Glycoprotein processing

Many of the effects of AGIs in biological systems are the result of changes which they bring about in the sequence of sugars in the oligosaccharide side-chains (glycans) of the so-called *N*-asparagine-linked glycoproteins. The glycans have numerous roles, including targeting information, prevention of proteolytic degradation, and facilitating the correct folding of the molecule and also its interaction with carbohydrate binding sites on other proteins. Alterations to glycan structure bring about changes in the immunologial and physico-chemical properties of the molecule. The assembly of the glycan chain on this type of glycoprotein is complex but almost identical in plants and animals (Elbein and Molyneux 1987; Faye *et al.* 1989). Firstly, a preformed glycan of structure $Glc_9Man_3GlcNAc_2$ is co-translationally transferred from a lipid carrier to specific asparagine residues on the nascent peptide chain. This is then trimmed ('processed') one residue at a time (but with the addition of two extra *N*-acetyl glucosamine residues) by a series of processing enzymes down to a smaller structure of composition $GlcNAc_2Man_3GlcNAc_2$. Sugar residues are then added to this by a series of glycosylations to produce a structure unique to the glycoprotein in question. Certain AGIs inhibit specific steps in the processing sequence. For example, castanospermine and DNJ both inhibit the first step, removal of the terminal glucose by so-called glucosidase I, and lead to the accumulation of high mannose structures (reviewed in Elbein and Molyneux 1987). DMJ and swainsonine inhibit mannosidase I (which cleaves α-1-2 mannoside links) and mannosidase II (which cleaves α-1-3 and α-1-6 mannosides) respectively, leading to the accumulation of hybrid structures. AGIs are being used to probe the structure–function relationship of glycoproteins, particularly in animal cells. Their potential in plant glycoprotein research, for instance in plant breeding mechanisms, has not been sufficiently explored (Elbein and Molyneux 1987; Faye *et al.* 1989).

Anti-viral effects

Castanospermine and DNJ inhibit the replication of the retroviruses HIV1 and HIV2 (which cause AIDS), and murine leukaemia virus at non-toxic concentrations: they also inhibit the herpes virus cytomegalovirus (reviewed in Fellows *et al.* 1989*a*). Although the mechanisms of their anti-viral action have not been fully elucidated, it is known that they result at least in part from the inhibition of glucosidase 1 and the resulting imperfect processing of viral glycoproteins. There is a possibility that derivatives of these two AGIs, 6-*O*-butyryl-castanospermine and *N*-butyl DNJ, both of which are more potent anti-viral agents than the parent compound, will find a role in the clinical management of AIDS, either alone or in combination with reverse transcriptase inhibitors such as AZT (Fleet *et al.* 1988; Johnson *et al.* 1989; Sunkara *et al.* 1989). Despite alterations in the glycoproteins of other viruses, there are few reports of reduced infectivity (reviewed in Fellows *et al.* 1989a).

Cancer and the immune response

There is considerable evidence that changes in glycoproteins accompany the transformation of normal cells to cancer cells, and that oligosaccharides on the surface of tumour cells are involved in metastatic colonization (Dennis *et al.* 1987). Some AGIs, notably swainsonine and castanospermine, have been shown to reduce the metastatic potential of some cancer cell lines in culture, and to cause a loss of the transformed phenotype in others (reviewed in Fellows *et al.* 1989a). Swainsonine also reduces the invasiveness of B16-F10 murine melanoma cells if the animal, as opposed to the cancer cells, is supplied with swainsonine in drinking water prior to inoculation with cancer cells; there is evidence that this results from the direct stimulation of the immune response and an increase in the activity of natural killer cells. Although these protective effects are lost if swainsonine treatment commences after injection of the cancer cells, it is restored by the treatment of the animal with cyclophosphamide, which blocks the action of suppressor T cells (Myc *et al.* 1989; Humphries *et al.* 1990). Swainsonine is therefore likely to prove a useful tool with which to probe the mechanism of the immune response to cancer. Recently castanospermine has also been shown to stimulate the immune system (Rothman *et al.* 1990).

Inhibition of other types of hydrolase

Some AGIs have been reported to inhibit hydrolases other than simple glycosidases. For example, castanospermine inhibits the hydrolysis of the cyanogenic β-glycoside vicianin from the fern *Davallia trichomaniodes* Blume. (Lizotte and Poulton 1988.) DMDP and castanospermine also inhibit the action of myrosinase (thioglucosidase) from mustard and the cabbage aphid, *Brevicoryne brassicae*. Alexine, DAB-1, and DNJ also have some effect on the myrosinase of the aphid. However, AGIs are considerably less potent as inhibitors of the activity of thioglucosidases than of mammalian and insect glucosidases (Scofield *et al.* 1990).

Synthesis and biosynthesis

The biosynthetic pathways to AGIs in higher plants are unknown. Nojirimycin is synthesized in *Streptomyces* from glucose with C1/C6 inversion (Inouye *et al.* 1968). The piperidine ring of swainsonine in the mould *Rhizoctonia leguminicola* derives from lysine via pipecolic acid. Two additional carbons are introduced to the side chain which cyclizes to give 1-hydroxyindolizidine and 1,2-indolizidine diol intermediates before being converted to swainsonine (Harris *et al.* 1988). All the naturally occurring AGIs have been synthesized, as have many unnatural structural variants (Fleet 1989). Some unnatural forms have potentially useful activity. For example, 1-deoxy-6,8a-*diepi*castanospermine has no effect on human glucosidases but is a powerful inhibitor of α-L-fucosidase. 2-*epi*swainsonine strongly inhibits α-glucosidase, but has no effect on any mannosidase tested (Elbein and Molyneux 1987; Winchester *et al.* 1990).

Ecological implications

AGIs have the potential to deter a wide range of predators from plants. Mixtures of AGIs with different inhibitory properties have frequently been found in the same plant species (Kite *et al.* 1988; Fellows *et al.* 1989a). The ratio of DMDP to DMJ varies considerably in species of the pan-tropical legume genus *Lonchocarpus* (Evans *et al.* 1985). It would be of interest to determine whether there is selection for variation in the levels of individual AGIs in response to changing patterns of predator pressure. AGIs are, however, by no means detrimental to all insects. Bruchid beetles of the genus *Ctenocolum* breed on seeds of *Lonchocarpus* species which may contain over 2 per cent dry weight of DMDP (Janzen 1980). It is possible that DMDP may be also be a nutrient to those species that can tolerate it. The day-flying moth *Urania fulgens* breeds on *Omphalea diandra* which contains DMDP, DMJ, and HNJ. Larvae of the moth contain DMDP and HNJ but not DMJ. Since the larval frass is relatively enriched with DMJ, it is possible that the moth selectively accumulates those AGIs from its diet which might protect it from predators, such as birds. HNJ is a powerful inhibitor of avian (chicken) digestive α-glucosidase. DMDP accumulates in *Urania* eggs and may serve to protect them from insect attack. DMJ is a poor inhibitor of glucosidases, and its accumulation would presumably serve no purpose (Fellows and Fleet 1988; Kite *et al.* 1990).

A species of night-flying moth from Australia, *Nyctalemon patroclus*, belonging to a genus related to *Urania*, breeds on another genus of Euphorbiaceae, *Endospermum*. Although *Endospermum* is not considered to be closely related to *Omphalea*, it has also been shown to contain similar levels of DMDP and HNJ which also accumulate in the larvae of *N. patroclus*. Many species of *Endospermum* are known to be hosts for ant colonies. It will be of interest to investigate the part played by AGIs in the interaction between host plant, ants and uraniid larvae (Monteith and Wood 1987; Kite *et al.* 1991).

Summary

Alkaloids which are glycosidase inhibitors have been isolated from five families of higher plants and preliminary work in this laboratory suggests that they are present in many others (Fellows *et al.* 1989*a*; Kite *et al.* 1988). They are proving useful tools in many areas of biochemical research, and are pointing the way to new drugs and pesticides. Despite this, to our knowledge no studies of their biosynthesis and metabolic fate in higher plants have been undertaken.

Acknowledgements

We thank Prof. W. M. Blaney, entomologist, for advice, and the Medical Research Council, UK, for financial support (R J N).

References

Campbell, B. C., Molyneux, R. J., and Jones, K. C. (1987). Differential inhibition by castanospermine of various insect disaccharidases. *Journal of Chemical Ecology*, **13**, 1759–70.

Dennis, J. W., Laferte, S., Waghorne, C., Breitman, M. L., and Kerbel, R. S. (1987). B1–6 branching of asparagine-linked oligosaccharides is directly associated with metastasis. *Science*, **236**, 582–5.

Elbein, A. D. and Molyneux, R. J. (1987). The chemistry and biochemistry of simple indolizidine and related polyhydroxy alkaloids. In *Alkaloids; chemical and biological perspectives* (ed. S. W. Pelletier), Vol. 5, pp. 1–54. Wiley, New York.

Evans, S. V., Fellows, L. E., Shing, T. K. M., and Fleet, G. W. J. (1985). Glycosidase inhibition by plant alkaloids which are structural analogues of monosaccharides. *Phytochemistry*, **24**, 1953–5.

Faye, L., Johnson, K. D., Sturm, A., and Chrispeels, M. J. (1989). Structure, biosynthesis and function of asparagine-linked glycans on plant glycoproteins. *Physiologia Plantarum*, **75**, 309–14.

Fellows, L. E. (1986). The biological activity of polyhydroxy alkaloids from plants. *Pesticide Science*, **17**, 602–6.

Fellows, L. E. and Fleet, G. W. J. (1988). Alkaloidal glycosidase inhibitors from plants. In *Natural products isolation* (ed. G. H. Wagman and R. Cooper), pp. 540–60. Elsevier, Amsterdam.

Fellows, L. E., Kite, G. C., Nash, R. J., Simmonds, M. S. J., and Scofield, A. M. (1989*a*). Castanospermine, swainsonine and related polyhydroxy alkaloids: structure, distribution and biological activity. In *Plant nitrogen metabolism* (ed. J. E. Poulton and E. E. Conn), pp. 395–427. Plenum, New York.

Fellows, L. E., Doherty, C. H., Horn, J. M., Kite, G. C., Nash, R. J., Romeo, J. T., Simmonds, M. S. J., and Scofield, A. M. (1989*b*). Distribution and biological activity of alkaloidal glycosidase inhibitors from plants. In *Swainsonine and* *related glycosidase inhibitors* (ed. L. F. James, A. D. Elbein, R. J. Molyneux, and C. D. Warren), pp. 396–416. Iowa State University Press, Ames.

Fleet, G. W. J., Karpas, A., Dwek, R. A., Fellows, L. E., Tyms, A. S., Petursson, S., Namgoong, S. K., Ramsden, N. G., Smith, P. W., Son, J. C., Wilson, F., Witty, D. R., Jacob, G. S., and Rademacher, T. W. (1988). Inhibition of HIV replication by amino-sugar derivatives. *FEBS Letters*, **237**, 128–32.

Fleet, G. W. J. (1989). Synthesis of highly functionalised homochiral compounds from sugars. *Chemistry in Britain*, **25**, 287–9.

Harris, C. M., Schneider, M. J., Ungemach, F. S., Hill, J. E., and Harris, T. M. (1988). Biosynthesis of the toxic indolizidine alkaloids slaframine and swainsonine in *Rhizoctonia leguminicola*: metabolism of 1-hydroxyindolizidines. *Journal of the American Chemical Society*, **110**, 940–9.

Humphries, M. J., Matsumoto, K., White, S. L., Molyneux, R. J., and Olden, K. (1990). An assessment of the effects of swainsonine on survival of mice injected with B16-F10 melanoma cells. *Clinical and Experimental Metastasis*, **8**, 89–102.

Inouye, S., Tsuruoka, T., Ito, T., Niida, T. (1968). Structure and synthesis of nojirimycin. *Tetrahedron*, **24**, 2125–44.

Janzen, D. H. (1980). Specificity of seed-attacking beetles in a Costa Rican deciduous forest. *Journal of Chemical Ecology*, **68**, 925–9.

Johnson, V. A., Walker, B. D., Barlow, M. A., Paradis, T. J., Chou, T. C., and Hirsch, M. S. (1989). Synergistic inhibition of human immunodeficiency virus type 1 and type 2 replication *in vitro* by castanospermine and 3'-azido-3'-deoxythymidine. *Antimicrobial Agents and Chemotherapy*, **33**, 53–7.

Kite, G. C., Fellows, L. E., Fleet, G. W. J., Liu, P. S., Scofield, A. M., and Smith, N. G. (1988). Alpha-homonojirimycin (2,6-dideoxy-2,6-imino-D-glycero-L-gulo-heptitol) from *Omphalea diandra* L.: isolation and glycosidase inhibition. *Tetrahedron Letters*, **29**, 6483 6.

Kite, G. C., Horn, J. M., Romeo, J. T., Fellows, L. E., Lees, D. C., Scofield, A. M., and Smith, N. G. (1990). Alpha-homonojirimycin and 2,5,-dihydroxymethyl-3,4-dihydroxypyrrolidine: alkaloidal glycosidase inhibitors in the moth *Urania fulgens*. *Phytochemistry*, **29**, 103–5.

Kite, G. C., Fellows, L. E., Lees, O. C., Kitchen, O., and Monteith, G. B. (1991). Alkaloidal glycosidase inhibitors in nocturnal and diurnal uraniine moths and their respective food plant genera, *Endospermum* and *Omphalea*. *Biochemical Systematics and Ecology*. (In press.)

Lizotte, P. A. and Poulton, J. E. (1988). Catabolism of cyanogenic glycosides by purified vicianin hydrolase from squirrel's foot fern (*Davallia trichomanoides* Blume). *Plant Physiology*, **86**, 322–4.

Monteith, B. B. and Wood, G. A. (1987). *Endospermum*, ants and uraniid moths in Australia. *Queensland Naturalist*, **28**, 35–41.

Myc, A., Kunicka, J. E., Melamed, M. R., Darzynkiewicz, Z. (1989). The effect of swainsonine on stimulation and cell cycle progression of human lymphocytes. *Cancer Research*, **49**, 2879–83.

Nash, R. J., Fellows, L. E., Dring, J. V., Fleet, G. W. J., Girdhar, A., Ramsden, N. G., Peach, J. M., Hegarty, M. P., and Scofield, A. M. (1990). Two alexines [3-hydroxymethyl-1, 2, 7-trihydroxypyrrolizidines] from *Castanospermum australe*. *Phytochemistry*, **29**, 111–14.

Rhinehart, B. L., Robinson, K. M., Liu, P. S., Payne, A. J., Wheatley, M. E., and Wagner, S. R. (1987). Inhibition of intestinal disaccharidases and the suppression of blood glucose by a new alpha-glucosidase inhibitor MDL 25,637. *Journal of Pharmacology and Experimental Therapeutics*, **241**, 915-19.

Rothman, R. J., Perussia, B., Herlyn, D., and Warren, L. (1990). Antibody-dependent cytotoxicity mediated by natural killer cells is enhanced by castanospermine induced alteration of IgG glycosylation. *Molecular Immunology*, **27**, 1113-24.

Scofield, A. M., Fellows, L. E., Nash, R. J., and Fleet, G. W. J. (1986). Inhibition of mammalian disaccharidases by polyhydroxy alkaloids. *Life Sciences*, **39**, 645-50.

Scofield, A. M., Rossiter, J. T., Witham, P., Kite, G. C., Nash, R. J., and Fellows, L. E. (1990). Inhibition of thioglucosidase-catalysed glucosinolate hydrolysis by castanospermine and related alkaloids. *Phytochemistry*, **29**, 107-9.

Shimada, I., Shiraishi, A., Kijima, H., and Morita, H. (1974). Separation of two receptor sites in a single libellar sugar receptor of the fleshfly by treatment with p-chloromercuribenzoate. *Journal of Insect Physiology*, **20**, 605-21.

Simmonds, M. S. J., Blaney, W. M., and Fellows, L. E. (1989). Wild plants as a source of novel anti-insect compounds: alkaloidal glycosidase inhibitors. In *New crops for food and industry* (ed. G. Wickens, N. Haq, and P. Day), pp. 365-77. Croom Helm, London.

Simmonds, M. S. J., Blaney W. M., and Fellows, L. E. (1990). Behavioral and electrophysiological study of the antifeedant mechanisms associated with polyhydroxy alkaloids. *Journal of Chemical Ecology*, **16**, 3167-96.

Sunkara, P. S., Taylor, D. L., Kang, M. S., Bowlin, T. L., Liu, P. S., Tyms, A. S., and Sjoerdsma, A. (1989). Anti-HIV activity of castanospermine analogues. *Lancet*, 27 May, 1206.

Winchester, B. G., Cenci di Bello, I., Richardson, A. C., Nash, R. J., Fellows, L. E., Ramsden, N. G., and Fleet, G. W. J. (1990). The structural basis of the inhibition of human glycosidases by castanospermine analogues. *Biochemical Journal*, **269**, 227-31.

Index of organisms

Subject index